Before the Backbone

Before the Backbone

Views on the origin of the vertebrates

Henry Gee

CHAPMAN & HALL
London · Weinheim · New York · Tokyo · Melbourne · Madras

Published by Chapman & Hall, 2–6 Boundary Row, London SE1 8HN, UK

Chapman & Hall, 2–6 Boundary Row, London SE1 8HN, UK

Chapman & Hall GmbH, Pappelallee 3, 69469 Weinheim, Germany

Chapman & Hall USA, 115 Fifth Avenue, New York, NY 10003, USA

Chapman & Hall Japan, ITP-Japan, Kyowa Building, 3F, 2-2-1 Hirakawacho, Chiyoda-ku, Tokyo 102, Japan

Chapman & Hall Australia, 102 Dodds Street, South Melbourne, Victoria 3205, Australia

Chapman & Hall India, R. Seshadri, 32 Second Main Road, CIT East, Madras 600 035, India

First edition 1996

Typeset in 10/12pt Palatino by Saxon Graphics Ltd, Derby

Printed in Great Britain by St Edmundsbury Press, Bury St Edmunds, Suffolk

ISBN 0 412 48300 9

A catalogue record for this book is available from the British Library

Library of Congress Catalog Card Number: 95-72203

Printed on permanent acid-free text paper, manufactured in accordance with ANSI/NISO Z39.48-1992 and ANSI/NISO Z39.48-1984 (Permanence of Paper).

To the memory of David Lanning, who first taught me which cranial nerve was which.

Contents

Preface

> We cannot catechise our stony ichthyolites, as did the necromantic lady of the *Arabian Nights* did the coloured fishes of the lake which had once been a city, when she touched their dead bodies with her wand, and they straightaway raised their heads and replied to her queries. We would have many a question to ask them if we could – questions never to be solved.
>
> Hugh Miller, *The Old Red Sandstone*

When I started this book in 1991, the subject of vertebrate origins was fusty and unfashionable. Early drafts for this preface read like an extended complaint at the lot of traditional morphologists, cast aside by the march of modern molecular biology.

But no longer – this book should reach you at a time of renewed interest in the origin of the vertebrates, our own particular corner of creation. For although the topic has excited interest for well over a century, molecular biology has only lately achieved the maturity necessary to test its predictions. As a legitimate field of study, it is fashionable again.

It has been a long time coming. Most people remember William Bateson as one of those who rehabilitated Mendelian inheritance at the turn of the century, an act that paved the way for the Modern Synthesis. They will also be familiar with the name of Thomas Hunt Morgan, a pioneer of the genetics of the fruit fly *Drosophila melanogaster*, and thus much of modern genetics. The unit of recombinational distance is named after him.

Less well known is that both men contributed to the debate on vertebrate origins through early studies on the acorn worm *Balanoglossus* (=*Saccoglossos*), a member of the group of animals known as hemichordates. The noted palaeontologist William King Gregory takes up the story (Gregory, 1946).

> Even Bateson (1884, 1885), in his younger, more liberal days at least, contributed through his studies on the embryology of the

> acorn worms (*Balanoglossus*) to the contemporary wide interest in the problem of the origin of the vertebrates. But after he had experienced the satisfactions of experimental science, Bateson (1914) formed a poor opinion of those who indulged in what he regarded as idle speculations and held them up to ridicule, comparing their attempts to explain 'adaptations' with the optimistic vaporings of Voltaire's Dr. Pangloss.

He continues:

> Bateson's American colleague T. H. Morgan, although he had also contributed to the knowledge of the development of *Balanoglossus* ... likewise encouraged a sceptical attitude toward supposedly outworn theories of the origin of the vertebrates and toward all a priori [*sic*] reasoning without benefit of experimental science ... In that period of brilliant and far-reaching successes in the genetic and experimental fields and in the application of statistical methods and mathematical analyses to the problems of growth and form, it was inevitable that the problem of the origin of the vertebrates, dealing with a very distant event by means of qualitative analysis of morphological patterns, should come to be regarded in many quarters as of less than academic importance. [1946: p348]

Bateson's reservations were justified given the once widespread misuse of Darwinian ideas to justify morphological just-so stories. Haeckel – an ardent fan of Darwin – perfected the idea of recapitulation. As Garstang (1922) showed, recapitulation has many good things to say provided that it is not taken too literally. Which is, of course, how it was generally taken.

Again, Gaskell's defence of his idea that vertebrates came from a crustacean-like ancestor (in Gaskell *et al.*, 1910) is less reprehensible for its morphological content than for its basis in a mode of evolutionary thought that owed more to pre-Darwinian ideas than natural selection. Morphologists in the Garstangian tradition, notably Berrill (1955), veered the other way in their promotion of a panselectionist view, in which absolutely everything could be explained by the agency of natural selection.

Bateson, Morgan and their colleagues were powerful people, and – as Gregory (1946) further shows – they used their influence actively to discourage the teaching of old-style morphological argument, replacing it with modern experimental science.

Matters have now come full circle, and, thanks to the experimental biology of the kind championed by Bateson and Morgan, the subject has been revitalized. Although a consensus about the relationships of vertebrates with invertebrates has yet to emerge, progress in molecular biology suggests that the answers will be with us soon.

When discussing the origin of vertebrates, it is often more correct to use the term 'craniate' than 'vertebrate' (Janvier, 1981). Chordates are those animals in which (among other things) the primary axis is supported by a stiffening, rod-like structure called a notochord. Craniates include those chordates in which the brain and organs of special sense are housed in a special protective box, or 'cranium'. Vertebrates, which are all craniates, include animals in which, in addition to the notochord, the spinal column is composed of a segmental series of vertebrae, which may supplant the notochord entirely. However, some animals – the hagfishes – contrive to be craniates, because they have skulls; but not vertebrates, because their notochords are not replaced by vertebral discs. Strictly speaking, then, hagfishes are craniates but not vertebrates. However, for the sake of not replacing a familiar term with an unfamiliar one, in this book I consider hagfishes as at least honorary vertebrates, at least when the term 'vertebrate' is used casually. So, most of the time it's safe to read 'craniate' when I write 'vertebrate', unless otherwise indicated.

In the conventional textbook scheme, as taught to generations of students, the closest living relatives of the craniates among the chordates are generally held to be the lancelets, technically known as the cephalochordates. Lancelets are small, superficially fish-like creatures which have a notochord and many other craniate-like features, but they lack organs of special sense and the cranium necessary to contain them (hence an alternative name for the group, the acraniates).

The sea-squirts, otherwise known as tunicates or urochordates, are also chordates. They are usually sessile creatures, in which the notochord is confined to a short-lived larval stage, and in sessile forms, atrophies on settlement. Some urochordates, though, retain the notochord and a propulsive tail throughout life. Urochordates are usually placed slightly more distant from the craniates and cephalochordates.

The craniates, cephalochordates and urochordates constitute a larger group, the phylum Chordata, on the basis of certain attributes (such as the notochord) which they share.

The textbook view of the position of hemichordates (Bateson and Morgan's *Balanoglossus* and its allies) is that they are probably the closest relatives of chordates, if they are not actually chordates themselves. Of more remote alliance are the echinoderms; the spiny-skinned starfish, sea-urchins, crinoids and their allies.

All these groups – the chordates, hemichordates and echinoderms – constitute a distinct branch of the animal kingdom, known as the deuterostomes, defined largely on embryonic features (Schaeffer, 1987). They are distinct from other major groups of animals, such as arthropods, segmented worms and molluscs, which share a different set of embryonic features.

Until the advent of the objective method of phylogenetic reconstruction known as cladistics, the justification of one phylogenetic scheme over another was no more amenable to test or scrutiny than a bedtime story. With little in the way of conclusive evidence, then, the case for favouring this particular scheme of vertebrate origins over another depended more on advocacy than experiment – hence Bateson's unease. Until surprisingly recently, the textbook scheme as outlined above achieved approbation largely through fashion and chance, reinforced by repetition, less than through experimental verification.

In recent years, though, the conventional wisdom has been subject to recent and continuing assault, not only from a renewed interest in comparative anatomy and embryology, but also from work in which phylogenies are reconstructed according to molecular phylogeny, the statistical assessment of the similarities and differences between the fine-scale nucleotide base-pair sequence of certain genes in different creatures.

In particular, comparative analysis of nucleotide base-pair sequences of the gene for the '18S' subunit of the ribosomal RNA (rRNA) molecule has shed new light on old phylogenetic questions. The 18S rRNA molecule is of profound structural importance by virtue of its central role in the process of genetic translation. Because it is so important, creatures harbouring mutant 18S rRNA genes would not survive. Changes in the base-pair sequence of the 18S rRNA gene, then, will be rare, and preserved long in evolution before being 'over-written' by other mutations. This 'long memory' makes comparative study of the 18S rRNA gene useful for untangling very ancient evolutionary events, and illuminating the relationships of major animal groups. Because of this property, molecular phylogeny based on this gene, and others like it, is sometimes claimed to be a surer yardstick for evolutionary change than divergence in anatomy or embryology.

More recently still, attempts have been made to assess relationships not by comparison of nucleotide base-pair sequences in single genes, such as the gene for the 18S rRNA subunit, but of the ordering of entire blocks of genes. Because the order in which genes occur on chromosomes changes much less frequently than the nucleotide base-pair sequence of any individual gene within the block, this approach is capable of resolving phylogenies even when the branching events happened hundreds of millions of years ago, with far higher statistical robustness than conventional single-gene phylogenies. This strategy promises to deliver answers to problems that have hampered and haunted zoologists for a century. Work on gene ordering in the mitochondrial genome has already helped untangle the vexed question of the higher-order phylogeny of arthropods (Boore *et al.*, 1995) and work on the duplication of clustered homeobox genes (more about them in Chapter 3) is beginning to resolve the origin of vertebrates, too (Ruddle *et al.*, 1994).

The latest work on vertebrate relationships is casting considerable doubt on the usual textbook picture. For example: the placement of urochordates as cousins to craniates and cephalochordates is more contentious than it seems at first. Claims based on fossil evidence (Jefferies, 1986) support a close relationship between urochordates and craniates which excludes cephalochordates. Molecular evidence supports a close link between craniates and cephalochordates: however, at the time of writing, molecular work has yet to settle the position of urochordates definitively, although work in a number of laboratories (P. W. H. Holland, H. Saiga, personal communications) suggests that resolution is not too far off. But some recent molecular work questions even this, suggesting that urochordates might be closer to echinoderms than craniates (Wada and Satoh, 1994; Halanych, 1995).

The traditional position of hemichordates as close cousins to chordates, with echinoderms standing further away, is supported by recent phylogenetic analyses based on anatomy (Peterson, 1994, 1995a) and embryology (Schaeffer, 1987). This, however, has been challenged by molecular evidence from ribosomal RNA gene sequences (Wada and Satoh, 1994; Halanych, 1995) which suggests a close relationship between hemichordates and echinoderms, exclusive of chordates. Interestingly, this revives the largely forgotten views of Metchnikoff (1881: cited in Gregory, 1946), who grouped hemichordates and echinoderms in a group called the Ambulacraria on the basis of shared embryonic features. The third of three possible resolutions – that the echinoderms are the closest to chordates, with hemichordates standing more distant, is adopted by Jefferies (1986) on the basis of fossil evidence.

Ribosomal RNA gene-sequence evidence tends to support the idea of the deuterostomes as a properly-constituted group (Wada and Satoh, 1994), although the embryological features used originally to recognize it are nowadays sometimes regarded as questionable (see for example Telford and Holland, 1993, and references therein).

Two areas, in particular, seem to be areas of dispute. The first is the position of urochordates with respect to cephalochordates and craniates. The second is the position of the hemichordates with respect to the echinoderms and chordates. I mention these now, because they recur at intervals throughout this book.

Given these multifarious challenges, then, there seems little reason to give much credence to traditional textbook tales. Less familiar schemes demand close vertebrate kinship with crustaceans, or spiders, or worms of some kind, and that urochordates and cephalochordates are not primitive chordates at all, but degenerate vertebrates. These ideas were taken seriously at the time of their proposal but are now almost forgotten. But if molecular biology is close to settling the issue, why should we keep their memory green?

The reason is that no idea, however odd, should ever fail simply because the reader finds it outrageous. Strangeness and novelty, after all, are attributes to be welcomed by science, for the oddball notion of today might be the orthodoxy of tomorrow. They also reflect more on the inexperience of the beholder than the object in question. Were one to look back far enough, the received wisdom found in textbooks started out on the borderlands of the bizarre. Even the most *outré* idea deserves serious consideration, and if it is found to fail, it will be by the standards of scientific investigation, not by the dictates of fashion or taste. Today, Gaskell's crustacean idea would be dismissed without thought as crackpot. Such dismissal would be wrong: it was once treated with due respect and only thrown out when its theoretical foundations – not its strange conclusions – were revealed as weak.

Second, a study of seemingly left-field ideas suggests new ways of looking at familiar turf, perhaps revealing interesting aspects, there all the time, masked by the comforting cloak of familiarity. Had Gaskell a better press (for all that he was wrong), the extensive similarities of developmental genetics now known to exist between arthropods and vertebrates (Manak and Scott, 1994; De Robertis and Sasai, 1996) might not seem so surprising.

Third, it could be that some of these ideas contain grains of truth that deserve further investigation in their own right. High time, then, for a survey of some of the attempts to search for vertebrate origins, with an examination of one or two of the more important ideas.

This, then, is my own effort. I have written it partly as a plea for the embrace of the unusual, but a qualified rather than a quixotic embrace, and to show that the investigation of vertebrate origins is far from the long-settled affair that textbook accounts are constrained to portray. Textbooks, of course, have space only for a brief nod before moving on, and vertebrate ancestry has the additional problem of falling between two stools. If not found as an appendix at the end of texts on invertebrate zoology, it must needs be *hors d'oeuvre* in a vertebrate zoology text.

As will now be evident, this book is not a textbook, nor is it meant to be used as such, and I have deliberately opted for opinion and for unanswered questions instead of encyclopaedic coverage. The object is to stimulate debate and discussion rather than solve problems, for which there can be no convenient answers to be found at the back of the book.

The latter part of the book concerns the so-called calcichordate theory as proposed by Jefferies and his associates over the past quarter of a century. This idea links vertebrates and the other chordates with the carpoids, a little-known and extinct group usually classified as primitive echinoderms.

Although I have some sympathy with the calcichordate theory, this is not the reason for its extensive coverage here. Rather, that the calcichordate theory has many of the attributes discussed above.

It is a scheme at first sight outlandish, and a refreshing alternative to the usual textbook story. For all that, it is based on a wealth of interpretive evidence and considerable erudition, encompasses an immense range of facts generally thought disparate, and makes testable predictions.

The fact that carpoids look like candidates for *The Alien*, rejected for being too weird, is no reason for discounting the theory that rests on their spiny and asymmetric forms – for it contains much of interest, and suggests new ways of looking at old problems.

It poses pertinent questions about the definition of chordates and echinoderms as groups, touches on important unsolved problems of phylogenetic reconstruction such as the use of fossils and the limits of parsimony, and exposes a number of explanatory gaps in the textbook story that would not be apparent without the significant shift in perspective it offers.

For example: is there a phylogenetic reason for the marked asymmetries in the development of the amphioxus, the best known cephalochordate? If so, is it connected with the asymmetries in echinoderm development? Where did vertebrates get their heads? Does the presence of phosphatic hard tissue preclude the simultaneous presence of calcitic tissue? Why do some cranial nerves penetrate remote regions of the viscera? Is the seriation of branchial slits in cephalochordates, vertebrates and hemichordates necessarily connected with the seriation of nerves or musculo-skeletal elements? Can an animal with a calcitic skeleton also have pharyngeal gill slits? If so, how does this affect the usual definition of echinoderms and chordates? The existence of animals with combinations of features not seen in extant animals raises questions about classifications constructed on the basis of modern character distributions: are such classifications adequate models of reality? If not, how can information from fossils be incorporated? And so on.

My sympathy for the calcichordate theory lies in the wealth and interest of questions it engenders, largely as a function of this shift in perspective – not because I advocate that it is 'right', or 'better' than the traditional textbook view, or any other view.

To begin at the beginning, Chapter 1 introduces Hennigian cladistics as the preferred method of phylogenetic reconstruction. In cladistics, organisms are grouped into hierarchies of relationship called 'cladograms', based on an analysis of features presumed to indicate shared common ancestry. In any analysis beyond the trivial, the number of possible cladograms generated by a given set of organisms and features is often very large, and convention selects as the provisional solution that which keeps the number of evolutionary-character-state changes (or 'steps') in the cladogram to a minimum.

This is the principle of parsimony: the problem is that there is no basis for supposing that evolution obeys it. Parsimony, therefore, does not necessarily present the 'correct' solution, but only the lower bound on what evolution can achieve. But amid an infinitude of potential solutions, parsimony offers a way to make systematics workable.

The origin of vertebrates is best appreciated in the context of their place among the larger grouping of the deuterostomes. To this end, I present a brief survey of deuterostomes (and some of their relatives), both living and extinct. Although echinoderms and chordates are the most visible extant deuterostomes, a cursory inspection turns up a surprising richness, particularly of fossil forms.

In Chapter 2 I look at some of the more important ideas about vertebrate ancestry, including Garstang's auricularia theory and its derivatives, Gaskell's plan to equate us all with remodelled crustaceans, Hubrecht's nemertean scheme, later taken up by Jensen and Willmer, and Gislén's views on the chordate affinities of carpoid echinoderms, work that presaged and inspired Jefferies' calcichordate theory. I have glossed over much, and probably omitted much more, but there should be enough here to set Garstang's ideas, the basis of the standard textbook model, into their proper context.

No theory of vertebrate origins would be complete without some treatment of the origin of the head, that most distinctive of vertebrate structures. The 'head problem' is the subject of Chapter 3. It is also the one problem in vertebrate origins which molecular genetics is singularly well-equipped to tackle, as demonstrated by its recent solution, at least with respect to the amphioxus (Garcia-Fernàndez and Holland, 1994; Gee, 1994).

The whole of Chapter 4 is devoted to Jefferies' calcichordate theory, and examines its persistence over 25 years, largely, it seems, through the inability of critics to find a refutation as sound as the theory itself. Even though firmly grounded in classical tradition, the calcichordate theory is in one sense ahead of its time – for those with the sufficiency to address it must have a good grounding in cladistics and molecular biology as well as classical morphology. Nevertheless, a new generation of morphologists, in the vanguard of renewed interest in morphological problems, has at last begun to test the predictions of the calcichordate theory. This is to be applauded: progress lies in the sober assessment of the merits of a theory (however odd it may seem) rather than standing on the sidelines and throwing rocks at it.

In Chapter 5 I return to one of the problems posed above, a source of one of the dramatic shifts of perspective wherein lies the appeal of the calcichordate theory. Features such as the calcite skeleton of echinoderms and the pharyngeal filtration system of chordates are useful because they are the features by which individuals belonging to these groups are generally recognized. No modern chordate has a calcite skeleton, and pharyngeal filter-feeding is foreign to modern echinoderms.

But what is one to make of a carpoid, in which, according to the calcichordate theory, both features are found in the same animals? Such combinations of characters make a nonsense of conventional classifications based on the extant fauna. Dismissing fossil evidence as uninformative is no help, especially as the diversity of form among deuterostomes is far greater among fossil forms than modern ones: and yet it is the modern fauna that dictates the character distributions on which classification is based. Is this valid? Should the classification of an entire group be based on the relict character distributions of an extant fauna known to be depauperate and (by implication) unrepresentative?

The reason, ultimately (and pragmatically), is that the question of vertebrate origins will be solved by molecular genetics, which by its nature can utilize information from extant groups only. Reliable information on the divergence of major lineages will soon be obtained by studies on highly conserved features, such as the large-scale arrangements of homeobox clusters or mitochondrial genes.

Fossils, however informative of phylogenetic information content, cannot contribute directly to a phylogenetic analysis based on molecular information, and will just have to be slotted in later. But does this necessarily follow? Can information from fossils overturn a phylogeny created on the basis of information, whether anatomical or molecular, from modern forms? This question lies at the heart of the argument about 'pattern cladistics' that exercised the systematics community during the 1980s, ever since the publication of a controversial paper (Rosen *et al.*, 1981) outlining its precepts. As far as I am aware, nobody has asked explicitly how this problem affects the search for the origin of vertebrates. The reason for this is, largely, because most people have addressed the problem using modern forms exclusively. Jefferies' calcichordate theory, on the other hand, is largely based on fossil evidence, with the assumption that fossils *can* and *do* bias phylogenetic reconstruction.

The challenge of these problems should not be grounds for despair. Far from that, it shows that the subject of vertebrate origins is very much alive, and worthy of consideration.

Acknowledgements

Many contributed to the writing of this book in ways both large and small. In particular I should like to thank R. J. Aldridge, R. McN. Alexander, Stuart Baldwin, Sandy Baker, Edoardo Boncinelli, Derek Briggs, Tony Carter, Euan Clarkson, Simon Conway Morris, Barry Cox, Tony Cripps, Paul Daley, Roberto Di Lauro, P. N. Dilly, Bertie Dorrington, Gina Douglas, Dan Fisher, Peter Forey, Bernd Fritzsch, John D. Gage, the late Humphry Greenwood, T. H. J. Gilmour, Brian K. Hall, James Hanken, N. D. Holland, Philippe Janvier, Robb Krumlauf, Thurston Lacalli, John Maisey, Charles Marshall, Chris Paul, Hidetoshi Saiga, Andrew B. Smith, Moya M. Smith, James Sprinkle, Max Telford, David Wake, Pat Willmer and J. Z. Young.

The following companies and institutions were kind enough to allow the use of copyright material: Cambridge University Press, Cell Press, the Linnean Society of London, Longman Education, Macmillan Magazines Ltd, the Trustees of the Natural History Museum, W. W. Norton and Company, Oxford University Press, the Palaeontological Association, Plenum Press, The Royal Society, Saunders College Publishing, the Systematics Association, and the Zoological Society of London.

In addition I thank most warmly R. P. S. Jefferies, P. W. H. Holland, Kevin Peterson, Per Erik Ahlberg and Barbara Cohen for taking the time to read, discuss and comment on all or part of the manuscript at various times: and Audrey Jefferies, the only person who managed to think up a livelier title for a book about the origin of the vertebrates than 'the origin of the vertebrates'.

Thanks are also due to my friends and colleagues at *Nature* in particular Editor Emeritus, Sir John Maddox, librarian Barbara Izdebska and artist Susanna Fox; and my friends at Chapman & Hall in particular Bob Carling and Ward Cooper.

The experience was greatly enriched by Demon Internet Services, Ltd, under whose auspices I made the acquaintance of many 'virtual' colleagues, too numerous to mention here, all of whom provided valuable information.

My wife Penny, ever the soul of patience, tolerated extempore expositions of the anatomy of crinoids and can now do quite a passable impression of *Uintacrinus*.

None of the above (or, indeed, any whose name has been omitted through oversight) are responsible for any errors that may have crept in. Nor should they be held responsible for any of the opinions expressed unless explicitly stated otherwise.

The Cranley is gone but not forgotten.

1

Introduction

> When we return home and our friends gleefully enquire, 'What then has been decided as to the Origin of Vertebrates?', so far we seem to have no reply ready, except that the disputants agreed on one single point, namely, that their opponents were all in the wrong.
>
> Stebbing in Gaskell *et al.* (1910, p. 45)

> With the data currently available, we are at an impasse regarding the origin of craniate autapomorphic tissues, and hence the origin of craniates themselves.
>
> (Peterson, 1994, p. 33)

1.1 OUR OWN CORNER OF CREATION

Vertebrates include many of the animals with which we are familiar. Fish, amphibians, reptiles, birds and mammals are vertebrates. But the vertebrates, as a group, comprise just one group in a larger, more inclusive group, the chordates. This is less familiar, as those chordates that are not vertebrates are not found every day. Invertebrate chordates include the urochordates, or tunicates, colloquially the sea-squirts, marine creatures that generally (but not always) spend their adult lives fixed in one place, as sponges and corals do; and the cephalochordates or lancelets – translucent, vaguely fishlike creatures which although in principle fully mobile throughout life, spend much of the time half-buried in the sand in inshore waters.

But the chordates are, again, just one group in a still larger assemblage of creatures, the deuterostomes (Figure 1.2). Non-chordate deuterostomes include the spiny-skinned echinoderms, a group of exclusively

Age	Stage/Epoch	Period	Era
		Quarternary	Cenozoic
2			
		Tertiary	
65			
		Cretaceous	Mesozoic
144			
		Jurassic	
213			
		Triassic	
248			
		Permian	Palaeozoic
286			
		Carboniferous	
360			
		Devonian	
408			
	Downtonian	Silurian	
	Ludlow		
	Wenlock		
	Llandovery		
438			
	Ashgill	Ordovician	
	Caradoc		
	Llandeilo		
	Llanvirn		
	Arenig		
	Tremadoc		
505			
	Merioneth	Cambrian	
	St David's		
	(other stages)		
approx 590			
	Ediacaran	Vendian	Proterozoic
	Varangian		
approx 670			
		Sturtian	

Figure 1.1 A chart of geological time to illustrate the relative ages of some of the fossil forms mentioned in this book. Ages are in millions of years before the present, and the uncertainty increases with age. Recent isotopic evidence suggests that the base of the Cambrian is perhaps no older than 540 million years (Bowring *et al.*, 1993). Stage names are given only for Periods in the lower Palaeozoic. (After various sources).

marine creatures that includes starfish, sea-urchins and sea-cucumbers as well as a host of other less well-known and extinct forms. But other, less familiar deuterostomes include the hemichordates and (less certainly) some or all of the so-called lophophorate phyla, the brachiopods and others, distinguished by the possession of a special feeding organ called the lophophore.

Looking still further afield, the deuterostomes form just one branch on the great tree of multicellular animals, the metazoa. Most non-deuterostome

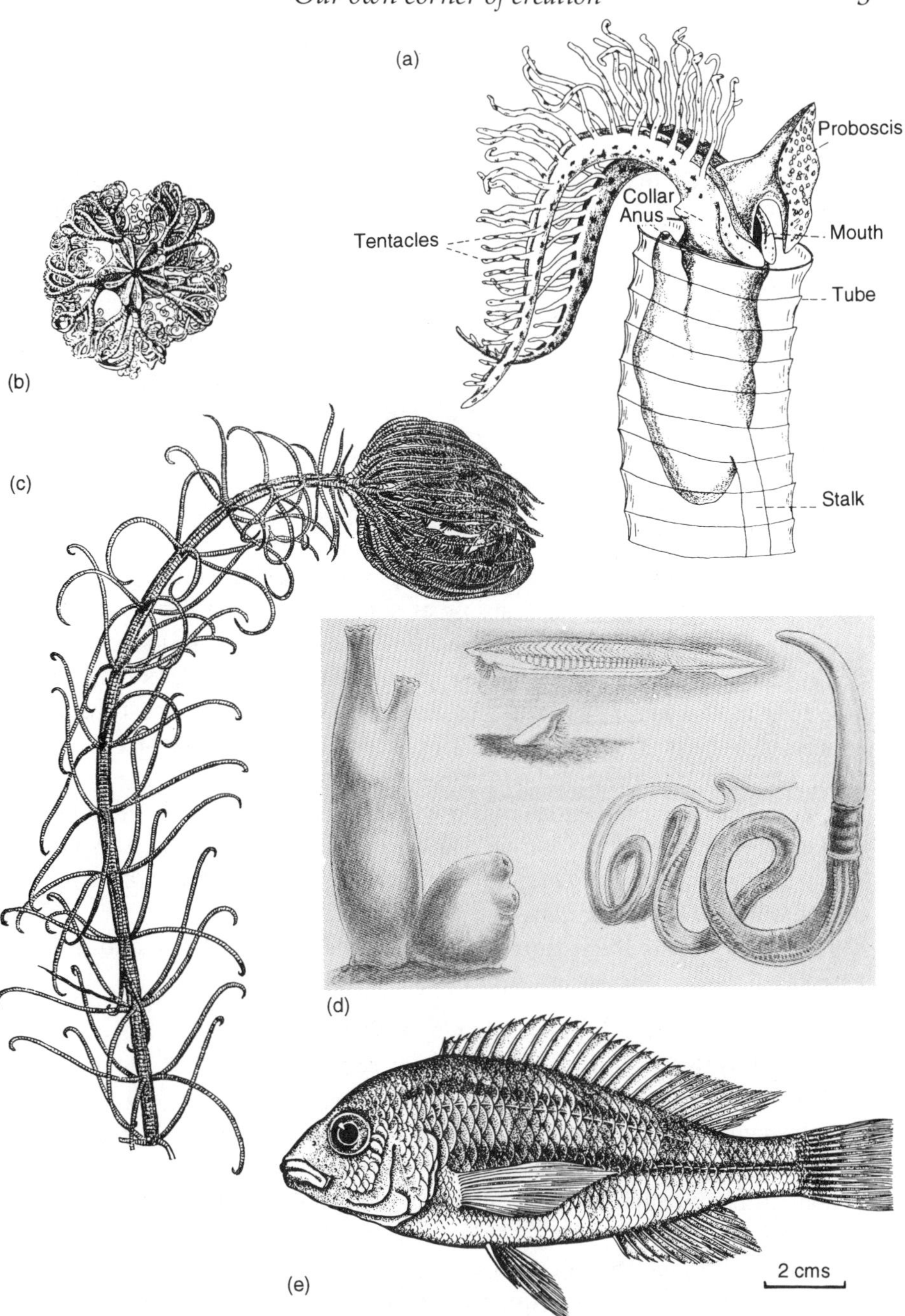

Figure 1.2 Deuterostome diversity. (a) A pterobranch hemichordate, *Rhabdopleura* (from Barnes, 1980, after Delage and Hérouard); (b) an eleutherozoan echinoderm, *Astrophyton* (from Hertwig, 1909, after Ludwig); (c) a pelmatozoan echinoderm, *Isocrinus* (from Bather, 1900, after Agassiz and Carpenter); (d) two ascidian tunicates, a cephalochordate and an enteropneust hemichordate (from Wells *et al.*, 1931); (e) a vertebrate chordate, *Platytaenoides* (from Greenwood, 1974, with permission of the Natural History Museum, London).

metazoa, including flatworms, molluscs, annelids, insects and all other arthropods and a host of other groups, form a large group called the protostomes.

But how are all these groups defined? Put another way, what are the attributes that define an animal as a member of one group rather than another? This is a big question because, at present, our understanding of metazoan interrelationships is somewhat fraught[1].

Traditionally, zoologists have used embryology to define different animal groups. In other words, they have supposed that such fundamental features of organization of a major animal group – its 'body plan' – will be defined at its earliest stages. Features that permit finer degrees of discrimination (such as between two different species) are assumed to be later additions in development[2].

1.2 MULTICELLS MAKE METAZOA

The metazoan condition is recognized by a number of features. That is, all metazoans possess these features at some part in the life cycle, and animals without them cannot be considered as metazoan. But more than being just criteria for similarity, the presence of a particular feature in any two metazoans is assumed to reflect shared inheritance through common descent. The latest common ancestor of all metazoans is presumed to have had these same features.

A group such as the metazoans, in which qualification for membership depends on shared common ancestry, is called monophyletic. Importantly, a monophyletic group includes not only the common ancestor but *all* its descendants.

The following are the features that are shared by all Metazoa and characterize the monophyletic group to which they belong.

- Metazoans are multicellular. That is, they are made of very many cells acting together in concert, rather than being constructed of a single cell.
- The sex cells are grown and fostered in special organs, and the sperm cells have whip-like tails or flagella.
- Each cell in the body of a multicellular animal is a specialist in a particular function, such as reproduction, feeding or transmission of nerve impulses.
- After the single-celled egg is fertilized by a sperm, it divides into two. These two cells divide again, repeatedly, until what is now the young embryo becomes a hollow ball of cells, called a blastula. The space within the ball is called the blastocoel, the presumptive body cavity.
- More controversially, all metazoans contain a distinctive cluster of so-called 'homeobox' genes which determine the spatial identity of body parts along the anterior–posterior axis (Slack *et al.*, 1993). More about homeobox genes in Chapter 3.

1.3 PROTOSTOMES AND DEUTEROSTOMES

Many of the features that define the Metazoa concern fertilization and very early development. To an extent, features of embryology are also used to define the subdivisions of the Metazoa. But first, a health warning. Many of the features used to distinguish protostomes and deuterostomes outlined in what follows are subject to considerable debate. I use them here to emphasize the systematic differences between the two groups. However, irrespective of the difference of opinion, the latest research picks out deuterostomes as a natural group, distinct from protostomes. For the latest thinking on the role of embryology in metazoan systematics, see Davidson *et al.*, 1995.

The blastula is a product of the repeated division of the single, fertilized egg cell. But the way this cell divides, in terms of the relative orientation of successive divisions, differs between animals. In protostomes, the first cells divide by a process termed spiral cleavage, and in some cases the exact fate of each embryonic cell can be traced all the way to adulthood: that is, cell fate is determinate[3].

In deuterostomes, on the other hand, the first cells divide by a process called radial cleavage, and the fates of individual cells are indeterminate, in that their ancestry cannot always be traced, exactly, to a progenitor cell in the blastula.

The blastula is one cell thick, and at some point in its life it crumples like a squashed football, usually by folding in on itself, to make a vase-shaped gastrula with two layers of cells, one on the inside (the endoderm), the other on the outside (the ectoderm). As a result of the infolding process, which occludes or abolishes the blastocoel, there is a hole or blastopore at one end of the embryo, connecting the space inside (the archenteron) with the outside world.

Some animals look very like this even as adults: superficially at least, jellyfish, sea-anemones and coral polyps (the coelenterates or cnidaria) are made of two layers of cells, a condition termed diploblastic. If there is a layer in between, it is filled with a jelly-like substance called the mesogloea. The archenteron becomes the gut, lined by the inner layer of cells, and the blastopore forms the only way to get into and out of the creature: there is no specific anus, as the blastopore continues to serve as mouth and anus combined. Creatures like this, then, are thought to be very primitive, representing an early stage in the history of the Metazoa[4].

In other animals, the triploblasts, a third layer of cells forms between the other two, partially or completely occluding the blastocoel. This third layer of cells is called the mesoderm. Generally speaking, the endoderm in metazoans forms the gut and associated structures such as the liver and other organs, the ectoderm forms the skin and nerves, and the mesoderm in triploblasts forms the circulatory system, the muscles and other organs.

The simplest triploblasts are the flatworms or platyhelminthes: a large and diverse group that includes free-living forms such as the planaria of ponds and streams, as well as parasitic forms such as flukes and tapeworms. Flatworms have mesoderm, but like cnidarians, they have no anus. They do, however, possess true bilateral symmetry. Very primitive animals indeed, the sponges have no symmetry at all, and cnidarians have radial symmetry: one can rotate a jellyfish about its centre, and it would look similar from a discrete number of angles, depending on the presence of tentacles and other structures. A featureless circle looks the same from any angle, no matter how much you rotate it (Figure 1.3). Flatworms, and all other metazoans have, at least primitively, a distinct front and rear end, and sides that are mirror-images of each other. The plane of symmetry runs from front to rear, and at right angles to the ground[5].

In bilaterally symmetrical animals, the determination of the front end of an animal (as opposed to the rear end) is set down early in embryogeny and differs from group to group. In protostomes, the mouth develops from, or near, the blastopore. But in deuterostomes, it is the anus that develops from, or near the blastopore, and the mouth forms secondarily some distance away. Hence the name for the group: 'deuterostome' means 'second mouth'.

Although protostomes and deuterostomes are all triploblasts, mesoderm originates and develops in a different way in each case. In protostomes, it generally forms when cells from the gut wall are shed into the blastocoel. In deuterostomes, it generally forms from the outpocketing of parts of the gut wall into the blastocoel. Apart from flatworms and a few other forms in which the mesoderm is more or less solid, the mesoderm becomes complicated by the appearance of spaces within it. These spaces are called coeloms, and animals (whether deuterostomes or protostomes)

(a)

that have coeloms are called coelomates. Flatworms have no coeloms and are called acoelomates[6].

Coelomic spaces are lined with sheets of mesoderm. Generally speaking, internal organs derived from any embryonic tissue abut either the inside of the body wall or the outside of the gut, and are excluded from direct contact with the coelom by a sheet of mesoderm called the peritoneum. Some animals have body cavities that are not, in anatomical terms, true coeloms, in that they are not lined with peritoneum, and the internal organs contact coelomic spaces directly. Such cavities are not coelomic, but are remnants of the blastocoel: animals with this style of bodily cavitation are called pseudocoelomates. Nematodes and rotifers are conventionally regarded as pseudocoelomates.

(b)

Figure 1.3 Patterns of symmetry. (a) (left) The baby brontosaurus is bilaterally symmetrical, in that a single, vertical plane of symmetry running from front to back, divides her into two halves, each a mirror image of the other. She has no other planes of symmetry. The toy football is a sphere, with rotational symmetry about an infinite number of axes of rotation (or bilateral symmetry about any number of planes). Closer examination reveals a complex radial symmetry based on the numbers five and six. (b) Although it appears to be bilaterally symmetrical with a plane of symmetry running up and down through the body and neck, this electric guitar (with penguin) is completely irregular, with no symmetry at all. It is reminiscent of the carpoid echinoderm *Cothurnocystis.* (Photographs by the author.)

In protostomes, the coelom generally forms from splitting of mesoderm into inner and outer portions, leaving a space between (a process called schizocoely). In deuterostomes, the coelom generally forms from outpocketings of the developing gut (enterocoely) as the mesoderm forms.

To sum up, there are a number of features that seem to define deuterostomes and protostomes (bearing in mind the health warning as described on p.5):

- Protostomes develop by spiral, determinate cleavage. The coelom, if present, develops by schizocoely. The mouth develops at or near the site of the blastopore.
- Deuterostomes develop by radial, indeterminate cleavage. The coelom, if present, develops by enterocoely. The anus develops at or near the site of the blastopore, and the mouth is a secondary perforation.

1.4 PROBLEMS WITH PROTOSTOMES

Now, the clear-cut scheme presented above is far neater in theory than in practice, and depends for its coherence on the assumption that several key features can be used as characteristics of monophyletic groups. In other words, that these features have appeared only once and can be used to define groups of shared common ancestry, and did not appear independently in several groups. These features are:

- the fate of the blastopore (protostomy or deuterostomy);
- the style of cleavage in the early embryo (spiral or radial);
- the fate of individual cells in the early embryo (determinate or indeterminate);
- similarities between various forms of larva, such as the distinctive trochophore of annelids and molluscs;
- the distinction between two and three layers of cells;
- the state of the body cavity (coelom or pseudocoelom);
- the manner of formation of the coelom (schizocoely or enterocoely).

It now seems clear that few of these assumptions are justified (see Willmer, 1990; Willmer and Holland, 1991; Telford and Holland, 1993). For example, it is often hard to tell in practice whether an animal is a deuterostome or a protostome, and close examination reveals a number of inconsistencies which blur this neat distinction. Several so-called protostome taxa do not form the mouth from the blastopore, and indeed protostomes display the whole range of possibilities from forming, from the blastopore, just the mouth (some molluscs); the mouth and anus (some annelids, nematodes, the obscure interstitial animals called gastrotrichs); or just the anus, as in *bona fide* deuterostomes (worms called nematomorphs, other annelids, other molluscs). The blastopore in the shellfish-like creatures called brachiopods – regarded either as protostomes, deuterostomes, or protostomes closely allied to deuterostomes – may form the mouth and anus, or just the anus (M. Telford, personal communication). More on brachiopods later.

Second, spiral cleavage may owe less to heritage than to a common solution to a problem of functional design. In spiral cleavage, the embryo is formed of tiers of cells such that the cells in each are offset by a small amount from those above and below. In the same way that offsets between bricks in adjacent layers in a wall make the wall much stronger than if the bricks were laid directly one atop another, spiral cleavage could reflect the fact that embryos are built for strength, rather than that animals in which embryos have spiral cleavage are related on that account alone.

Nevertheless, style of cleavage as a character may be saved by determination of cells. In all animals in which determinate cleavage can be demonstrated, the mesoderm arises from one particular cell in the blastula (known as '4d') for which the correspondents can be found in any embryo that divides in this way (see Barnes, 1980; Willmer, 1990 for discussion and illustrations, and Schaeffer, 1987 for a more recent review and citations). As far as is known, animals with spiral cleavage have deterministic cell fate, which suggests that the two features are linked, and that spiral cleavage can be used as a character (if only inasmuch as it correlates with cell fate).

Third, the same functional argument can be used to question whether resemblances between larvae, such as between those of molluscs and annelids, really reveal anything about ancestry: larvae are as subject to natural selection as are adults, so their forms may reveal no more about past history than present necessity.

Fourth, the distinction between two and three layers of cells is also harder to draw on a second look. The mesogloea of the nominally diploblastic cnidarians and their relatives, the comb-jellies or ctenophores, may be filled with cells to a greater or lesser degree so that the distinction between mesogloea and mesoderm is moot.

Fifth, the formation of the coelom is more complicated than simple schizocoely or enterocoely. It may be a mixture of both, and vary greatly even between supposedly closely related animals, depending on the circumstances. This lessens the utility of the style of coelom formation as a useful feature for resolving ancestry.

These problems cast some doubt about whether the traditional embryological definition of a deuterostome (an animal with deuterostomy, radial and indeterminate cleavage, and an enterocoelous coelom) really does define a monophyletic group. Nevertheless, there is some evidence from adult anatomy as well as molecular phylogeny (Wada and Satoh, 1994) to support the monophyly of a group containing those animals traditionally regarded as deuterostomes, even if the embryological features originally used to define the group are not consistent in every case.

Matters are not so simple for the protostomes, which include creatures both with and without coeloms, and with a wide variety of adult and larval structures. The protostomes do not seem to form a natural, monophyletic group. In fact, they can only be defined negatively, on the basis of what the group does *not* include: the protostomes seem to comprise all triploblastic, bilaterally symmetrical metazoans that are *not* deuterostomes.

Now, we have reason to think that the Metazoa are a monophyletic group. It is likely that metazoans with bilateral symmetry form a smaller, more select monophyletic group, nested within the Metazoa. Given that both protostomes and deuterostomes are all bilateral, triploblastic metazoans, one must assume that they share a common ancestry (Figure 1.4).

What was this ancestor like? It is fair to assume that it resembled a flatworm, a basic, no-frills bilaterally symmetrical triploblast without any extras such as a coelom, segmentation, anus and so on. However, acoelomate flatworms tend to have spiral cleavage, and the mesoderm of sev-

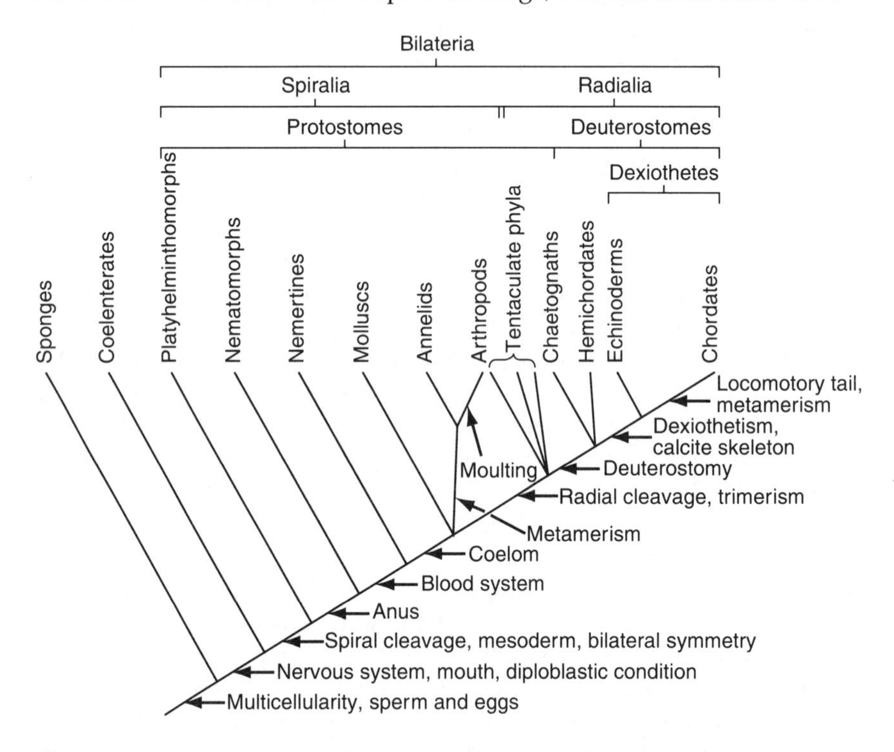

Figure 1.4 A tentative cladogram of the Metazoa, illustrating how a cladogram may be composed of nested sets of monophyletic groups, or clades. Note the several unresolved polychotomies (from Jefferies, 1986, with permission of the Natural History Museum, London).

eral kinds of worm develops from cell 4d. And given that they have no anus, they can almost be considered protostomes by definition.

If so, then the protostomes retained these ancestral, primitive features, whereas deuterostomes, which must have evolved from protostome stock, replaced them with new ones that appeared only once in evolution, circumscribing deuterostome monophyly.

So, although the deuterostomes form a monophyletic group, the protostomes do not.

Because deuterostomes evolved from protostomes, the protostomes as a group do not include *all* the descendants of the common ancestor, thus violating one of the criteria for monophyly. To put it another way, the protostomes can only be grouped together on the basis of shared, *primitive* characters, found in the common ancestor of protostomes *and* deuterostomes (spiral, determinate cleavage, protostomy), whereas the deuterostomes can be united on the basis of shared, derived features to exclude protostomes. In conclusion, the protostomes do not form a natural group: they are simply those coelomate triploblasts left over once deuterostomes are excluded. Such remnant groups are termed paraphyletic[7].

An important corollary of the difference between monophyly and paraphyly is as follows: because of their shared common ancestry, any member of a monophyletic group is as closely related to a creature *outside* the group (that is, in a more inclusive group) as is any other member of that same monophyletic group. But this does not work from the outside, looking in. Individual members of a paraphyletic group can be more or less closely related to members of the exclusive monophyletic group.

Therefore, one can imagine a 'stem' lineage, leading to an apical monophyletic 'crown' group, within which members of the paraphyletic stem groups are more or less closely related to members of the crown group. The degree of relationship is determined by the order of acquisition of key characters along the stem lineage (Figure 1.5)[8]. The aim of modern phylogenetic reconstruction is to express relatedness in terms of shared common ancestry by resolving relationships into nested sets of monophyletic groups.

1.5 WILLI HENNIG'S SYSTEMATICS

These terms – monophyletic, paraphyletic, stem group, crown group – are occasionally found in traditional discussions of systematic zoology[9]. However, they came to prominence as part of an objective method of systematic reconstruction developed in the 1960s by German entomologist Willi Hennig.

The traditional approach to reconstructing phylogeny is to pull together all the evidence (embryological, anatomical and palaeontological) and

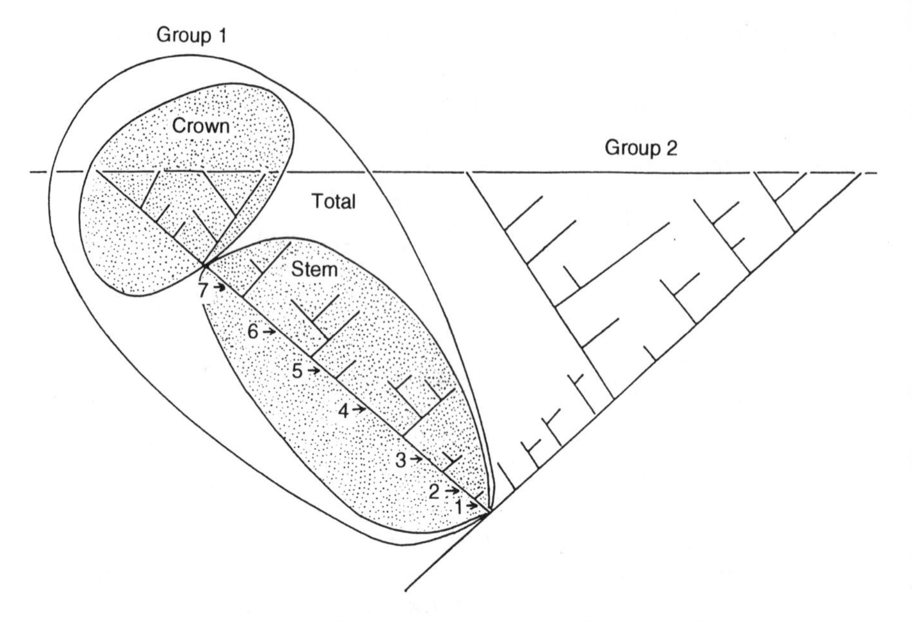

Figure 1.5 A cladogram to illustrate the concepts of stem and crown groups, paraphyly and monophyly. Group 1 – the 'total' group – is monophyletic because the lineages of all its members can be traced to a single common ancestor. The shaded 'stem' group within this group is paraphyletic, because it does not include forms (in the 'crown' group) that share its common ancestor. Note the stem lineage running through the stem group, with numbers denoting points of acquisition of evolutionary novelties (from Jefferies, 1986, with permission of the Natural History Museum, London).

weave it into a consistent account or 'scenario' of evolutionary history (Figure 1.6).

As an aim, this is admirable: but because these phylogenies mix palaeontological data with information of extant animals, there is a sense in which a preconceived idea of the course of evolution is 'built in' to the phylogeny. Therefore, such phylogenies are inappropriate models for the investigation of questions about evolutionary pattern, because they are already biased towards one particular outcome.

Second, again because traditional systematics mixes information from living and extinct animals, it is easy to make unwarranted assumptions about ancestor–descendant relationships, and to define groups on the basis of primitive characters. As we have seen, such groups tend to be paraphyletic and do not reflect 'natural' groups.

Another problem with the traditional approach is that it is hard, if not impossible, to judge the relative merits of alternative schemes based on the same data without a formal, independent method of testing. Often, choice boils down to subjective judgements about plausibility that cannot be supported objectively.

(a)

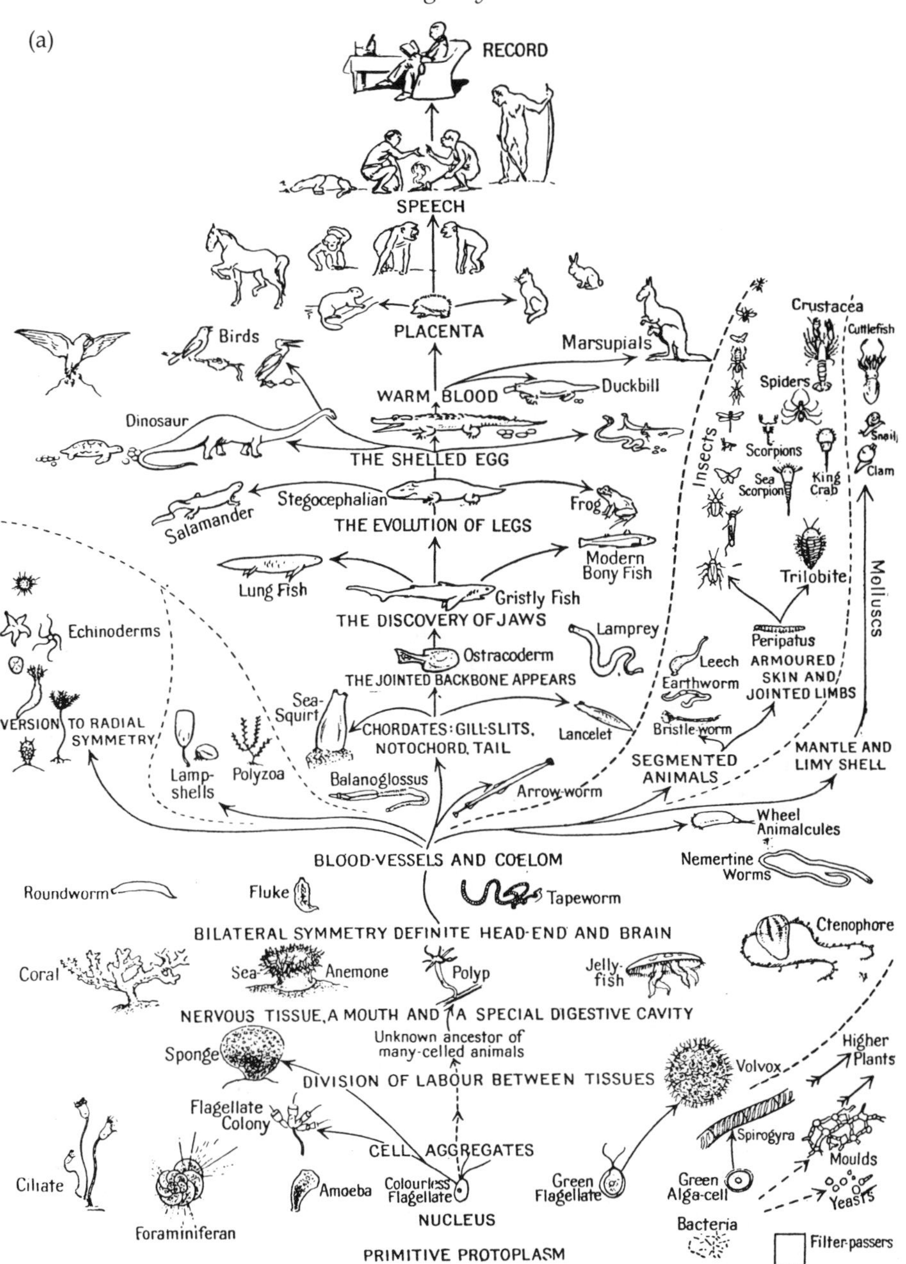

Figure 1.6 Cladograms and story-telling: in the traditional, didactic story-telling phylogeny (a), ancestry and descent are clearly determined (from Wells *et al.*, 1931). The cladogram (b) is free from such prior assumptions, and all groups are shown as terminal twigs (from Ahlberg and Milner, 1994). Although both are networks described by the acquisition of novelties, only the second is free from prior assumptions about ancestry and descent.

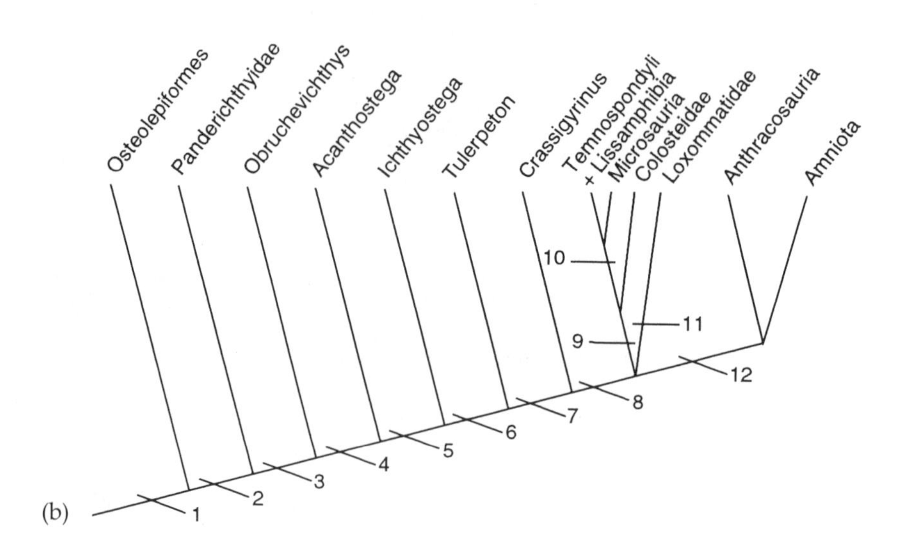

Hennig did two things to address these difficulties. First, he made the distinction between shared primitive and shared derived characters. From this concept the ideas of paraphyletic and monophyletic groups arise quite naturally. Groups defined on the basis of shared, primitive characters can be paraphyletic, whereas those associated on the basis of shared, derived characters are monophyletic.

Second, he did away with ancestor–descendant relationships, and developed a kind of diagram of relationship based on nesting sets of the kind discussed above. In contrast with traditional family trees, taxa in Hennigian diagrams are all shown at the ends of branches – on the same time plane, as it were. With this simple device, Hennig transcended time, and thus evolutionary preconception.

In these branching diagrams, or cladograms, species are grouped not according to prejudged or subjective ideas about ancestry and descent, but by the objective analysis of derived characters that they share. In this way, species are arranged into nesting sets, joined by lines representing the acquisition of derived characters.

Rather than 'descending' from 'ancestors', species in cladograms are 'related' through common-ancestral 'nodes' that do not represent 'real' ancestral species, but sets of characters that one might expect to be present in a common ancestor, were one ever found. Cladograms are not phylogenies as such, but graphic summaries of data from which phylogenies may be reconstructed later.

The immediate problem with cladograms is deciding which characters are primitive and which derived, with respect to the set of organisms in which one is interested. Establishing such polarity is important, because monophyletic groups are defined solely on the basis of derived characters, and not on primitive ones (an important distinction between cladistics and traditional methods of phylogenetic reconstruction).

The problem is solved by finding a species known by independent criteria to lie outside the assemblage of organisms under test. Because such a creature is held to lie outside a monophyletic group comprising *all* the other organisms under test (irrespective of their internal interrelationships), it may be supposed that it is primitive with respect to all the essential features in the test assemblage. Thus, the monophyly of deuterostomes is tested with reference to a creature that we know is *not* a deuterostome, such as a flatworm or cnidarian, which we can assume to be primitive with respect to deuterostomes in all the features in which we are interested. This 'reference' species is called an 'outgroup'. The process is called outgroup comparison, and is a vital part of reconstructing phylogenies on Hennigian principles.

It is important to note that in cladistics, 'primitiveness' is relative, and does not imply simplicity of organization or crudeness of adaptation. Because cladistics is strictly independent of time, it is easier to view organisms as adapted to the environments in which they find themselves – not as, as so often seems to creep into such discussions, orthogenetic staging posts between one type and another. In cladistics, there is no such thing as a 'missing link'. Or if there is, one can never presume to have found it.

1.6 THE PRINCIPLE OF PARSIMONY

A problem shared by both cladograms and more traditional family trees is convergence of adaptation. Does the sharing of a character by a pair of species reflect convergent evolution or real common ancestry? This problem can be broken down into a number of distinct situations.

First, a character shared by two species may reflect a shared heritage, in the sense that both species diverged from a common ancestral (or nodal) form in which that character was also present. But it is also possible that the two species each acquired this character independently: in which case the shared presence of the character says nothing about relationship.

Second, a character in one species may not be present in another. This could mean either that the character was gained by the first species, in which it is a derived feature. Alternatively, the *loss* of the feature by the second species could be the derived condition: in which case, the presence of this character is a primitive feature. Characters may have been gained and lost more than once in each or any lineage.

From this it is evident that it is impossible to retrieve the 'true' phylogeny straight off, for even with an outgroup to set the initial polarities of the characters, an unknown number of character acquisitions or reversals could have happened along the way.

Nevertheless, it is always possible to calculate the smallest possible number of character-state changes in a cladogram, as a kind of lower bound, and propose this most economical cladogram as representative of the true phylogeny. This rule-of-thumb, an application of Occam's Razor, is called the principle of parsimony.

Convergence besets phylogenetic reconstruction of any kind, cladistic or not, but it is easier to deal with in a Hennigian context because each character is essentially broken down to its fundamental, irreducible digital state. As nodes in cladograms summarize character states in potential (but not necessarily actual) ancestors, systematists are free to juggle them about until the most parsimonious solution is reached. Only at that point need one make the transition between cladogram and proposed evolutionary tree.

Ideally, each node should connect no more than two species distally, in which case the cladogram is described as fully resolved. This does not often happen in practice (Figure 1.7), but the reason for this failure lies often in insufficiency of data rather than inadequacy of method (for example, see Philippe *et al.*, 1994).

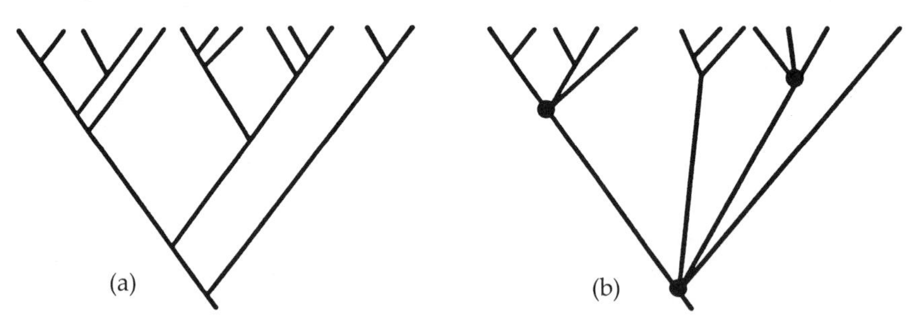

Figure 1.7 A diagram showing two cladograms (a) fully resolved, with dichotomous branching throughout, and (b) with unresolved tri- and polychotomies.

1.7 CHORDATE CHARACTERISTICS

Bearing all this in mind, we can be fairly confident that the chordates form a monophyletic group nested within the larger monophyletic group of the deuterostomes. But what are the derived characters that all chordates share, and which can be used to distinguish them from other animals, even other deuterostomes?

- All chordates have, at some point in their life cycle, a stiffening rod of mesodermally derived tissue that runs along the body axis. This rod is called the notochord.
- Connected with this, the mesoderm that forms the muscles of the

body (the somitic mesoderm) is divided into segments like the slices of a sausage. These segments are called somites. (However, the presence of somites in urochordates is debatable).

- The main nerve trunk in chordates runs along the dorsal side of the body, and is hollow. It develops as a tubular infolding of the ectoderm of the bilateral embryo. In other animals, the main nerve trunk generally runs along the ventral side of the body and is solid. (Again, the position of enteropneusts with respect to the nerve cord is under review).
- The body continues backwards dorsally, past the anus, into a tail that consists almost entirely of muscle, nerves, notochord and other structures connected with movement, but no viscera. In other animals, the body terminates with the anus or, if there are postanal parts to the body, they bear a variety of structures not necessarily connected with movement.
- The throat region or 'pharynx' of a chordate is, at some part of its life cycle, perforated by a number of paired gill-slits. These slits are lined with sheets (epithelia) of cells bearing hair-like protrusions called cilia. In adults or free-living larvae in which this arrangement persists, currents of water generated by organized ciliary beating drive water through the pharynx, trapping food particles in sheets and strings of mucus draped across the gills. Gill-slits are also found in deuterostomes called enteropneusts and pterobranchs, collectively the hemichordates, the status of which is unclear. Some researchers regard them as chordates (Young, 1981), in which case pharyngeal gills can be used as a definitive feature, but others (Barnes, 1980; Jefferies, 1986) do not, in which case they cannot.
- Running along the floor of the pharynx is a groove filled with ciliated and glandular strips. This structure, called the endostyle, secretes mucus and is rich in iodine-containing compounds. Neither enteropneusts nor pterobranchs have an endostyle.

These, then, are the main characteristics of chordates. Craniates and lancelets bear all these features as adults. Tunicates, though, have lost most of them by the time they reach sedentary adulthood. Although they have serially repeated pharyngeal slits, adult tunicates show no trace of somitic segmentation, and it is moot whether the groups of muscle cells in the microscopic tadpole larvae (Lankester, 1882; Crowther and Whittaker, 1994) can be regarded as true somites.

Most hemichordates have pharyngeal slits but lack the other components that make up the distinctive chordate pharynx, such as the endostyle. Neither do they have a notochord or dorsal tubular nerve chord at any stage of development (but see Peterson, 1994).

1.8 WHAT MAKES A CRANIATE?

In this section, if in no other, the distinction between vertebrates and craniates becomes important.

The craniates almost certainly form a monophyletic group within the chordates, defined by the following features (Figure 1.8):

- Craniates have a true 'head'. Much of the head develops from a special tissue called neural crest, which as its name suggests develops at the edges of the developing dorsal nerve tube along its length. Associated with the neural crest at the head end of the embryonic craniate are knots of ectodermal tissue called epidermal placodes. These placodes mark the sites of the special sense organs: the ears, nose and eyes.
- The neural crest starts out as a pair of longitudinal strips of ectoderm on either side of the infolding dorsal nerve tube. These strips are folded inwards as the tube closes, whereupon the cells that comprise it migrate to other parts of the body. Together, the neural crest and epidermal placodes form much of the head, including the eyes, nose and ears, the cartilaginous capsules that protect them and the nerves that connect them to the brain, as well as parts of the pharyngeal gill arches. Neural crest cells also form the sympathetic nervous system, the spinal ganglia (absent, for example, in non-craniate chordates such as lancelets) and parts of the adrenal gland. Neural crest-derived and epidermal placode tissues have been regarded as the embodiment of most of the things that are characteristically craniate (Gans and Northcutt, 1983).
- The pharyngeal arches are modified for ventilation rather than filter-feeding[10]. Unlike other chordates, craniate pharyngeal clefts lack ciliated epithelia (that is, they have been secondarily lost).
- There is a well-developed and regionated brain, housed in its own cartilaginous box, the cranium. The braincase is essentially derived from mesoderm, but much of the outside of the skull, derived as it is from neural crest, is ectodermal in origin, or derives from the interactions between ectoderm and mesoderm (see Hanken and Thorogood, 1993, for a recent review).
- In craniates, the notochord is unrestricted. In vertebrates, the notochord is bounded by, ringed with or replaced by blocks of cartilagi-

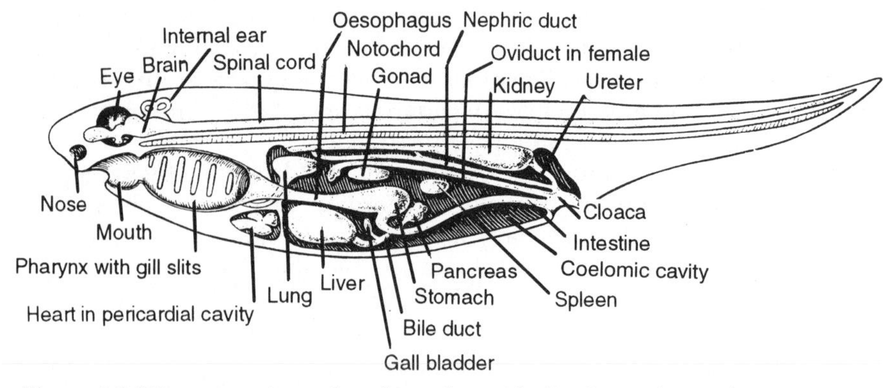

Figure 1.8 Diagrammatic section through an idealized vertebrate (from Romer, 1970).

nous tissue, the vertebrae. This distinction explains why hagfishes (which have an unrestricted notochord) are invertebrate craniates.

- The ventral part of the mesoderm in the pharyngeal region is muscular, forming much of the throat musculature as well as that of the heart (Gans and Northcutt, 1983).
- The nerves are sheathed by a fatty substance called myelin that insulates them thereby improving their efficiency.
- A downward growth of the brain called the infundibulum meets an upward outpocketing of the pharynx. The fruit of this union is the pituitary body.
- The endostyle is transformed from an open ditch-like structure into a closed gland called the thyroid which, like the endostyles of other chordates, sequesters iodine. The growth hormone thyroxine, the principal export of the thyroid, is rich in iodine.
- At least primitively, craniates contain no hard tissues of any kind, whether mineralized cartilage, bone, dentine or enamel. The presence of hard tissues is thought to be a feature of vertebrates (Halstead, 1987). Lampreys, vertebrates that contain no hard tissues, are believed to have descended from forms clothed in bony armour, whereas the ancestors of the hagfish were always naked (see Janvier, 1981; Forey and Janvier, 1993).
- Craniates have longitudinal tail fins. Vertebrates often have, in addition, lateral paired fins derived from lateral folds of ectodermal and mesodermal tissues.
- The blood system is highly developed, and the blood contains haemoglobin for the transport of oxygen. Vertebrates also have an 'immune system' whereby cells in the blood are able to distinguish between the tissue of the body and interlopers such as micro-organisms.
- In vertebrates, the body fluids are generally less concentrated in dissolved salts than is seawater. This is thought to be an advance on the primitive craniate condition, exemplified by the hagfish, in which the body fluids are isotonic with seawater. Maintaining osmotic balance is the job of the kidneys, likewise distinctively craniate structures, which in hagfish serve an excretory function.

1.9 DEUTEROSTOMES: *DRAMATIS PERSONAE*

Now that I have set out the characters that define vertebrates and chordates as monophyletic groups, I have something for which to aim. The invertebrate relatives of the chordates can now be sought as groups, each one progressively further away from the monophyletic group of the chordates, each helping to define a more inclusive monophyletic group.

As I discussed above, the chordates seem to be allied with a number of other groups by virtue of a set of features that define a monophyletic group, the deuterostomes.

To recapitulate, the cells in deuterostome embryos tend to cleave radially and (in most cases) have indeterminate fates. The anus, if present, develops from the blastopore or near it, and the mouth generally forms as a perforation elsewhere in the gastrula. The coelom typically develops by enterocoely, from pockets in the mesoderm as the mesoderm balloons into the blastocoel from the primitive gut.

In addition, the coelom in deuterostomes tends to be divided into three separate compartments, often paired, comprising (from front to back) left and right protocoels, left and right mesocoels, and left and right metacoels (Figure 1.9).

In this section I shall introduce the living and extinct animals that are conventionally regarded as deuterostomes, and which are therefore the most likely candidates for the closest relatives of chordates and vertebrates. I shall also take a brief look at chordates and vertebrates to see how they fit into the general plan. Brief, because detailed descriptions of

Figure 1.9 The three-paired coeloms of deuterostomes, as represented by this culinary still-life. The two miniature pots of jam represent the right (peach) and left (apricot) protocoels. The tins of Indian spices are right (haldi) and left mesocoels, and the large tins of ratatouille at the back are the right and left metacoels. (From the author's kitchen.)

the anatomies and lifestyles of all the animals are readily available elsewhere: the textbooks already mentioned provide excellent places to start. Apart from these, Jefferies (1986) fills in a few gaps and refers to many sources that may not otherwise be readily available.

1.9.1 Echinoderms: five way stretch

Apart from the chordates, the echinoderms comprise the largest and most successful group of deuterostomes, with about 6000 known species today[11]. The anatomy of the adult is characterized by a radial symmetry based on the number five. This is most evident from the familiar starfish, which has five 'arms', and closer examination typically reveals a fivefold multiplication of the internal organs as well.

Echinoderms generally have an internal skeleton of calcite plates. The shapes and sizes of these plates are quite variable: in sea-urchins or echinoids they meet to form a complete test, but in most sea-cucumbers or holothurians they have all but disappeared, reduced to small needle-like spicules. But whatever its final shape, each plate is made from a single crystal of calcite.

Another distinctive feature of echinoderms is the water–vascular system, derived from the left mesocoel. A doughnut-shaped 'water ring' around the central mouth is connected by a tube called the stone canal (on account of its impregnation with calcite in some species) to the surface of the animal through an opening called the hydropore, which perforates a plate-like structure, the madreporite. From the water-ring, tubes extend into (typically five) radial fields or ambulacra, which extend along the arms or, if the animal has no arms as such, over the surface of the body.

As they extend, the tubes throw off side-branches each of which terminates in a tube-foot or podium, which resembles an inflated and animated condom. These coelomic outpocketings extend to the outside (although they are not in themselves perforated), and in many cases are strengthened with muscles. The ambulacra are (translated literally from the Latin) 'avenues' of podia that appear to perforate the surface. Sometimes the ambulacrum has a superficial radial groove running between the rows of podia.

Passage of water into and out of the podia from the rest of the system is assisted by associated sac-like muscular outgrowths from the radial tubes called ampullae. The arrays of flexible podia constitute a single, integrated hydraulic system that can be used for defence, locomotion, feeding or combinations of all three.

Echinoderm skin consists of a thin cuticle, secreted by a thin layer of epidermis, which may be ciliated. Gland cells coat the skin with a layer of mucus. Beneath the outermost layer of skin is a fine mesh of nerves, which constitutes the main nervous system of the animal. The calcite plates grow in the underlying muscular dermis.

The skeleton of echinoderms, no matter how superficial it seems, is mesodermal in origin and thus in developmental terms very much an internal skeleton rather than an external suit of armour. The skin may bear spines that may or may not be connected to or articulated with the skeleton beneath, and the body wall as a whole may be perforated with coelomic outgrowths called papulae. These are used for respiration and for voiding coelomic cells called coelomocytes, which perform many of the functions of blood cells in vertebrates. The skin of some starfish and sea urchins may also bear scissor-like organs called pedicellariae, each a pair of shaped calcitic 'jaws' sprung with muscle. These are used for clearing the skin of debris and parasites or, in some cases, catching prey. Some starfish use them to catch small fish.

Altogether, the skins of echinoderms are unusual organs, unique in the animal kingdom and different in many ways from the skins of chordates (D. Nichols, personal communication). All echinoderms live in the sea and, as far as we know, have never left it.

The echinoderm nervous system is diffuse, and apart from a nerve ring around the digestive tract is largely associated with the epidermis. There is no real brain, central nervous system, eyes, ears or other organs of special sense. For all these apparent deficits, echinoderms are conspicuous and successful residents of the sea from rockpools on the beach to the deepest ocean trench, and have been so for at least 600 million years.

1.9.2 The lives and times of modern echinoderms

Five classes of echinoderm exist today. In addition to the holothurians and echinoids already mentioned, there are the true starfishes or asteroids, the brittle-stars or ophiuroids and the sea-lilies and feather-stars, collectively the crinoids. These five classes are conventionally grouped together as the Pelmatozoa (the crinoids) and Eleutherozoa (everything else).

(a) Starfish

Asteroids typically have five arms surrounding a central disc. The mouth is found in the middle of the lower, or oral surface, and from it radiate the podia-fringed ambulacral grooves, along the oral surface of each of these arms. The upper, or aboral surface is perforated by the madreporite and the anus. The water–vascular system is very well-developed and starfishes use their tube feet not only for locomotion, but also for prizing open the shells of clams – a habit that has persisted for at least 400 million years, judging by the intimate association of fossil starfishes with bivalves (Clarkson, 1986; p.249). Typically, the skeleton is a meshwork of articulated calcite plates somewhat like chain mail. Canals, gland cells and papulae communicate with the outside through gaps between the plates.

Starfishes fossilize relatively rarely as the calcite skeleton tends to disintegrate after death. The earliest-known starfishes are Tremadocian in age (Lowest Ordovician, around 500 million years ago), and by the Middle Ordovician, starfishes very like the living forms were present. About 1600 species are known today.

(b) Brittle-stars

Ophiuroids are only superficially similar to starfishes. The arms are well demarcated from the central disk, in which is contained most of the viscera: the arms are largely taken up by articulating calcitic ossicles called vertebrae, which obliterate most of the coelomic spaces. The vertebrae augment an already well-developed dermal skeleton and accessory spines. The ambulacral grooves are covered by calcite plates and so are effectively internal tubes rather than external channels: the podia protrude from gaps between the plates. Whereas asteroids generally crawl about using their podia, ophiuroids use sinuous movements of their long, serpentine arms for movement, which may be relatively rapid: the podia play a small part in locomotion, and there are no ampullae. The madreporite is found on the oral rather than the aboral surface.

Unlike the often carnivorous asteroids, ophiuroids are scavengers or filter-feeders. The digestive system is simple, and there is no anus. As with the asteroids, the fossil record of the ophiuroids goes back to the Lowest Ordovician. Although today's ophiuroids look very like their early fossil relatives, this conservatism has not been maintained at the expense of success. With 2000 living species, the ophiuroids constitute the most speciose modern echinoderm group.

(c) Sea-urchins

Echinoids, or sea-urchins, have a skeleton in which all the calcite plates have fused to make a rigid, box-like test, covered in a thin skin forested with tweezer-like organs called pedicellariae and spines of various kinds. Five ambulacra radiate from the mouth in the middle of the lower, oral surface, and run longitudinally around the animal towards the aboral 'pole' – the ambulacra are not borne on arms. The regions of the test between the ambulacra are called interambulacra. As in asteroids, the anus and madreporite are found on the upper surface.

The podia and the water–vascular system in general are well-developed in many of the 900 species. The animals use a combination of spines and podia to move about, and largely live on detritus and plant material scraped from hard surfaces using a sophisticated arrangement of teeth, jaws and muscles called 'Aristotle's Lantern'.

Although many echinoids have the distinctive, radially symmetrical, almost spherical shape of the familiar sea urchin, some have adopted a kind of secondary bilateral symmetry. They have a preferred direction of movement, and the mouth and anus have become displaced to the front and backs respectively. Some of these 'irregular' echinoids are burrowers, and others, like the 'sand dollars', have become very flat in shape.

Echinoids date back to the Late Ordovician, but did not become common until the Triassic Period. The first echinoids looked similar to extinct kinds of holothurian or an obscure fossil group called the ophiocystioids.

Unlike many modern forms, the earliest forms generally had big tests in which the plates were poorly articulated. The small, rigid tests of modern forms appeared later on.

The group only just survived the mass-extinction event that wiped out more than 90% of marine genera at the end of the Permian, about 250 million years ago: but for the fortune that kept a single genus, *Miocidaris*, grimly hanging on into the Triassic, the echinoids would have become extinct (Erwin, 1994). Today, they are arguably as abundant now as they ever were.

(d) Sea-cucumbers

Holothurians, of which there are around 900 species, are similar to echinoids in that they have no arms, and the ambulacra run around the body surface from mouth to anus. There the similarity ends: the body is greatly elongated into a cucumber shape (hence their vernacular name) and is flexible – in all but a few primitively armoured species, the calcite skeleton consists of nothing more than vestigial, microscopic spicules in the skin.

The podia around the mouth are developed into long, branching tentacles. The elongation of the body means that holothurians tend to lie on their sides, and in many species one particular 'side' is developed into the 'sole', the three ambulacra of which are used for locomotion. The whole animal creeps along like one arm of a starfish. The water–vascular system is well-developed except for the fact that the madreporite hangs loose in the coelom rather than connecting up with the outside.

Although one usually thinks of holothurians as living on the sea floor, some are worm-like burrowers and others are swimmers. All, though, feed on detritus in the sediment or the water that they trap on the mucus of their tentacles. The animals feed in a way horribly reminiscent of newly weaned infants: they shove their mucus-slimed, food-laden tentacles into their mouths, scraping the food off as the tentacles are drawn out again.

The fossil record of holothurians is at best meagre, consisting almost exclusively of isolated spicules. Although the earliest undoubted holothurian remains are Devonian, some armoured echinoderms dating from as early as the Ordovician and occasionally described as early echinoids may be close to holothurians of the more primitive, plated kind that still exist today. However, if the Burgess Shales fossil *Eldonia* is in fact a holothurian (Durham, 1974; Duncan Friend, personal communication) rather than a representative of some other group (Gould, 1989), then forms of a modern and indeed specialized type had appeared even earlier, by the middle of the Cambrian period.

(e) Sea-lilies and feather stars

The 600 or so species of crinoid represent the last of what was once a large and very diverse group. They include the only modern echinoderms that are stalked in adult life, the sea-lilies. These are attached to the seafloor by a long, flexible stalk based on articulating, cylindrical ossicles. However, most stalked crinoids can release their hold on the substratum and attach themselves elsewhere, so they are not helplessly sessile. Indeed, most modern crinoids have all but dispensed with their stalks entirely – these are the free-living feather-stars (Figure 1.10).

Although the heritage of the crinoids is no more ancient than that of any other modern group of echinoderm, the relative paucity of living forms set against the richness and diversity of fossil crinoids fosters the view that the living forms are somehow 'living fossils' themselves, primitive relics of times gone by. This impression is misleading. No living species of crinoid is known as a fossil from before the Pleistocene, and the unfamiliarity may be a symptom of the fact that the approximately 80 sea-lilies known today live at depths greater than 100 metres. The 550 species of feather-star are highly successful residents of most marine habitats.

The crinoid body is based around the cup-shaped calyx, which is shaped somewhat like a soup tureen. The bowl of the tureen is made of stiff, rigid plates laid around the bowl in courses of five. In unstalked forms such as the modern feather-star *Antedon* the base of the calyx is made of a single plate, the remnants of the stalk (feather stars are stalked in early life, as sea-lilies are stalked into adulthood). Although stalkless,

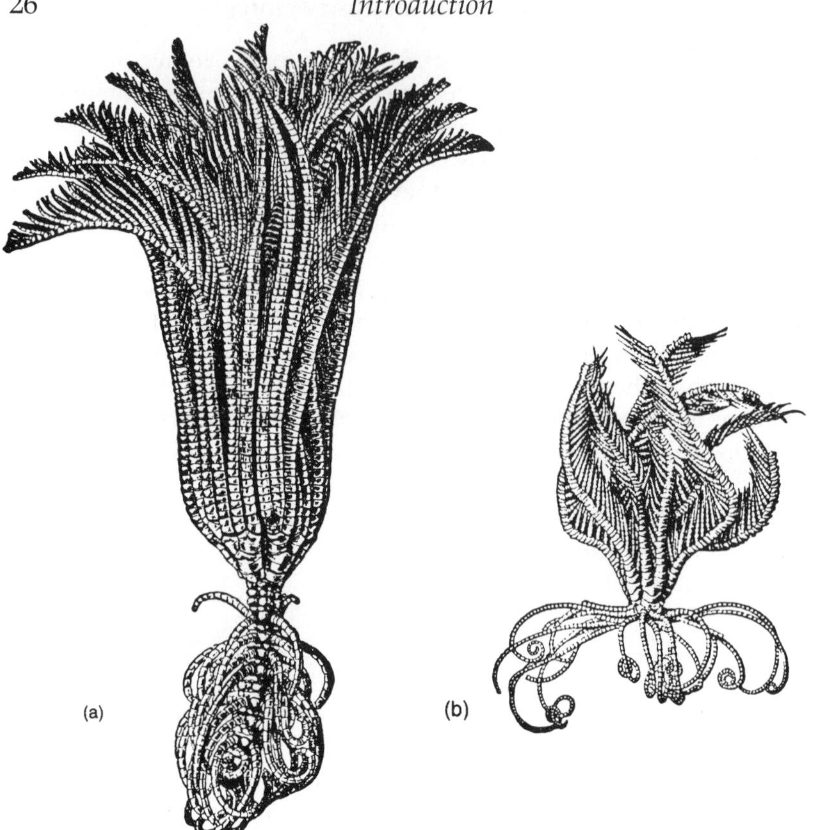

Figure 1.10 (a) Sea lily *Pentacrinus macleayanus* (from Hertwig, 1909, after Wyville Thompson) and (b) feather-star *Antedon macronema* (from Hertwig, 1909, after Carpenter).

the base of the calyx sprouts a number of flexible, plated tentacles or 'cirri' that are used for crawling about or for holding onto nearby objects. The stalks of sea-lilies also often bear cirri.

The 'lid' of the tureen is flatter and more flexible, and is called the 'tegmen'. This bears the anus (often on a special anal 'cone' or 'pyramid'), and the mouth. There is no madreporite as such, but a large number of radial stone canals that debouch into the coelom from the ring canal. The coelom is joined to the outside through a system of hundreds or even thousands of tiny ciliated funnels. This system, therefore, connects the water–vascular system to the outside in similar fashion to the madreporite and stone canal of other groups, but in a more roundabout way.

Five food grooves run from the mouth across the top of the tegmen (dividing it into ambulacral and interambulacral areas) and into the plated arms. In *Antedon*, as in many other crinoids, each of the five arms bifurcates as soon as it has left the tegmen, making a total of ten arms, but some crinoids have as many as 200 arms. The arms give off jointed pinnules on either side, rather like the feathers of birds. The water–vascular system runs into every last pinnule, and the ambulacra can be covered by flexible cover-plates.

Crinoids, whether stalked or free-living, are suspension feeders, and spread out their arrangement of arms and pinnules like a net, often into the prevailing current, to trap passing detritus and plankton. Because of the large numbers of arms and pinnules, the effective length of the ambulacra may be many tens of metres long, even in relatively small species, which is all the better for catching food. The catch is trapped, usually on the downstream side of the pinnules or arms, on mucus-laden podia, which toss it along the food grooves towards the mouth. The coelom is reduced by the volume of skeleton and muscle, but there are coelomic spaces in the arms, and in sea-lilies five tubular coeloms bunch together to make the 'chambered organ' that runs down the stalk, inside the ossicles.

Thus, crinoids differ from other living echinoderms in that the mouth faces upwards, rather than downwards. Whether this is especially significant is doubtful: after all, the madreporite of ophiuroids is found on the oral surface, as in crinoids but not asteroids or echinoids. Nevertheless, the upwardly directed mouth, fringed by an elaborate food-gathering system echoes the upwardly-directed mouths of lophophorates, fringed by a food-gathering system, so it is perhaps not unfair to regard the crinoid lifestyle as primitive in this respect.

The crinoid nervous system is rather more centralized than in other modern echinoderms. In addition to a sensory epidermal plexus similar to the nervous system of other echinoderms, there is a separate system centred on a mass of nervous tissue, the aboral ganglion, in the base of the calyx. In stalked forms, the aboral ganglion gives off a neural 'sheath', the peduncular nerve, that surrounds the chambered organ like the skin of a sausage. The peduncular nerve supplies side branches to the cirri, if present. The aboral ganglion also supplies nerves to all the arms and pinnules.

Like most of the other modern echinoderm groups, crinoids came to prominence in the Ordovician. There have been four major groups of crinoid: the camerates and flexibilians, which became extinct in the Permian; the inadunates, which just made it into the Triassic before they too became extinct, and the articulates, which survive today, and include both the stalked and stalkless forms. Although most modern articulates are stalkless, the unattached mode of life is not a modern innovation. Even in the Palaeozoic, some crinoids appear to have been stalkless, and the Mesozoic stalkless crinoids *Uintacrinus* and *Marsupites* are common as fossils.

1.9.3 Echinoderm development: two into five will go

Echinoderms have adopted a wide range of forms and lifestyles, but there seems no escape from the fundamentally pentamerous symmetry of the adult. But this symmetry is secondary and arises during the metamorphosis of the bilaterally symmetrical larva, a complex process that I shall outline here (Figure 1.11).

After gastrulation and the formation of the mesoderm and coeloms by enterocoely, the larva is bilaterally symmetrical. The mouth and anus are both situated on the midline of the undersurface, the mouth in front of the anus, and are connected by a simple, untwisted gut.

As is the way in deuterostomes, the coelom is divided into six, arranged as three pairs of two. In front are the two axocoels (protocoels), arranged left and right; behind these, the hydrocoels (mesocoels), likewise left and right, and bringing up the rear the large left and right somatocoels (metacoels). But this bilaterality is not complete: the left axocoel (but not the right) meets the outside through a canal and an opening, the hydropore. In turn, the left axocoel (but not the right) is connected internally with the left hydrocoel (but not the right) by a short tube. The right axocoel and right hydrocoel are neither connected to each other nor to the outside. In some species the right hydrocoel becomes a structure called the dorsal sac, but in others the right axocoel and right hydrocoel never actually form at all.

Bilaterality is further compromised by movements of the gut. The anus migrates forwards along the ventral midline, so the gut becomes a U-shape, but the mouth sidesteps leftwards, so that it emerges somewhat on the left-hand side above but slightly behind the anus, in front of the left somatocoel and close to the left hydrocoel. The coelomic compartments stay as they were, except that the left hydrocoel snakes itself round the oesophagus until it joins itself to form a hollow ring around it, the water ring – the germ of the entire water–vascular system. The connection between the left hydrocoel through the left axocoel to the hydropore becomes the stone canal.

Then, the entire posterior half of the body is twisted a quarter turn to the left, relative to the axocoels. The mouth retires to its former ventral position, taking the encircling left hydrocoel with it. The stone canal thus stretches across the inside of the body from the mid-ventral surface to the anterior dorsal surface. The anus shifts leftwards to accommodate this, and the changes in the position of the gut also force both somatocoels to rotate a quarter turn to the left, so that the left somatocoel comes to rest in the ventral (now the oral) half of the body, the right somatocoel in the dorsal (now the aboral) half.

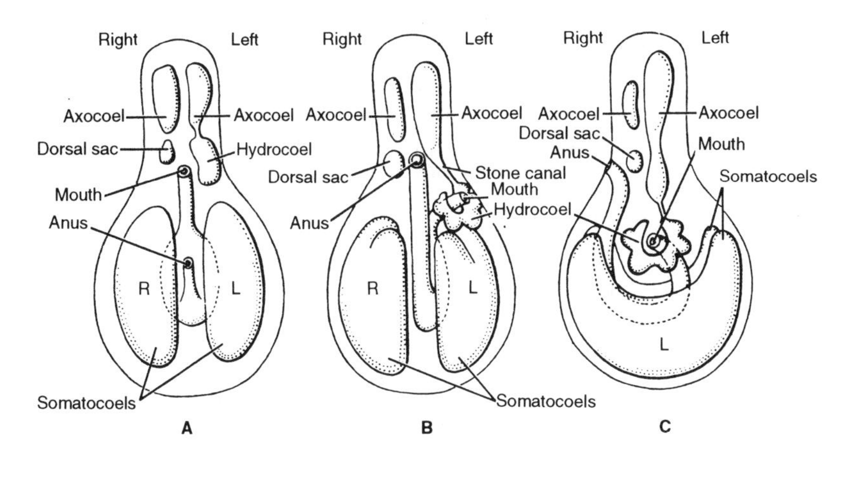

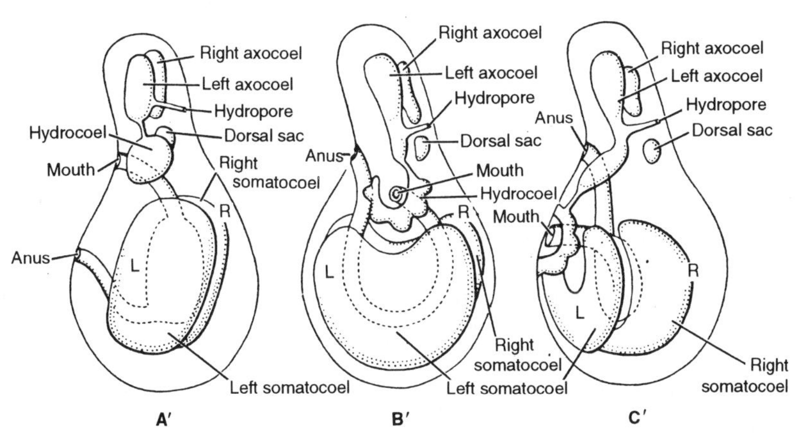

Figure 1.11 Above: a generalized view of echinoderm metamorphosis, top row (A–C) in ventral view; bottom row (A′–C′) in lateral view (from Barnes, 1980, after Ubaghs). Below: the hypothetical dipleurula ancestor of the echinoderms (from Bather in Lankester, 1900).

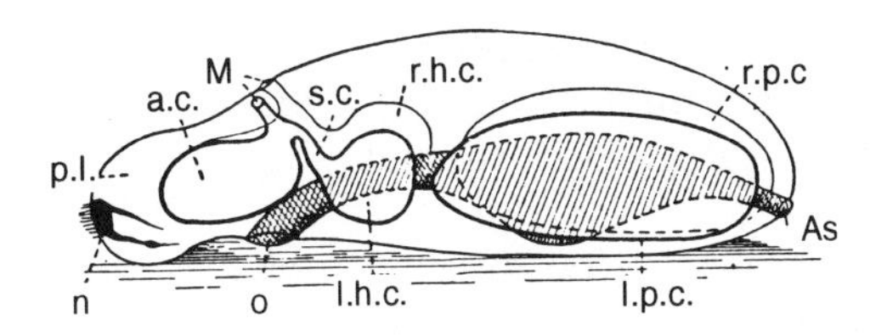

The left hydrocoel, now the water ring, develops to form the entire water–vascular system, which is of course pentameral. Because the water–vascular system pervades the entire body, the left hydrocoel alone effectively determines the (pentameral) symmetry of what was once a (bilaterally) symmetrical deuterostome based on three pairs of coelomic compartments. The axocoels and the right hydrocoel do not amount to much in the adult, and the left (oral) and especially the right (aboral) somatocoels form the bulk of the body cavity. For example, five outpocketings of the right somatocoel form the chambered organ in the crinoid stalk.

1.9.4 Echinoderm development: brothers in arms

This, of course, is a very generalized scheme, and the particular way in which an echinoderm larva reaches adulthood depends, ultimately, on its specific heritage and circumstances.

In starfish, sea-cucumbers, crinoids and brittle-stars, the coelom arises as an outpocketing of the advancing end of the archenteron, and the three pairs of coeloms are formed in a complicated process in which the axocoel and hydrocoel never quite separate. The conjoined compartment is called the axohydrocoel, and in sea-cucumbers and crinoids, the right axohydrocoel does not form at all.

The larvae of sea-urchins are often used in embryology experiments, so the details of echinoid development are known rather well. In sea urchins, the enterocoely of the mesoderm is preceded by a few cells of 'mesenchyme' that slough from the inner surface of the blastula. At first the larva is covered in fine cilia, but these get grouped together into a ciliated band that snakes around the larva like the seam round a tennis ball. The band is formed of two lateral halves, right and left, that meet in front, just ahead of the mouth, and behind, just in front of the anus. The loop in front of the mouth may detach to form a separate loop, or form separately to start with. The larva then sprouts special arms on which the locomotor bands are borne. These arms have nothing to do with the arms of the adult, but serve to spread the bands of cilia more widely. At this point in the embryogeny of an asteroid, the larva is called a bipinnaria, but with the appearance of yet more arms it graduates into a brachiolaria.

The auricularia larva of sea-cucumbers looks a lot like a bipinnaria, and has a similar arrangement of bands. It develops into a barrel-shaped doliolaria, with five latitudinal ciliated bands. Some sea-cucumbers have an armless (and therefore non-feeding), doliolaria-like ciliated larva called a vitellaria instead, which resembles the larvae of crinoids and a few brittle-stars.

Early crinoid larvae develop directly to vitellariae, and do not pass through a stage equivalent to an auricularia. The larvae of most brittle-stars, though, bear arms but look rather different from auriculariae or bipinnariae – more similar, in fact, to the echinopluteus larvae of sea-urchins – and each one is called an ophiopluteus. Barnes (1980) points out that of all 2000 species of ophiuroid, the larvae of only 71 have been recorded, and 55 of these brood their young rather than rely on external fertilization. So it is probably true to say that the 'typical' ophiuroid larva has not been discovered.

The left hydrocoel starts to develop into the adult water–vascular system at about this time. In starfish, the larva attaches itself to some substrate by its front end, but brittle-stars manage to metamorphose and stay swimming at the same time. Generally, the front end and its coelom (the remaining, left, axocoel) atrophy. The mouth, anus and much of the gut disappear along with the front half, and are rebuilt. The adult is formed from the back half, and essentially contains three coeloms: the left hydrocoel, left (oral) somatocoel and right (aboral) somatocoel. Sea-urchin larvae sink to the bottom under the weight of the developing skeleton and metamorphose very rapidly, without attachment. Sea-cucumbers change rather gradually from doliolaria to adult with neither attachment nor drastic remodelling of the anatomy.

Attachment, though, is characteristic of crinoids: the vitellaria up-ends itself on the substrate, gripping with an 'adhesive pit' near the anterior end. There then follows a long, involved metamorphosis (summarized by Jefferies, 1986, pp.47–51), the result of which (at least in the feather-star *Antedon*) is a tiny stalked pentacrinoid larva that looks like a sea-lily only three millimetres tall.

The complexity of development in *Antedon* is related to some extent to its passing through a stalked phase, but it is no different from other echinoderms in that the right somatocoel becomes the aboral and the left somatocoel oral in position, and the water vascular system originates from the left hydrocoel. Eventually, the young crinoid breaks free.

The search in echinoderms for genes involved in development, such as homeobox genes, has been largely confined to echinoids. Homeobox genes for which the vertebrate homologues are expressed in the posterior parts of the animal appear to be present, but those expressed in anterior regions have not been found (Ruddle *et al.*, 1994). It may be the case that they await discovery, and work with tunicates suggests that this might be the case (Di Gregorio *et al.*, 1995; Hidetoshi Saiga, personal communication). But echinoids are highly derived, and one might legitimately ask how the expression of developmentally important genes in echinoderms has altered through time (though perhaps with little hope of an answer). It would be interesting to learn about genes involved specifically in the elaboration of the hydrocoel to form the water–vascular system, and in the use of the *left* hydrocoel for this purpose.

1.9.5 Fossil echinoderms

The five groups of echinoderm alive today are just the survivors. In the Palaeozoic Era, they were joined by 15 others (depending on the classification scheme one adopts); the edrioasteroids, diploporite cystoids, rhombiferan cystoids, blastoids, parablastoids, eocrinoids, paracrinoids, cinctans, solutes, cornutes, mitrates, ctenocystoids, ophiocystioids, cyclocystoids and helicoplacoids (Figure 1.12). The cinctans, solutes, cornutes and mitrates are collectively known as carpoids. One may also see cornutes and mitrates referred to, collectively, as stylophora. As discussed below, Jefferies (1986) regards some carpoids, more particularly stylophora, as chordates rather than echinoderms.

All echinoderm groups, living or fossil, appeared as part of the 'Cambrian Explosion' (Conway Morris, 1993), and all had appeared by the mid-Ordovician. That was the heyday of echinoderm higher-order diversity, when 17 classes (compared with today's five) existed all at once – representatives of as many as 13 having been recovered from a single site. Yet in terms of numbers of species (rather than classes), echinoderms are probably as diverse today as they have ever been (Sprinkle, 1983).

Every so often somebody finds a fossil believed to represent a hitherto unknown group of echinoderm. The helicoplacoids, for example, were described just 30 years ago (Durham and Caster, 1963). Many of the now-extinct forms had appeared by the Lower Cambrian and some are only known from that period, but the last of these had gone by the end of the Permian, leaving the five we have today.

Despite their variety, all these creatures had skeletons formed of single crystals of calcite and most had some kind of water–vascular system, so they are clearly recognizable as echinoderms in the sense that they share these characters with the echinoderms we know today.

There were, though, some intriguing differences between ancient and modern forms. For example, Palaeozoic echinoderms were generally stalked and sessile, and had their mouths directed upwards, which is not the case for modern forms except crinoids.

Second, although even all modern echinoderms have five ambulacra apiece, some Palaeozoic echinoderms did not have pentameral symmetry. Some appear to have adopted a three-way symmetry, which may have been a precursor of pentamerism: some early pentameral animals, such as the Cambrian edrioasteroid *Stromatocystites*, show signs of a '2+1+2' symmetry, in which the larger two of the three primary ambulacra bifurcate, to give five.

The cornutes, cinctans, solutes and mitrates – collectively the carpoids – are of particular interest in the context of chordate affinities, as I will show later on. Cinctans and mitrates seem bilaterally symmetrical at first glance, but closer inspection reveals that this bilaterality is imposed on an awkward irregularity. Cinctans have just two ambulacra, and solutes only one (Jefferies, 1990).

Cornutes, though, are thoroughly irregular, and whether cornutes and mitrates possessed ambulacra is a matter of debate. However, cornutes seem to have a row of apertures seen nowhere else (at least, not as clearly) that resemble the gill slits of hemichordates or chordates. This was first pointed out by Bather (1913), if only to caution readers that this feature was convergent. Nevertheless, the idea appealed to Matsumoto (1929), who wrote that the carpoids 'may probably stand at the base of the Chordata and be allied with the Urochorda in a certain way', but who neither elaborates nor gives reasons – perhaps these were assumed to have been obvious to the reader. The idea was taken up by Gislén (1930) (discussed in Chapter 2) and, of course, by Jefferies (discussed in detail in Chapter 4).

The classification of echinoderms has always been contentious, and the reader would do well to consult the paper by Smith (1984) before looking any further. Smith (1984) despairs of traditional echinoderm classifications, in which the search for relationships through patterns of character distribution have become 'progressively divorced from the production of classification schemes'. He starts as he means to go on – in Hennigian fashion – throwing out data from fossils as things to be fitted in later, after the adoption of a scheme based on the embryology and comparative anatomy of living forms. Smith, therefore, offers a cladistic classification of Recent forms (Figure 1.13a). Note the union of echinoids with holothurians; the progressively less related ophiuroids and asteroids; and the crinoids, set apart as 'pelmatozoa'.

He then uses the concept of monophyletic crown- and paraphyletic stem-groups to slot the plethora of extinct echinoderms into this scheme. Fossil forms are arranged as extinct stem groups (the 'plesions' of Patterson and Rosen, 1977) according to the distribution of shared, derived characters that define the Recent crown-groups:

> Neontologists search for the most narrowly defined crown group whereas palaeontologists attempt to discover the unique stem group that each fossil belongs to. (p.454)

Not surprisingly, given their abundance and variety, the fossil echinoderms dominate the picture. It is easy to see why traditional systematists are swamped: cladistic principles are needed, if only as a guide to teasing from a patchwork of fossils which features are primitive, and which derived. Figure 1.13b shows what Smith's grand echinoderm cladogram looks like.

However, given that all modern echinoderms share (by definition) the key echinoderm features of calcite endoskeleton, ambulacra (indicating a water–vascular system) and pentaradial symmetry, it is hard to decide the order in which these features were acquired in echinoderm history.

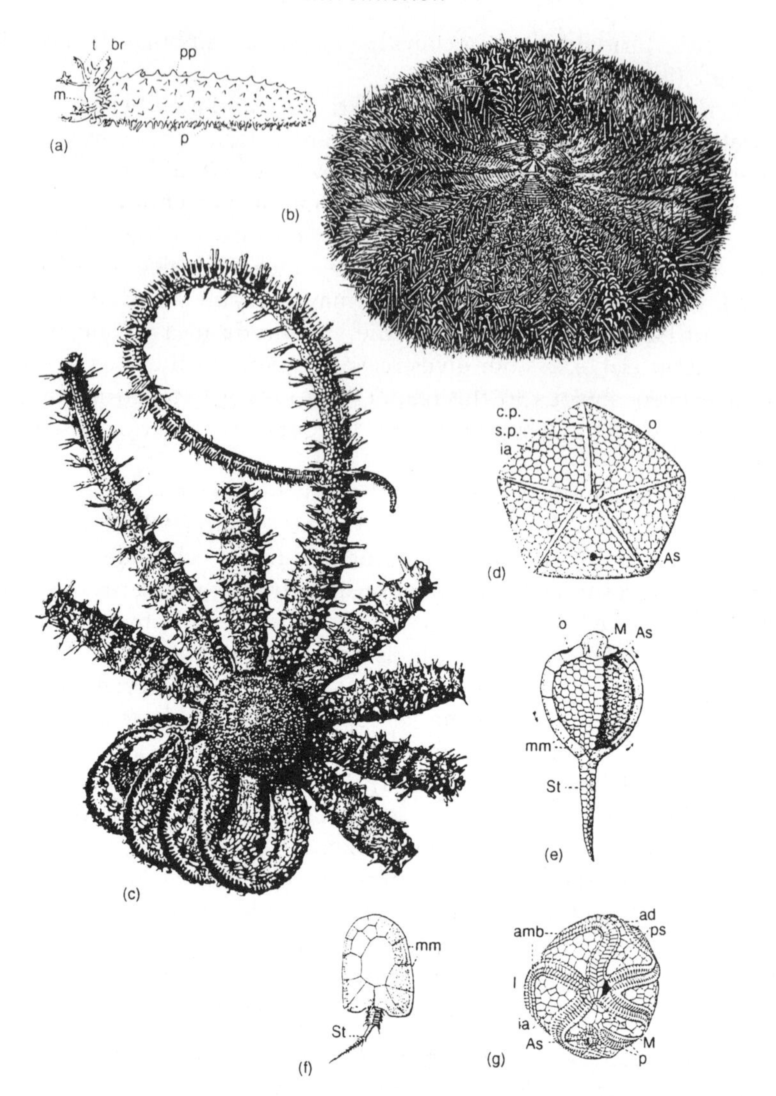

Figure 1.12 Diversity of Recent and fossil echinoderms. (a) A holothurian *Holothuria forskali,* (b) an echinoid *Asthenosoma hystrix,* (c) an asteroid *Brisinga coronata,* (d) an edrioasteroid *Stromatocystites,* (e) the carpoids *Trochocystis bohemicus* and (f) *Mitrocystis mitra,* (g) the edrioasteroid *Edrioaster bigsbyi* (all from Bather in Lankester, 1900). (h) The enigmatic *Eldonia ludwigi* from the Burgess Shales, possibly a holothurian (photograph courtesy of Duncan Friend, University of Cambridge). (i) A tableau of stalked Palaeozoic echinoderms. The fourth from the left is a solute (from Wells *et al.,* 1931). (j) Evolutionary diversification of early Palaeozoic echinoderms interpreted in terms of feeding strategy and habitat. Stipple indicates unconsolidated substratum, diagonal hatching, hard substratum (from Smith, 1990, by permission of the author and the Systematics Association).

(h)

(i)

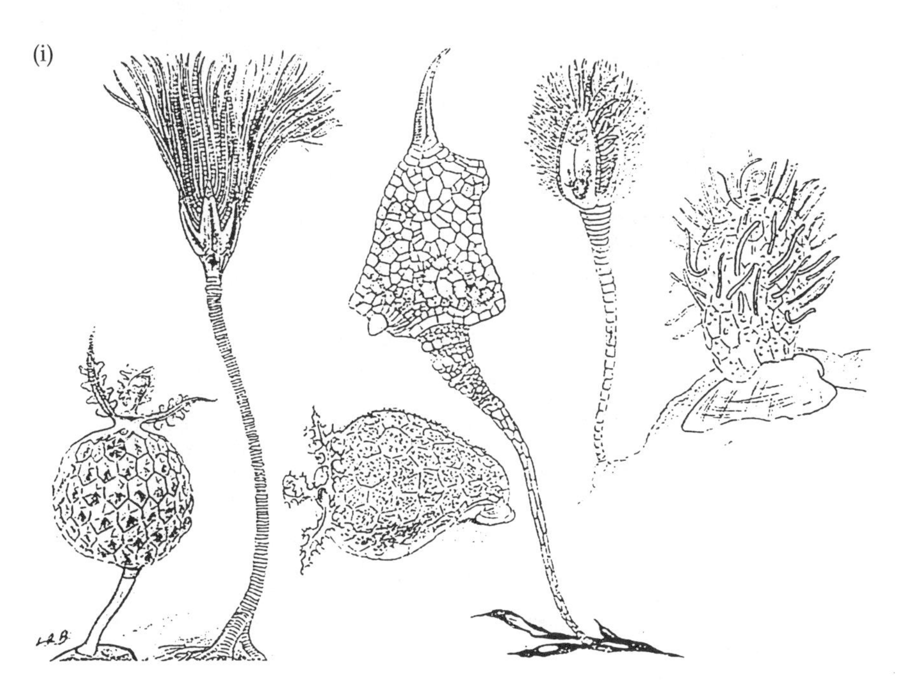

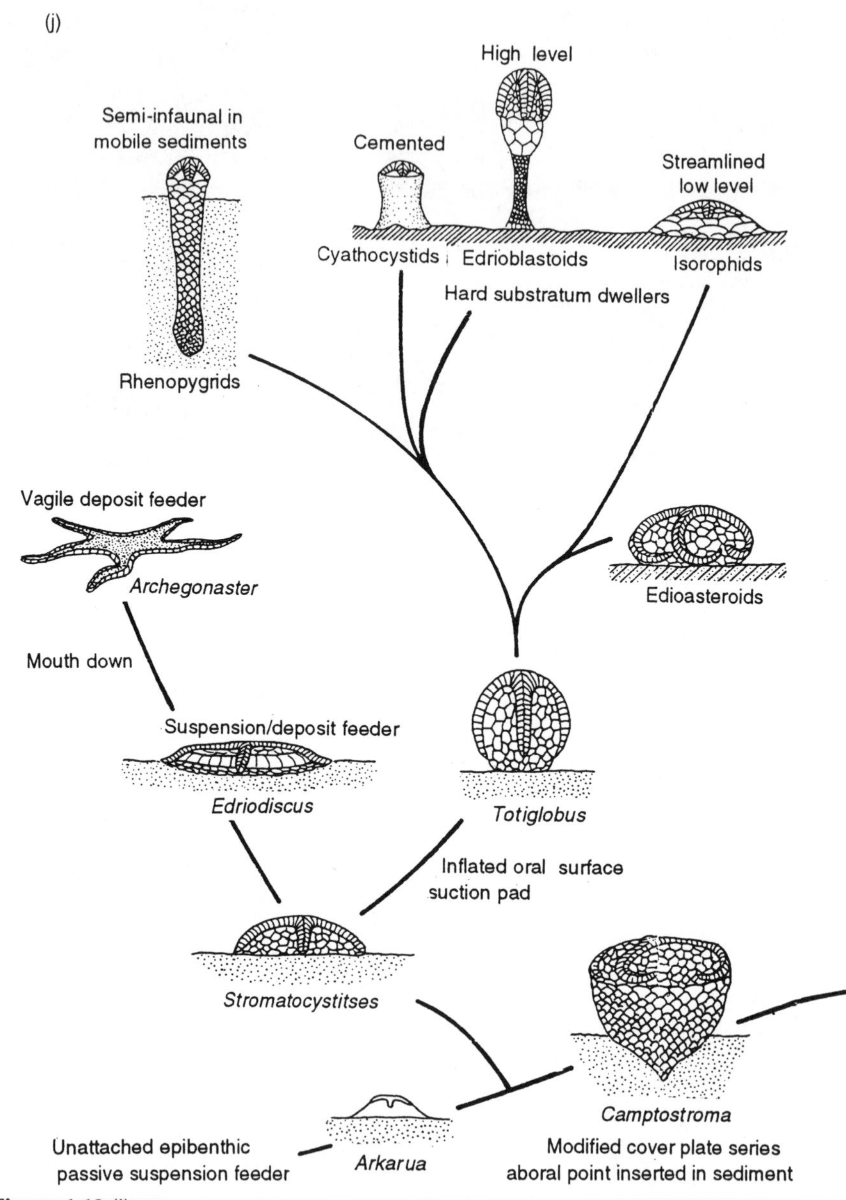
(j)
High level
Semi-infaunal in
mobile sediments
Cemented
Streamlined
low level
Cyathocystids
Edrioblastoids
Isorophids
Hard substratum dwellers
Rhenopygrids
Vagile deposit feeder
Archegonaster
Edioasteroids
Mouth down
Suspension/deposit feeder
Edriodiscus
Totiglobus
Inflated oral surface
suction pad
Stromatocystitses
Camptostroma
Unattached epibenthic
passive suspension feeder
Arkarua
Modified cover plate series
aboral point inserted in sediment

Figure 1.12 (j)

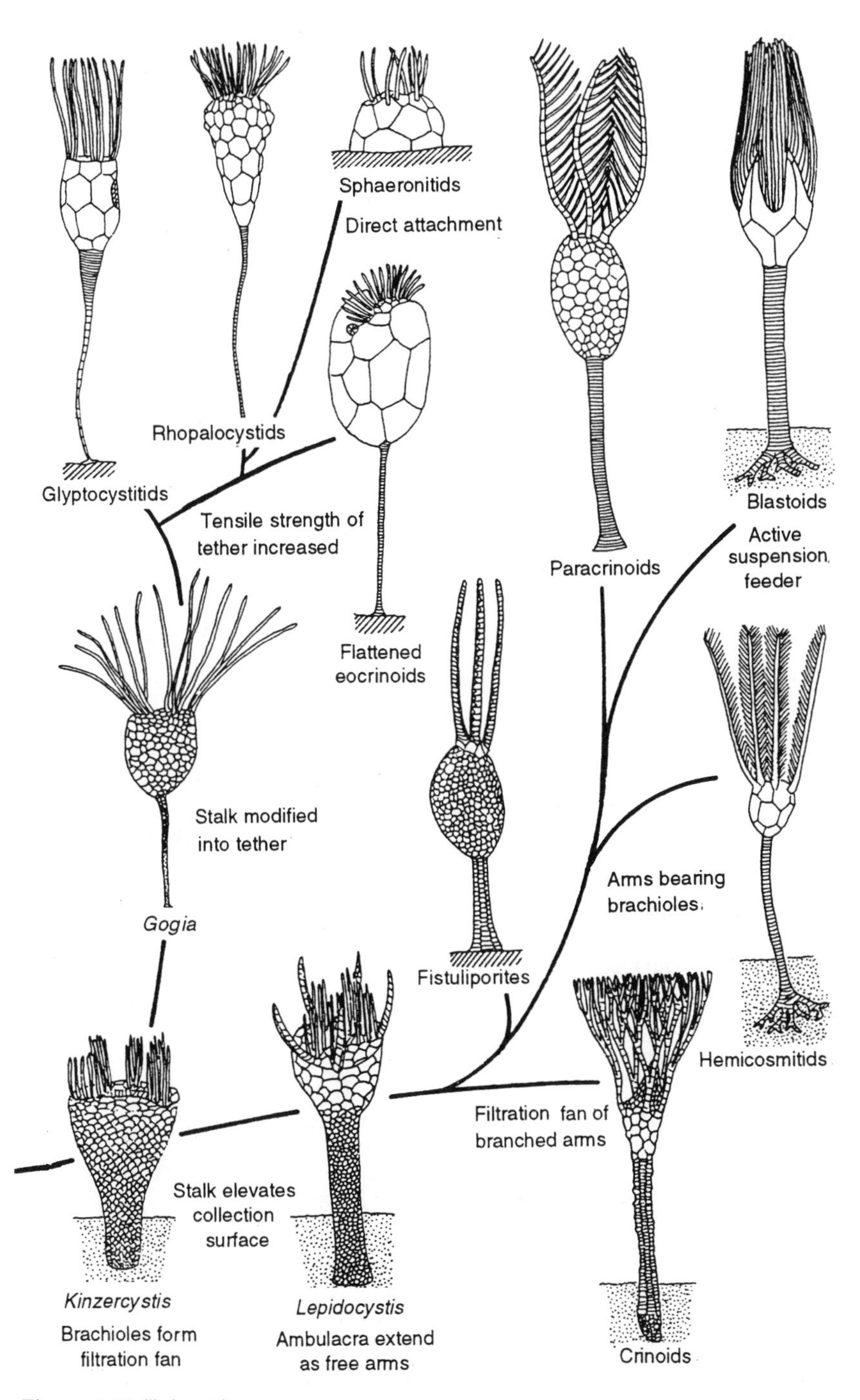

Figure 1.12 (j) *(cont.)*

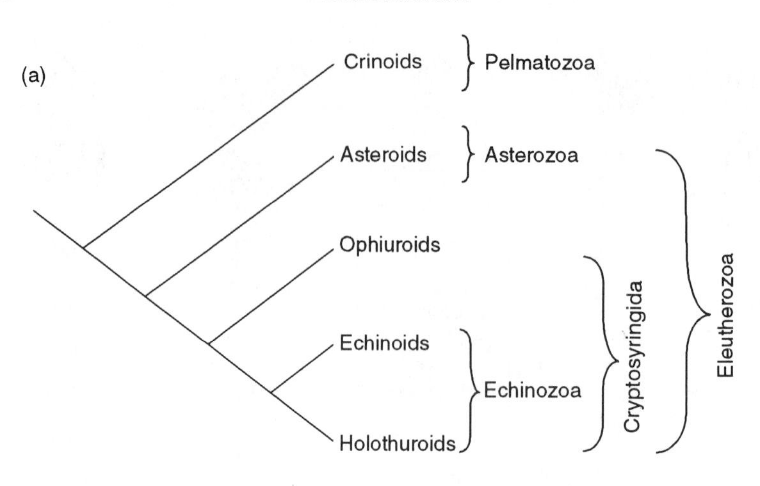

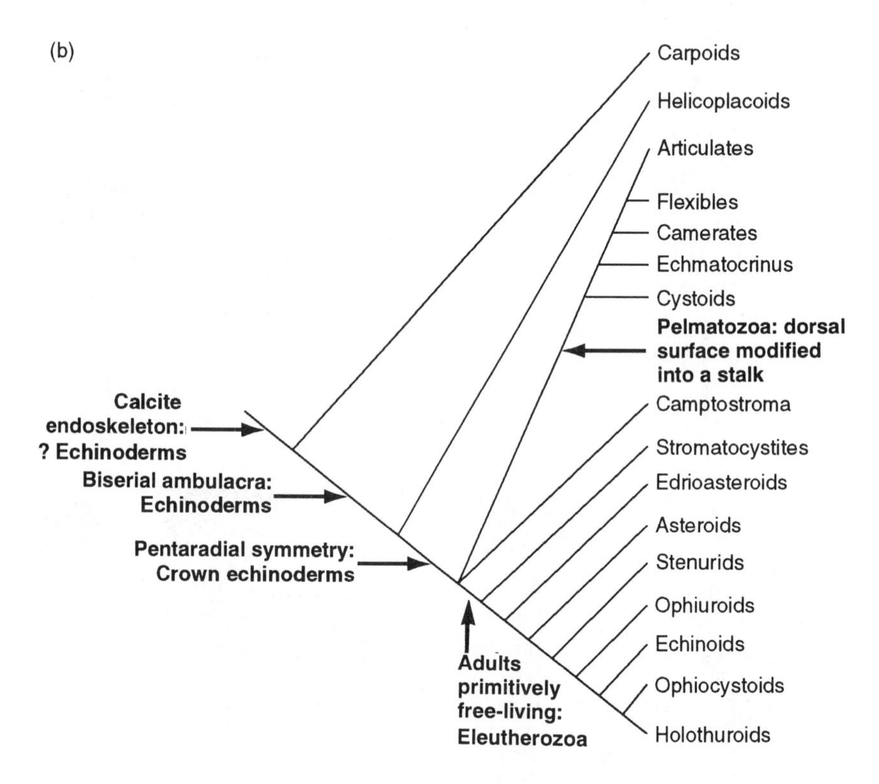

Figure 1.13 Smith's cladograms of (a) living and (b) living and fossil echinoderms. Note the position of carpoids as a sister-taxon to all other echinoderms. Redrawn by Sue Fox from Smith (1984), with the permission of the author.

The calcite endoskeleton is a logical choice for the first such feature to have been acquired, given that there is one group, the carpoids, that shares just this feature – and no others – with all other echinoderms, living or fossil:

> Carpoids are all basically asymmetric, without a trace of radial symmetry and either lack ambulacra or have a single exothecal appendage. They all have a single feature, their calcite endoskeleton, which they share with crown group echinoderms. (p.455)

The carpoids, therefore, become the first offshoot from Smith's cladogram. This poses a problem, because carpoids have some decidedly un-echinoderm features all their own. He continues:

> However, Jefferies (1981)[12] believes that Stylophora [cornutes + mitrates] show evidence of gill slits and a post-anal tail and should therefore be classified as stem chordates. If this proves to be correct then the other carpoids may be stem chordates, stem echinoderms, or stem (chordates plus echinoderms). Further work is required to resolve the phylogenetic position of these groups and I shall not consider them further. (p.455)

Smith's doubt seems awkward, as the position of the carpoids calls into question the utility of the calcite endoskeleton as a shared, derived feature of echinoderms, and, with that, echinoderm monophyly.

For if some carpoids are chordates, then the calcite skeleton must have been acquired by the common ancestor of both chordates *and* echinoderms, and subsequently lost by chordates. In which case, the calcite endoskeleton defines echinoderms only as a paraphyletic group – the remainder once chordates have been taken away.

1.9.6 Enteropneusts: coming to terms with worms

In contrast with echinoderms, the enteropneusts (Figure 1.14a–d) are wormlike creatures that burrow in marine muds or under stones and among weeds in shallow seas. There are about 70 known species, in 11 genera, and they range in size from about two centimetres to a hundred times this length. Unlike echinoderms, they have no skeleton and are rather flaccid. Some enteropneusts have a ciliated planktonic larva called a tornaria that is similar to the larvae of starfishes.

The coelom is divided into three, although the protocoel – unlike the axocoel – is unpaired. Again, in contrast with echinoderms, the body of the adult is bilaterally symmetrical and the external appearance reflects

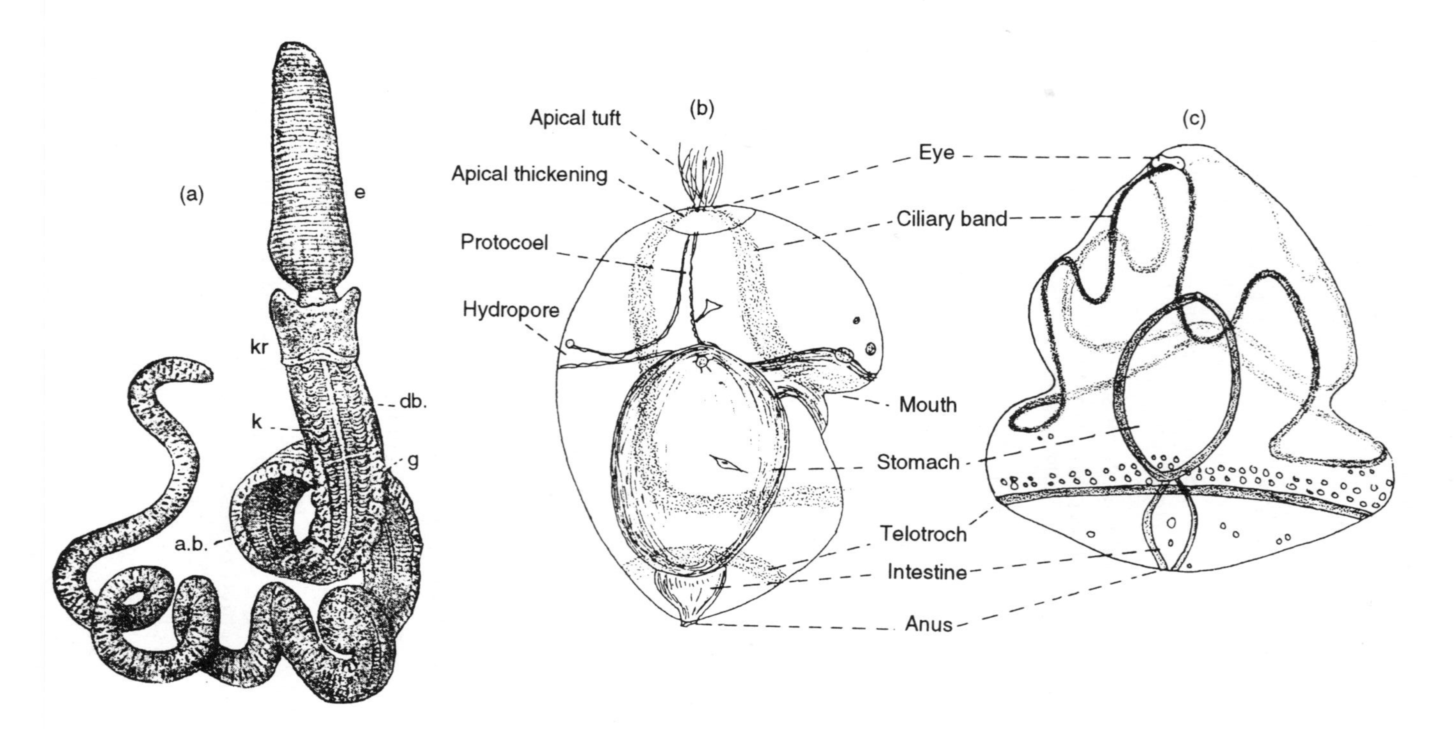

Figure 1.14 (a) The enteropneust *Balanoglossus kowalewskii* (from Hertwig, 1909, after Korschelt-Heider and Agassiz), (b) early tornaria of *Balanoglossus clavigerus* in lateral view, and (c) fully developed tornaria in lateral view (from Barnes, 1968, after Morgan from Hyman). (d) Diagrammatic section of the front end of *Balanoglossus* (from Young, 1981, after Spengel). (e) Sagittal section of a *Cephalodiscus* zooid (from Jefferies, 1986 (after Barrington and Schepotieff), with permission of the Natural History Museum, London).

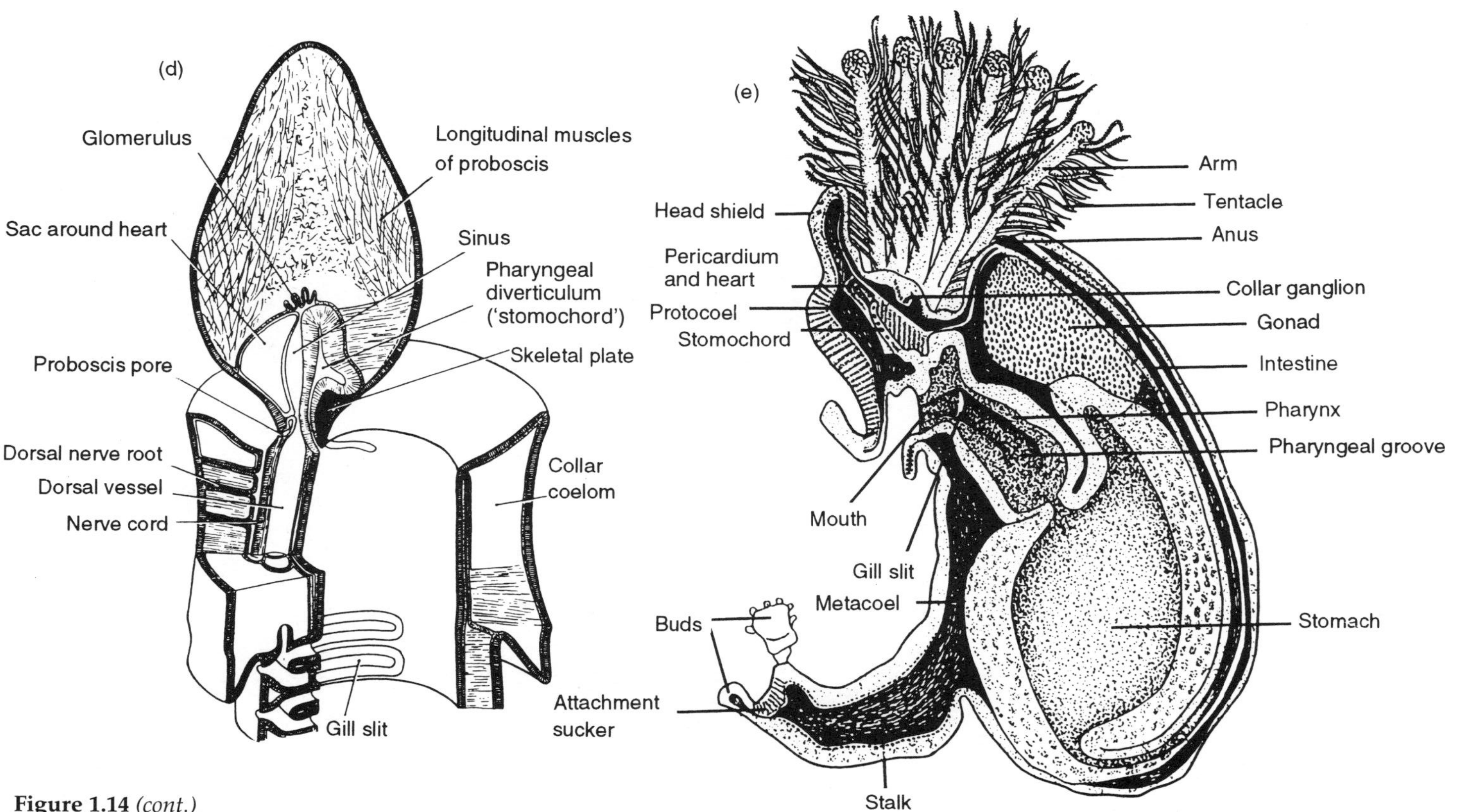

Figure 1.14 *(cont.)*

the internal coelomic divisions. The protocoel inhabits a distinctive preoral proboscis, a flexible all-purpose burrowing and food-gathering organ covered in mucus-laden cilia that conveys food to the mouth.

Inside the proboscis is the 'heart' – a pulsatile vesicle in the generally open circulatory system – as well as an outgrowth of the pharynx called the stomochord. Slight resemblance between this structure and the chordate notochord suggests a link between enteropneusts and chordates (Bateson, 1886). Nowadays, the two structures are not usually thought to be homologous (Komai, 1951; but see Peterson, 1995b).

The proboscis is attached by a short stalk to the 'collar' region, which contains the paired mesocoels. The collar appears to support the proboscis as a cup supports an acorn (hence the popular name for the enteropneusts of 'acorn worms'). The mouth perforates the body wall in the mid-ventral line within the collar, and the rest of the long wormlike body bounds the metacoels.

The body wall is perforated by a series of pharyngeal slits on left and right. These are very like chordate gill slits in structure (indeed, the resemblances with the amphioxus have often excited comment) and in their formation, and may be secondarily divided by vertical bars called tongue bars. Unlike chordate gills they appear to be used for gas exchange rather than for filter feeding (a function performed by the ciliated proboscis).

The nervous system is similar to that of echinoderms in that it is generalized and epidermal, without any particular brain. A local thickening in the collar called the 'collar cord' runs along the dorsal midline and is hollow. It was, therefore, once thought suggestive of the chordate dorsal tubular nerve cord (again, see Bateson, 1886) especially considering its proximity to the hollow supposedly notochord-like stomochord – but the enteropneust body contains a number of longitudinal nerve trunks besides this.

Modes of development vary greatly in the group. In some species development is direct, the animal hatching out as a miniature version of the adult. In others, it proceeds via a ciliated planktonic larva called a tornaria. Like echinoderm larvae, this tends to have a tuft of cilia at the anterior end, and a complicated band of cilia which snakes round the animal. There is, also, another band of cilia running latitudinally around the posterior end like a grass skirt. The protocoel separates from the archenteron quite early and sends out a canal that opens on the surface, the future protocoel pore.

The mode of origin of the five coelomic compartments in different enteropneust species likewise shows an astonishing variety. They may all originate as separate evaginations from the archenteron, or only the protocoel evaginates in this way with the other four compartments originating by splits in mesenchyme (rather like schizocoely in protostomes), or some other scheme between these two extremes.

Bateson (1886) reports that the tornaria was first described in 1849 as something akin to an echinoderm larva, but until the link between the two was demonstrated by Metschnikoff in 1880, adult enteropneusts languished as 'worms of some kind'. The discovery of tornariae elevated them to the status of aberrant bilateral echinoderms, thanks solely to the structure of the larva. The adult was at that time poorly known, but known well enough that Metschnikoff

> was led to suppose that the gill-slits of Balanoglossus [*sic*] were mere amplifications of the water-vascular system of Echinoderms, which could hardly have been suggested had it not been felt that no other solution was possible. (Bateson, 1886, p.552)

On this basis Metschnikoff united enteropneusts with echinoderms in a single group, the Ambulacraria (Metschnikoff, 1881: cited in Gregory, 1946).

The 1880s ushered in a new understanding of enteropneusts as animals which, although having some connections with echinoderms, occupied a position more akin to chordates. Today, they are seen as close relatives of pterobranchs, but the doubt cast on the stomochord and neural architecture as homologues of chordate structures means that the position of enteropneusts with respect to other major deuterostome groups remains equivocal.

A preliminary study of ribosomal RNA genes of the enteropneust *Saccoglossus* (= *Balanoglossus*) *cambrensis* reaffirms the kinship of enteropneusts with chordates, rather than with echinoderms (Holland *et al.* 1991). However, more recent evidence from 18S rRNA sequence data (Wada and Satoh, 1994; Turbeville *et al.*, 1994; Halanych, 1995) suggests that they might in fact be closer to echinoderms, lending some validity to Metchnikoff's 'Ambulacraria' concept. If so, then one cannot use the presence of pharyngeal slits to recognize hemichordates and chordates to the exclusion of echinoderms. It follows from this that the echinoderm stem-lineage included animals with pharyngeal slits, which were subsequently lost.

The fossil heritage of enteropneusts is exiguous, but such records as exist are discussed by Beuton (1993) and Conway Morris (1993). If the enigmatic fossil *Yunnanozoon* is a hemichordate (Shu *et al.*, 1996), then the record stretches back to the Cambrian.

1.9.7 Pterobranchs: old colonials

The pterobranchs are small, mainly tube-dwelling, almost bilaterally symmetrical, occasionally colonial animals that live on the ocean floor, often at great depths. Only three genera are known, *Cephalodiscus*, *Rhabdopleura* and *Atubaria*, of which the biology of *Cephalodiscus* (Figure 1.14e) is probably best known.

Ciliated all over, the bulbous body of *Cephalodiscus* bears a lophophore, twin plumes of tentacle-bearing arms, separated by a flexible 'head-shield'. The arrangement of arms and tentacles is used for filtering food material from the water, and is similar in structure and function to the lophophore found in brachiopods, bryozoa and phoronids (see below) and the vascular system of echinoderms.

The body contains a very large stomach: the gut is U-shaped, looping down from a pharynx (perforated by a single pair of gill clefts) into the stomach and up again, so that the anus opens dorsally just behind the lophophore. At the other end of the body is a long, flexible and prehensile stalk with which the animal attaches to the inside of the tube in which it lives or, if it emerges, uses as a temporary attachment as it crawls about, leech-like, using its head shield as an anterior sucker. The distal end of the stalk can also be used as a sucker.

Although they look quite different, pterobranchs are traditionally grouped together with the enteropneusts in a group called the hemichordata. This may be a natural group, although many of the features held in common probably represent generalized deuterostome – or even lophophorate – characteristics. Nevertheless, pterobranchs and enteropneusts have similar structure, regionated into a head shield (proboscis), mesosome (collar region) and body, and share general features of organization of the nervous and blood systems. Development in both groups is similar, at least in general and in the early stages: both show radial cleavage, and gastrulation in both pterobranchs and enteropneusts resembles that in echinoderms, although these features could be characteristic of deuterostomes in general rather than indicative of a close relationship between echinoderms and hemichordates (see Schaeffer, 1987, for a recent review.) However, molecular evidence from 18S rRNA sequences suggests that pterobranchs and enteropneusts are close relatives (Halanych, 1995).

Like enteropneusts, *Cephalodiscus* has pharyngeal gills, although just one pair rather than many, and a generalized, epidermal nervous system.

As in enteropneusts, the body of *Cephalodiscus* consists of five coelomic compartments. The unpaired anterior protocoel connects to the outside by a pair of pores, and contains the heart as well as a stomochord. Each of the paired (left and right) mesocoels bear a pore as well as one half each of the lophophore. Enteropneusts have the pores but lack the lophophore. The large, paired (left and right) metacoels do not connect to the outside. Although the metacoels are unrestricted, they are separated from each other by a membrane or mesentery, except in the stalk. The mesocoels are also demarcated by a mesentery, but are not completely separated by it.

The early embryogeny of *Cephalodiscus* is heavily influenced by the yolkiness of the eggs, and the product is a ciliated larva that looks like the planula of a cnidarian (see Gilchrist, 1917 and citations therein): there is no suggestion of an echinoderm–like larva as seen in enteropneusts. The animal settles on the bottom and gradually metamorphoses into an adult.

The gills appear relatively late in development, contrasting with chordates such as the amphioxus. Indeed, *Rhabdopleura* has no gills at all (Hyman, 1959)[13]. The coeloms originate by outpocketings of the archenteron, although the order in which they develop may differ between species. In any case, the protocoel is always a single, unpaired structure.

Cephalodiscus usually (but not always) lives in deep water, mainly in the southern hemisphere. However, the smaller, more obligately colonial and somewhat asymmetrical *Rhabdopleura* lives worldwide, and indeed was first found in arctic waters off the Lofoten Islands in northern Norway (see Stebbing, 1970, and citations therein). The lack of gill slits in *Rhabdopleura* is perhaps the most significant difference between the two. All known examples of *Atubaria* come from a single catch dredged off Japan in 1935. This solitary form could be a non-colonial form of *Cephalodiscus*.

The fossil record of pterobranchs is patchy but not as poor as one might expect, as their horny tubes preserve quite well as carbon films on certain kinds of rock. Fossil pterobranch tubes that look like *Rhabdopleura* colonies are known from the Middle Cambrian of Sweden (Bengtson and Urbanek, 1986) and other records of pterobranchs are scattered throughout the geological column (see Benton, 1993, and citations therein.)

The fossil history of pterobranchs may be more illustrious, though, if the remarkable fossils called graptolites are added to their number (see below).

1.9.8 A descent into the abyss

Concrete confirmation of the link between pterobranchs and enteropneusts may come with the lophenteropneusts, little-known residents of the deep-sea benthos that look like enteropneusts, but with a collar-region fringed with lophophore-like tentacles (see Gage and Tyler, 1991, for pictures and citations).

1.9.9 Graptolites

Graptolites were tiny marine animals that lived in colonies encased in protein, preserved in rocks as carbonaceous films. In contrast to the two-dimensional colonies of corals or bryozoa, graptolite individuals were arranged in lines along one or both sides of the blade-like stem or stipe. One or more stipes joined at an apex, so the whole colony looked like a bunch of hacksaw blades (Figure 1.15).

The first graptolites to appear, in the mid-Cambrian – and the last to disappear, in the Carboniferous – were the bushy dendroids, which lived rooted to the seafloor. More abundant as fossils are the true graptolites or graptoloids, which floated freely in the plankton of the world's oceans from the Ordovician to the Devonian.

Although the structure of graptolite skeletons is known in microscopic detail (Palmer and Rickards, 1991), the identities and affinities of the animals that inhabited the colonies are rather less certain (Kozlowski, 1947; Urbanek, 1986; Palmer and Rickards, 1991). Current opinion favours an affinity with pterobranchs, as the colonies are in some ways similar to those of *Rhabdopleura*. This view is supported by the discovery of the pterobranch *Cephalodiscus graptoloides* (Dilly, 1993) with a colonial structure very similar to that of graptolites. This may be regarded as a 'living fossil' (Rigby, 1993), strongly suggestive of a phylogenetic connection with pterobranchs.

1.9.10 A look at the lophophorates

The so-called 'lophophorate' phyla comprise three relatively small animal groups that resemble deuterostomes in that they tend to have radial cleavage and some suggestion of a tripartite coelom plan (strongly so in phoronids). They resemble pterobranchs in their possession of a lophophore, a convoluted, crescentic tentacle-bearing feeding structure derived from the paired mesocoels. If the lophophore is homologous with the water–vascular system of echinoderms by virtue of common coelomic derivation, the homology is, strictly, between the *left* lophophore and the *entire* water–vascular system, as the right hydrocoel (mesocoel) of echinoderms invariably atrophies in embryogeny, if it develops at all. All lophophorates are sessile filter-feeders.

The ten or so species of phoronid (Figure 1.16a) are tube-dwelling marine worms. Unlike protostomes in general, cleavage of the blastula is radial and the mesoderm enterocoelous: it is these features that suggest some affiliation with deuterostomes. Phoronids have no hard parts, but fossil evidence suggests existence since at least Devonian times (Emig, 1982).

Better known are the moss animals or bryozoa (ectoprocta), with 4000 marine and freshwater species (Figure 1.16b). Bryozoa are colonial, and individuals are very small, highly specialized and live in small calcitic boxes. Nevertheless, the feeding individuals in bryozoan colonies use their lophophores for feeding, just like phoronids and pterobranchs. The colonies are usually found as encrustations on rocks or other surfaces such as seaweeds. Cleavage appears to be radial, at least in marine

(a)

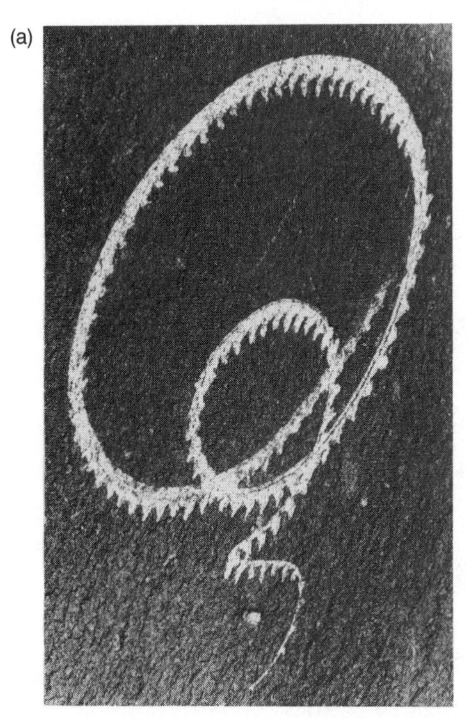

(b)

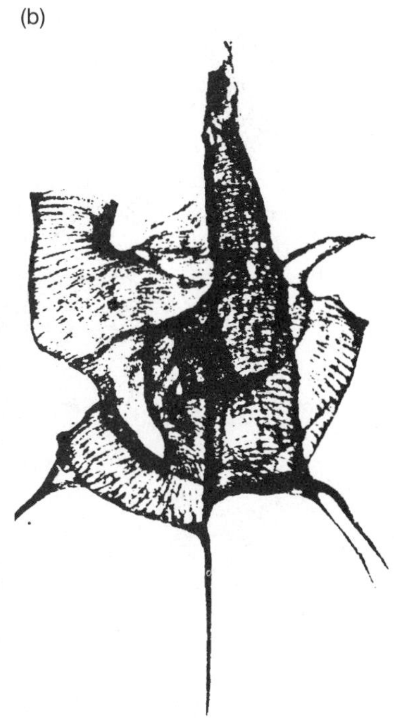

(c)

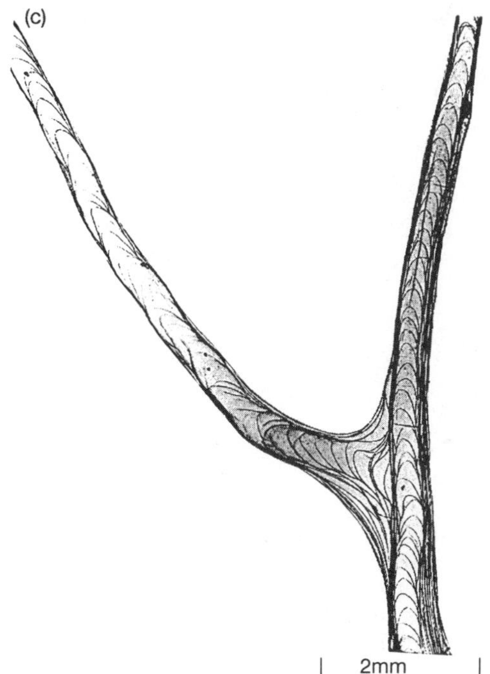

Figure 1.15 Graptolites.
(a) *Monograptus* and (b) *Amplexograptus* (from Palmer and Rickards, 1991), (c) a bifid spine secreted by *Cephalodiscus graptolitoides,* a possible extant graptolite (from Dilly, 1993).

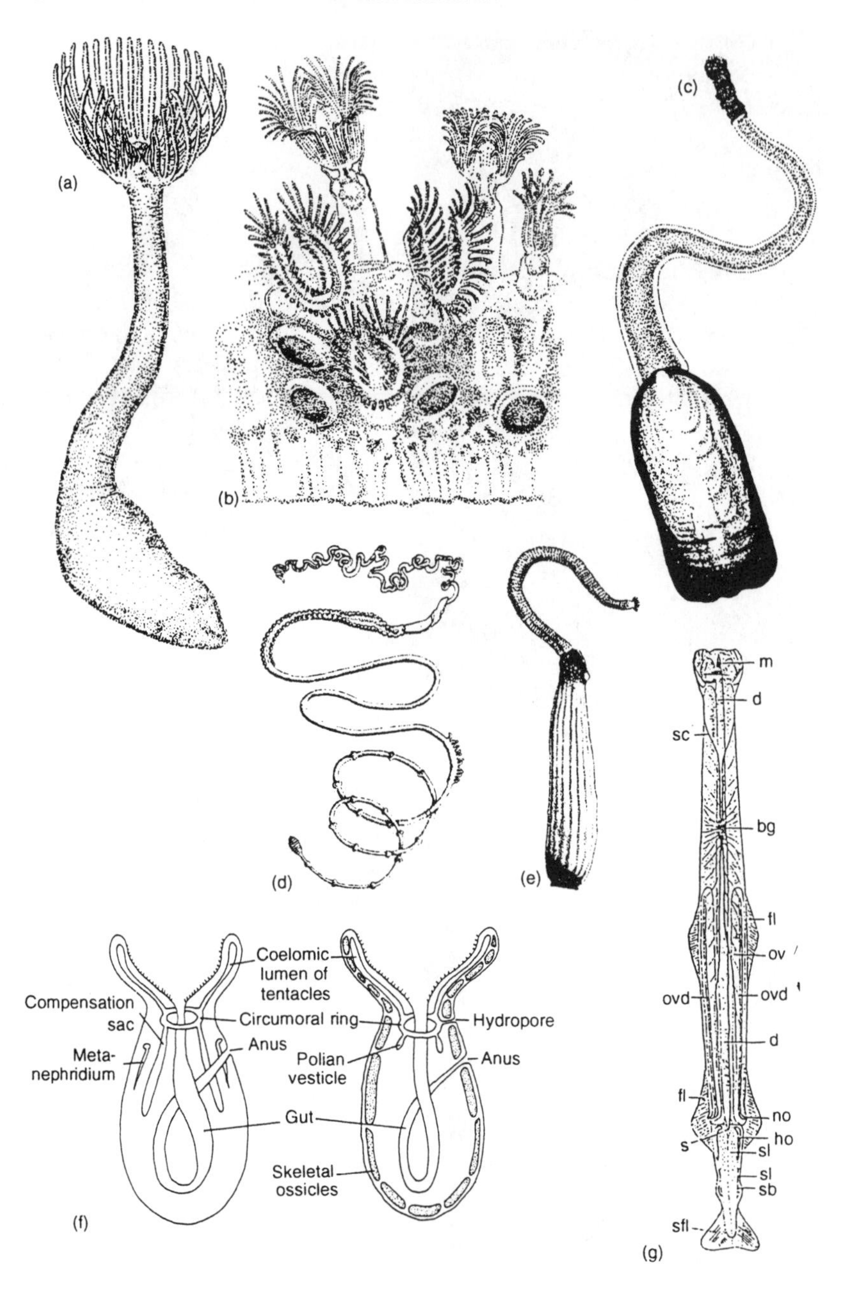

Figure 1.16 The shadowy borderlands between protostomes and deuterostomes. (a) A phoronid, (b) bryozoans, (c) a brachiopod, (d) a pogonophore and (e) a sipunculid (from Willmer, 1990). (f) The supposed homologies between sipunculids (left) and echinoderms (right) after Nichols' theories (from Willmer, 1990). (g) a chaetognath, *Sagitta* (from Hertwig, 1909).

species. Because of their calcitic exoskeletons, bryozoans are extremely well-known as fossils. About 15 000 fossil species have been named, and the group dates back to the Ordovician.

Although about 30 000 species of fossil brachiopod have been described, dating back to the early Cambrian, barely 280 species survive today (Figure 1.16c). At first sight they look like bivalve molluscs, as the body is encased within a bivalved shell. Unlike bivalves, though, the shells encase the top and bottom of the body, rather than the left and right sides, and are of unequal size and shape rather than mirror-images. The brachiopod shell once opened reveals a lophophore, something which, of course, is never found inside a bivalve.

There are two kinds of brachiopod; 'articulate' forms in which the shells are hinged, and may be calcareous or phosphatic, and 'inarticulate' forms in which the phosphatic shells are quite separate. *Lingula*, perhaps the most familiar extant brachiopod genus, is inarticulate.

Cleavage in brachiopods is radial, and mesoderm and coelom formation appear to be enterocoelous. These attributes suggest an affinity with the deuterostomes, although the exact details vary from one species to another.

However, brachiopods also bear hair-like structures or 'setae' that are characteristic of annelids among the protostomes, and an increasing body of evidence suggests that brachiopods, as well as the other lophophorates, are protostomes closely related to molluscs and annelids. Cambrian armoured worms known as wiwaxiids, similar in some ways to annelids, appear to have had setae. A slightly older group, the halkieriids, had body armour similar to wiwaxiids, and also sported brachiopod-like shells at each end of its body (Conway Morris, 1993, 1995; Conway Morris and Peel, 1995). Molecular evidence based on partial sequences of 18S rRNA genes also indicates a coelomate protostome affinity for brachiopods (Field *et al.*, 1988; Lake, 1990). This has since been amplified by work on complete sequences (Halanych *et al.*, 1995), a particularly interesting study in that it considers bryozoa and phoronids as well as both articulate and inarticulate brachiopods. The analysis places bryozoa as an outgroup to a group of protostomes containing brachiopods, phoronids, molluscs and annelids. Interestingly, the phoronids come out as a group of shell-less brachiopods, being closely related to articulate brachiopods at the expense of inarticulate forms.

An abiding problem with all this work is the phylogenetic significance of the lophophore. Such an organ is unlikely to have evolved twice – in protostome lophophorates and deuterostome pterobranchs – yet Halanych *et al.* (1995) do not consider this inconsistency. However, complete 18S rRNA sequence data now exist for pterobranchs, and sequence comparison confirms a close relationship with enteropneusts (Halanych, 1995). Until sequences from pterobranchs and lophophorates are compared in the same analysis, therefore, the phylogenetic significance of the lophophore remains untested (Gee, 1995a).

1.9.11 Chaetognaths

The chaetognaths (Figure 1.16g), or arrow-worms, are a small group of about 50 species of marine animals, traditionally regarded as relatives (albeit remote) of deuterostomes. The following is largely drawn from Barnes (1980). They are bilaterally symmetrical animals with straight and more or less rigid bodies, bounded by fins, and are active predators. There seem to be two rather than three coelomic compartments, but the coelom may in fact be a pseudocoelom, as the presence of a pentoneum is not obvious (but see Shinn, 1994). Indeed, chaetognaths resemble giant rotifers in their general appearance.

Development appears to be deuterostomatous, with radial cleavage and an aberrant form of enterocoely. However, much recent anatomical work (Telford, personal communication) shows that chaetognaths are almost certainly not deuterostomes. Comparative analysis of 18S rRNA sequences puts the chaetognath *Sagitta* outside the coelomates, more primitive than either protostomes or deuterostomes (Telford and Holland, 1993). Although this supports the view that chaetognaths are pseudocoelomates rather than true coelomates, other recent ultrastructural work places them as coelomates (Shinn and Roberts, 1994), but with a form of schizocoely rather than enterocoely. Clearly, the debate about the affinities of chaetognaths is now concerned with their affinities with coelomates in general, rather than deuterostomes in particular. For a review of bodily cavitation in the 'Aschelminth' phyla, see Rupert, 1991.

Fossil chaetognaths are known from the Middle Cambrian Burgess Shales of Canada, and also from the Carboniferous Mazon Creek fauna of Illinois. All these fossil forms are reported to look like modern chaetognaths in all important respects, so reveal little about the relationships of chaetognaths in general (Telford, personal communication). The fossil *Amiskwia sagittiformis* from the Burgess Shales also looks something like a chaetognath, but close inspection suggests that this is a passing resemblance only. Recent work links this form with nemertean worms (Conway Morris, 1993).

1.9.12 Pogonophores and sipunculids: distant dreams

Two other groups deserve a mention here, although their affinities with deuterostomes are arguably even more tenuous than those of the lophophorates.

The pogonophores (Figure 1.16d) constitute a small group of about 80 worm-like animals, known for barely a century on account of their exclusively deep-sea habitat. Interestingly, adult pogonophores have neither digestive tract nor anus, and appear to absorb food through the body wall. The characteristic anterior mop of tentacles and a collar-like forepart are reminiscent of pterobranchs, but pogonophore embryogeny is teasingly ambiguous: there is no blastocoel, and cleavage could be a modified form of either spiral or radial.

Until 1964 it was suggested that the body reflected a tripartite condition, like deuterostomes – only then it was discovered that the creatures had a multisegmented hind region bearing annelid-like setae. This part had always been absent from the dredged specimens examined before this date. Pogonophores are now regarded as closer to annelids than deuterostomes (see Southward, 1975, and references therein).

The 300 species of marine worm-like sipunculids are similarly thought to be allied to annelids. Cleavage is spiral and the larva looks like a trochophore along annelid lines. And yet they lack segmentation and have an anterior fringe of tentacles somewhat reminiscent in some ways of a lophophore, and an internal hydraulic system so reminiscent of the distinctive water–vascular system that Nichols (1967a) posited a phylogenetic link between the two phyla (Figure 1.16e,f).

A comparative analysis of ribosomal RNA sequences (Lake, 1990) groups pogonophores and sipunculids with annelids and molluscs, not with deuterostomes.

1.9.13 Tunicates

Tunicates (Figure 1.17) are true chordates, in the sense that they display those features described above as diagnostically chordate, at least in some part of the life cycle. All tunicates are filter-feeders: a current of water is drawn by ciliary action through the mouth and a sieve-like multitude of fine gill slits (subdivided by tongue bars and cross-struts called synapticulae), arranged in a cylindrical basket-shape, and is then expelled through an exhalant siphon.

Particles of food are trapped in a filter of sticky mucus draped over the gills. This mucus is secreted in a gutter that runs along the ventral edge of the gill basket: the endostyle. Once secreted, the mucus is drawn by cilia upwards along the gill bars and across the spaces between them. At the top edge of the gill basket, opposite the endostyle, the upwardly mobile sheets of mucus are rolled up with the help of a curved plate-like structure called the dorsal lamina. The roll of mucus and food is then passed into the oesophagus and digested within the viscera.

The whole filter-feeding apparatus between mouth and oesophagus constitutes the pharynx: it seems to correspond with the gill-bearing regions in *Cephalodiscus* and enteropneusts, although these creatures strain food from sea water using structures outside the pharynx: respectively the lophophore and the ciliated proboscis. Echinoderms have no structure that corresponds with the filter-feeding pharynx of tunicates.

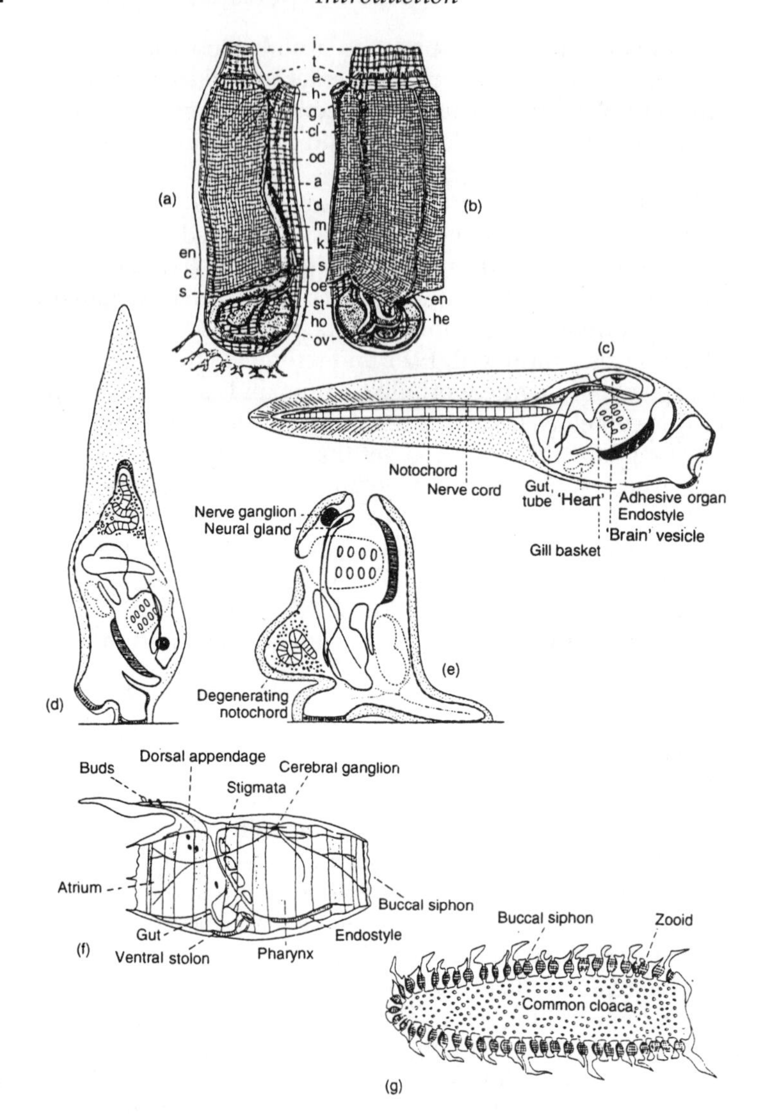

Figure 1.17 Tunicates. The ascidian *Ciona intestinalis* from (a) the left side with tunic removed and (b) from the right side with tunic removed and pharynx opened up. a = anus, c = cellulose tunic, cl = cloacal (exhalant) siphon, d = rectum, e = exhalant atrial opening, en = endostyle ending above in peripharyngeal band, g = ganglion, h = mouth of 'hypophysis' (subneural gland), hr = heart with pericardium, ho = gonads, i = inhalant opening, k = branchial basket, m = muscular sac, oe = oesophagus, od = oviduct, ov = ovary, s = partition between atrium and body cavity, st = stomach, t = tentacular crown (from Hertwig, 1909). (c–e) Metamorphosis of a solitary tunicate. (c) A free-swimming larva. (d) and (e) Settlement and metamorphosis is accompanied by the degeneration of larval structures such as the notochord (from Romer, 1970, after Dawydoff). (f) An oozoid of the solitary thaliacean *Doliolum* and (g) the colonial thaliacian *Pyrosoma* in longitudinal section (from Barnes, 1968).

Another difference between tunicates and hemichordates is that the delicate, much-subdivided tunicate gill-basket is enclosed within an 'extra' body wall or tunic, the structure that gives its name to the group. This is secreted by the epidermis, and although external to it, contains many cells of mesodermal origin, including blood cells. It also contains cellulose, a compound usually associated with the cell walls of plants. This is laid down in the tunic by a special kind of blood cell called a vanadocyte, so-called because it sequesters trace amounts of the metal vanadium, used in the substances the cells need to produce the cellulose. Tunicates also gather other elements found as traces in sea water, notably iodine[14].

Most tunicates belong to a group called the Ascidiacea, which numbers about 1000 species worldwide. These are bottom living, occasionally colonial and mainly hermaphrodite animals. Although sessile as adults, they are dispersed through their motile larval stage, the so-called tadpole larva.

Tunicate development starts much as in other deuterostomes in that cleavage is radial, although development is deterministic up to at least the 64-cell stage (after which gastrulation starts). The archenteron obliterates the entire blastocoel. The blastopore forms the presumptive anus.

After that, events take a rather odd turn: the mesoderm does not originate by enterocoely, and the coelem appears restricted to the pericardium. The mesoderm is made of cells shaved from the inner surface of the archenteron.

The larva extends longitudinally, and the top edge of the archenteron pinches off along its length to form the (mesodermal) notochord: above the notochord, which is largely confined to the tail region, is a hollow nerve cord.

The formation of the nerve cord and the central nervous system depends on interactions between tissues, in particular the presumptive nerve-tissue cells and those cells that give rise to the notochord and the roof of the archenteron. Thus, for all its oddity, tunicate development has at least the essentials of neural induction, stimulated by chordamesoderm and/or endoderm, seen in cephalochordates and vertebrates. The notochord and nerve chord are characteristic features of chordates, and together form the 'tail' of the tadpole.

The apparent strangeness of tunicate early development should be seen in context. Even within vertebrates, the early development of chicks, say, and mice differ in important respects from one another. Yet by the time of neural induction and the formation of axial structures – the so-called 'phylotypic' stage – many common features characteristic of chordates emerge. It could be that the early ontogeny of a given species reflects local cellular conditions and mechanical constraints (such as the abundance of yolk) rather than heritage, the signal of which becomes apparent later on (see Hall, 1992 for detailed discussion of this point).

Such mechanical constraints have profound effects on the form of the gastrula, and yet in molecular terms the gastrulae of vertebrates in general have much in common (De Robertis *et al.*, 1994), and gastrulation appears to be continuous with the formation and extension of the tail. In tunicates, however, gastrulation happens at such an early stage relative to the number of cell divisions, that possible tissue movements and interactions are limited by the small number of cells (Berrill, 1955).

Signs of segmentation are scarce in adults as well as larval tunicates. The muscles of the tadpole tail are sometimes clumped into groups of fibres but are not divided into discrete somites. Whether this lack of segmentation is a primitive (Schaeffer, 1987) or degenerate (Hubrecht, 1883; Jollie, 1982) condition remains unresolved. Recently, Crowther and Whittaker (1994) reported a serial repetition of ten equidistantly placed cilia pairs, arranged in two rows immediately opposite to each other mid-dorsally and mid-ventrally, embedded in the matrix of the test that forms the flattened tail-fin in the larva of *Ciona intestinalis*. These immobile cilia originate from cell bodies in mid-dorsal and mid-ventral peripheral nerves running beneath the epidermis of the tail. Crowther and Whittaker (1994) speculate that this serial repetition could be related to segmentation as found in vertebrates. Some genetic evidence supports this view: a homeobox gene recently isolated from tunicates is, remarkably, expressed in a 'segmental' pattern in the neural tube (Di Lauro, personal communication).

The 'head' of the tadpole contains the germs of the adult organs, surrounded in an epidermally secreted tunic with adhesive suckers at the front: the larva does not feed. Adult tunicates have no organs of special sense, but the tadpole larva has a single, simple eye and a gravity-sensing organ called a statolith.

The function of the larva seems to be to locate a suitable place for settlement and metamorphosis into an adult, as quickly as possible (Berrill, 1955). After a larval life that may be as short as a few minutes, the tadpole sticks itself head-first to the substrate. The tail atrophies, and along with it the notochord and nerve cord, and the animal turns into the sessile adult. The lack of segmentation in the tail is consistent with the tadpole as a short-lived phase: if it lived longer, segmented muscle blocks as seen in cephalochordates and vertebrates might have appeared (Schaeffer, 1987) although the fact that the larvacean tunicates lack muscle-blocks speaks against this.

The tadpole, then, is a specialized dispersal mechanism for otherwise sessile animals living in varied seafloor-scapes (Berrill, 1955). The lives of ascidians that live on fairly monotonous sandy or muddy seafloors seem to bear this out. As one place is very much like another, these species tend to lack the tadpole stage entirely and rely on passive dispersal.

Some tunicates have adopted a free-swimming, planktonic lifestyle, a dynamism embodied in the shapes of the salps or thaliacea: cylindrical and reminiscent of the engine pods of jumbo jets. Unlike the samovar-shaped ascidians, the inhalant water current of a salp passes through the front end, straight through the pharynx and out the back. Gas exchange and a kind of jet propulsion are added to filter-feeding as a pharyngeal function. Some salps are colonial, and have sex lives of great delicacy and complexity: the salp formed from the egg has the sole function of budding off other individuals asexually, these secondary individuals being capable of sexual reproduction.

Stranger still are the larvacea or appendicularians, so called because the adults seem to retain many of the characteristics of the tadpole larva, in particular the long notochord-supported tail (Figure 1.18). The presence of only one gill slit per side and the lack of a tunic are other distinctive features. Larvaceans are very small and float freely in the surface waters of the sea, occasionally in great numbers.

This sociability is illusory, though, as each animal is solitary, each living in a mucous structure called a 'house', secreted by the animal and continually replaced. Unlike the relatively snug fit of the tunic, the house is very much bigger than the animal. For example, *Oikopleura* is hardly bigger than a grain of rice, yet its house is the size of a walnut. Larvacean houses are often complicated in structure, with a number of passageways as well as incurrent and excurrent openings. Sitting at the centre of its house like a spider in its web, the larvacean sets up water currents with its tail that drive seawater through the house.

The animal lives on tiny planktonic organisms carried in on the currents and trapped in a series of fine meshwork screens. The food particles reach the mouth of the animal on a conveyor belt of mucus and, once inside the creature itself, are ensnared in mucus produced by the endostyle. Perpetually on the move, the tadpole larva of a larvacean develops directly into the adult, as one might expect[15].

Recently, a fourth kind of tunicate, was discovered, the sorberacea. Exclusively benthic, these creatures lack branchial sacs and are carnivorous. They capture prey (up to and including small worms and crustaceans) by means of finger-like extensions of the oral siphons (reviewed in Gage and Tyler, 1991)[16].

The discussion of the origins of vertebrates – indeed, the development of evolutionary theory in general – is intimately connected with research into the affinities of tunicates. In the mid-nineteenth century they were grouped with polyzoa (bryozoa and ectoprocts), but the general view was that they were molluscs (see Katz, 1983 for a short history and references). One scheme had it that the barrel-shaped pharynx of tunicates formed from the midline fusion of the gill-plates of a lamellibranch-like ancestor.

Tunicates have been regarded as chordates only since 1866, when Kowalevsky drew the comparison between the notochord in the

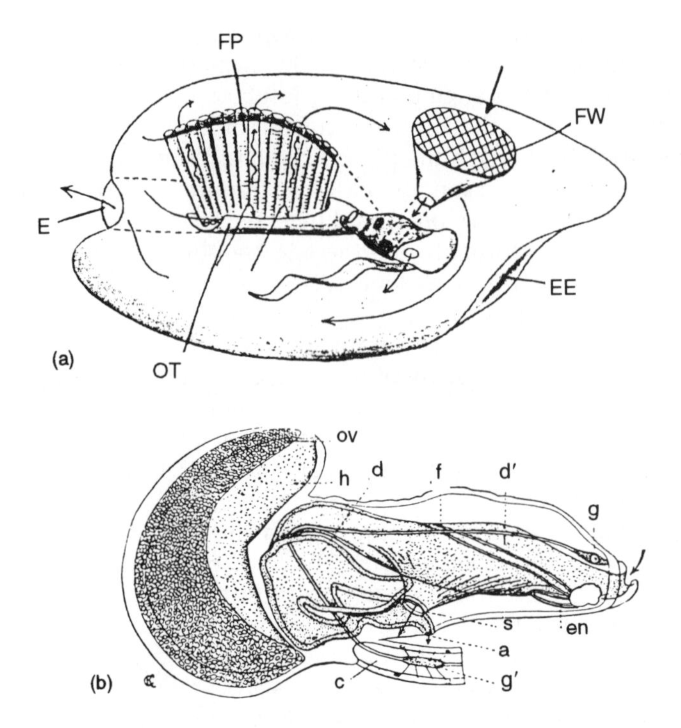

Figure 1.18 The larvacean or appendicularian *Oikopleura* (a) at home, in the middle of its house (from Garstang, 1928). The arrows indicate direction of water flux. (b) The whole animal removed from its house, in side view with the base of the tail (note that it is seen from the side obverse to that above); a = anus, c = notochord, d′ = branchial region, d = stomach, en = endostyle, f = ciliated peripharyngeal bands, g, g′ = brain and first tail ganglion, h = testis, m = mouth, ov = ovary, s = gill slits (from Hertwig, 1909).

amphioxus and the 'axis' in the tail of the tunicate tadpole (Kowalevsky, 1866a, p.13)[17]. This discovery is largely responsible for sparking off the modern debate about the origin of vertebrates. (Indeed, Haeckel used it as the basis for recapitulation.)

The isolation of tunicate genes (such as homeobox genes) involved in developmental regulation is at an early stage compared with progress in other chordates and many other organisms (Holland *et al.*, 1994b; Ruddle *et al.*, 1994). However, enough has been found out to suggest that tunicates are highly specialized and derived creatures. Those genes which have been isolated are divergent from homologues in other animals (explaining why they have been hard to isolate). However, such technical difficulties appear to have been overcome, and many interesting tunicate homeobox genes have now been identified (Di Gregorio *et al.*, 1995; Saiga, personal communication).

1.9.14 Cephalochordates: shaved fish

The most familiar cephalochordate, or acraniate, is the lancelet or amphioxus, *Branchiostoma lanceolatum*, although there are around 25 species in all, divided into two genera, the other one being *Epigonichthys* (=*Asymmetron*) (Poss, personal communication). Most spend their time buried in sandy seafloors, although the possibly paedomorphic form *Amphioxides* is pelagic. To describe these creatures as shaved fish seems most apt – preserved specimens look like anchovy fillets.

They are leaf-shaped, laterally compressed, translucent and streamlined animals a few centimetres in length (Figure 1.19). Although superficially fish-like, they have no scales, eyes, jaws or paired fins. They do, however, have a notochord and overlying dorsal nerve cord, and a well-developed segmental arrangement of chevron-shaped muscle blocks (Figures 1.20 and 1.21). These features mark cephalochordates as true chordates.

Nevertheless, they share a filter-feeding habit with tunicates. Although notionally free-living, adults tend to remain buried tail-first in the sand in nearshore waters. There they use ciliary action to draw water currents into the mouth and into the pharynx (Orton, 1913). Like tunicates, the pharyngeal floor is scored by the endostyle, which secretes mucus. This mucus is spread over the gill slits (much subdivided by tongue bars similar to those of enteropneusts). The water currents pass through the gills and into an atrium. Water escapes the animal from a ventral median atriopore. As in tunicates, the atrium is an accessory structure that protects the extensively perforated pharynx from damage.

On the other hand, the bilaterally symmetrical, mobile cephalochordates look similar to vertebrates in their general organization and in detail. Even several features long thought to be unique to cephalochordates, setting them apart from vertebrates, now turn out to be illusory or mistaken.

For example, the notochord extends from the very tip of the tail to the most anterior tip of the head. In vertebrates, the anterior end of the notochord is to be found in the middle of the head. But despite every appearance of their being headless, recent morphological (Lacalli *et al.*, 1994)

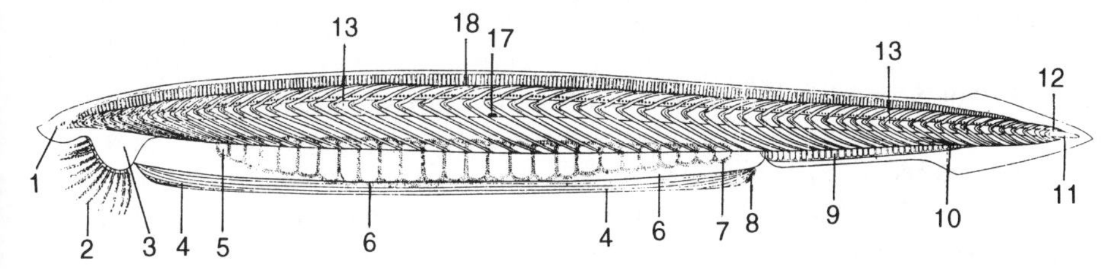

Figure 1.19 Amphioxus adult, from the left side (from Sedgwick, 1905, after Lankester).

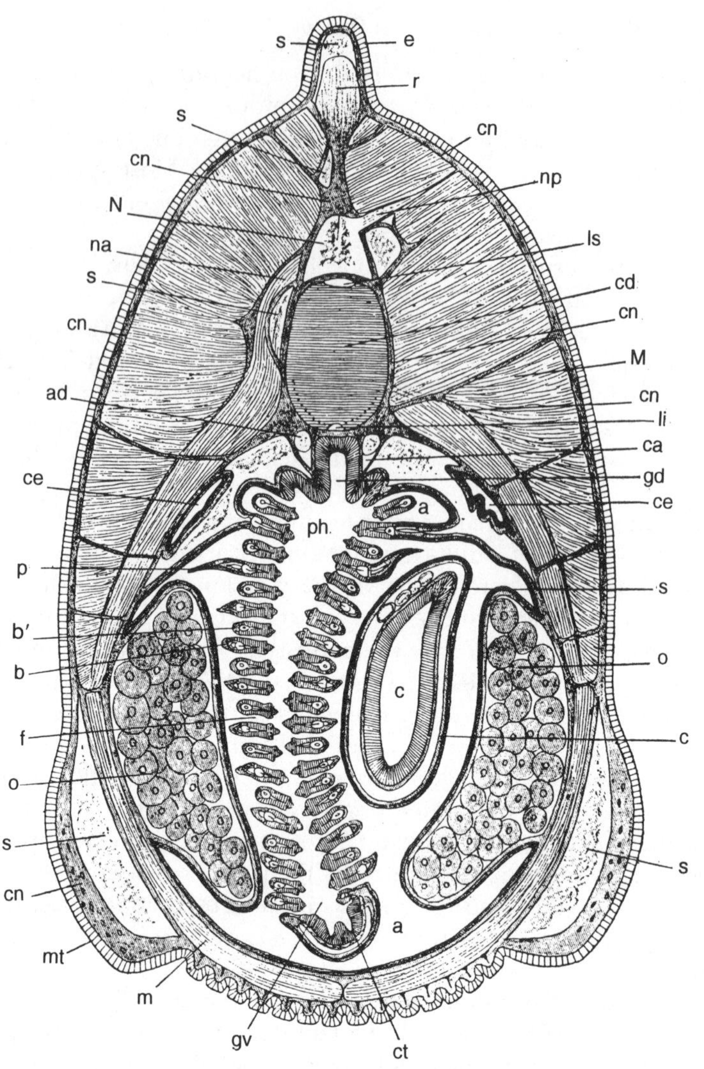

Figure 1.20 Amphioxus in transverse section through the hinder part of the pharyngeal region (from Sedgwick, 1905, after Lankester and Perrier). cd = notochord, N = nerve cord, ph = pharynx, b = gill bar, ct = endostyle.

and genetic work (Holland *et al.*, 1992a; Garcia-Fernàndez and Holland, 1994; Gee, 1994) shows that much of the anterior region of the animal is homologous with the vertebrate head.

Indeed, the tiny cerebral vesicle, bearing a single eyespot, seems to be homologous with the diencephalon, the part of the vertebrate forebrain from which the paired eyes originate, and much of the neural tube corresponds with the vertebrate hindbrain (Lacalli *et al.*, 1994).

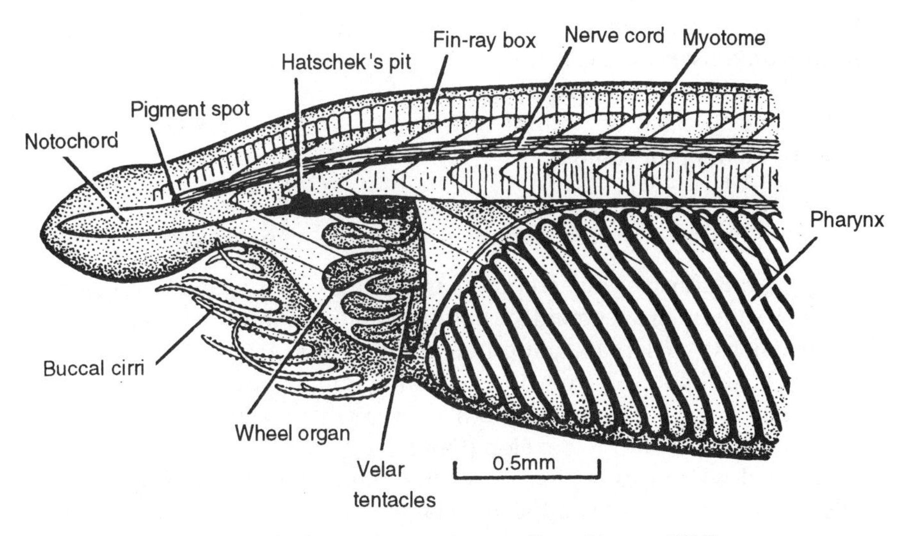

Figure 1.21 Anterior end of a young amphioxus (from Young, 1981).

Second, the excretory system – long thought to be made of a loose network of individual cells and canals with a superficial resemblance to the 'flame cells' of flatworms – is now thought to be more similar to vertebrate glomeruli, with modifications that one might expect in compensation for small size (Ruppert, 1994).

What does seem to set craniates apart from cephalochordates in particular (and everything else in general) is the presence in the former of the 'neural crest' from which much that is distinctive in the vertebrate head, nervous system and skin is derived (Gans and Northcutt, 1983; Gans, 1987). The amphioxus differs from vertebrates mainly in the lack of neural crest.

There are other features of the anatomy, though, that set cephalochordates apart. One is the lack of spinal ganglia. Moreover, instead of neurons innervating the muscle blocks, the muscles send 'tails' that make direct contact with the nerve cord (Flood, 1966). This style of innervation is the exact opposite of what happens in vertebrates, in which nerves send branches that innervate muscles.

Perhaps the most interesting features of cephalochordates are to be found in their development[18], for it contains many surprising and unique phenomena for which, even after more than a century of study, there remain no clear explanations.

The earliest stages follow the deuterostome pattern, and bear few surprises, being broadly similar to that in vertebrates, but for the lack of neural crest. Neural induction depends on the presence of notochord cells, and during enterocoely the somitic mesoderm develops through the interaction of notochord and endoderm, so that each somite as it forms contains its own coelomic compartment.

In their account of amphioxus development, Lankester and Willey (1890) add a touch of natural history that is almost poetic. The early development of the animal

> ... always commences at dusk – between the hours of seven and eight – and goes on very rapidly through the night. (p.447)

Early in this furtive and crepuscular process, both somitic and lateral plate mesoderm are segmented, but the somites of the lateral plate mesoderm (and its coelomic compartments) merge, to become (secondarily) continuous, separated from the overlying somites by horizontal partitions. The result is a planktic, fish-like larva with a distinct 'head' region, in which the gills start to develop (Figures 1.22 and 1.23).

The real weirdness begins somewhat later, when the young creature adopts for a mouth what appears to be the greatly enlarged first anterior left gill slit. Even stranger, a second gill slit appears not on right or left but in the ventral midline, whence it moves to lie on the right side of the body.

Once the animal becomes a free larva, 12–15 unpaired slits appear in anterior–posterior sequence, one after the other, following the lead of the second slit – starting ventrally and swinging up to the right. By this time, the longitudinal ridges that mark the rudiments of the accessory body wall, the atrium, have appeared. According to Lankester and Willey (1890) again, 'in this stage the larva rests habitually on one side at the bottom of the [laboratory] vessel and does not bury itself in sand or mud' (p.448), after which the animal undergoes a process of recovering a semblance of the basic symmetry lost during the *outré* gill-forming episode.

The importance of the asymmetric phase in the development of the amphioxus cannot be stressed enough. Lankester and Willey's century-old account – really a commentary on earlier work by Kowalevsky (1866b) – cannot be bettered for clarity, and is worth quoting at length:

> According to that account [i.e. Kovalevsky (1866b)], after some dozen gill-slits have taken up their position on the animal's right side – having moved into that position from the median line – a new and startling change occurs. The whole series moves downwards across the median line and up the left side of the pharynx, so that the primitive right-side gill-slits become the left-side series; and in the meanwhile a new series corresponding in number make their appearance not one by one, but all together, in the right side of the pharynx, occupying, as it were, the position deserted by the rotated primitive series. This movement of growth appears to be a general one affecting the whole pharynx, for, simultaneously, with

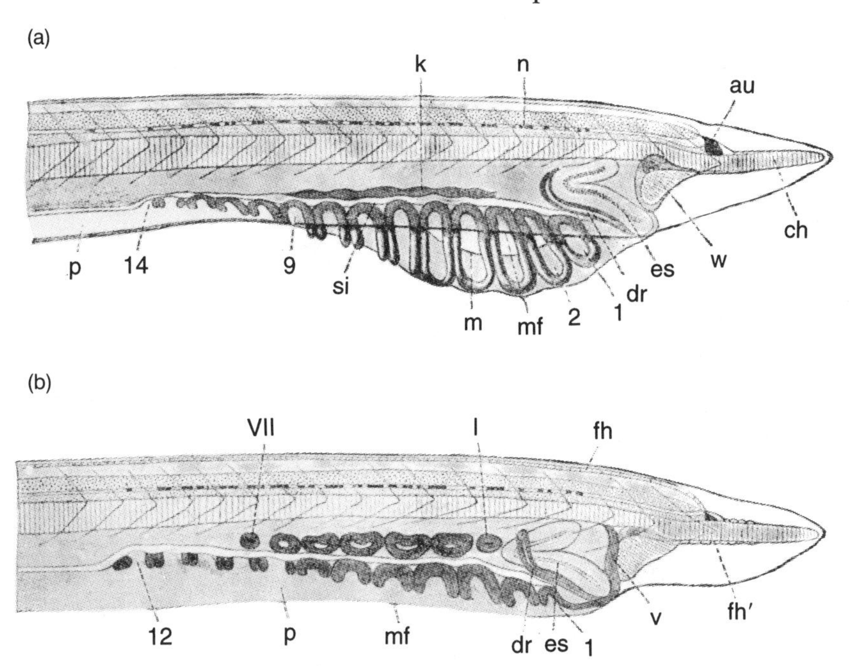

Figure 1.22 The origin of gills in the amphioxus. The left series appears in the mid ventral line, and moves across the mid-ventral line to the right side. The right-hand series develops on the right above this series, after which both series swing leftwards to rest in their adult positions. (a, b) Two larval stages of the amphioxus in right-lateral view, (a) later than (b), showing this process at work (from Sedgwick, 1909, after Korschelt and Heider, after Willey). (a) shows the future left series on the right side of the animal, before the right-hand set has started to form (their rudiments are marked *k*). In (b), the right-hand set has started to form above the left-hand set, which is now rotating leftwards round the ventral margin.

> the translation of the primitive gill slits from right to left, the great larval mouth moves from its extraordinary position on the animal's left side, and, becoming relatively much smaller, takes up its permanent position as an anterior median orifice whilst its hood and tentacles appear. We have not, we regret to say, at present been able to study any larvae in which these remarkable changes are in progress (p.460)

They continue:

> ... it is a curious fact that the morphologically median plane of the pharynx of the young larva becomes the left side of the adult, whilst the relations of the mouth to median plane, in adult and larva respectively, are even more curiously divergent. It is probable

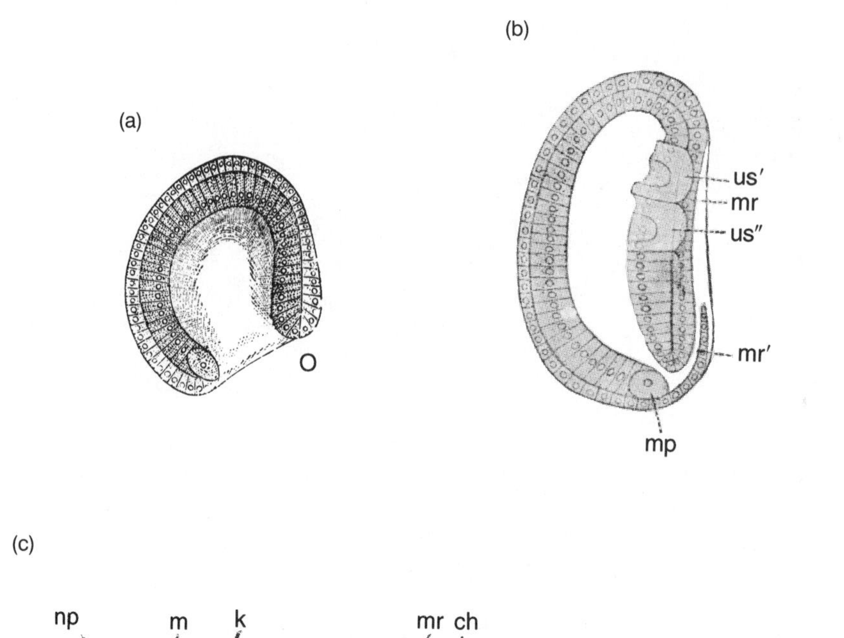

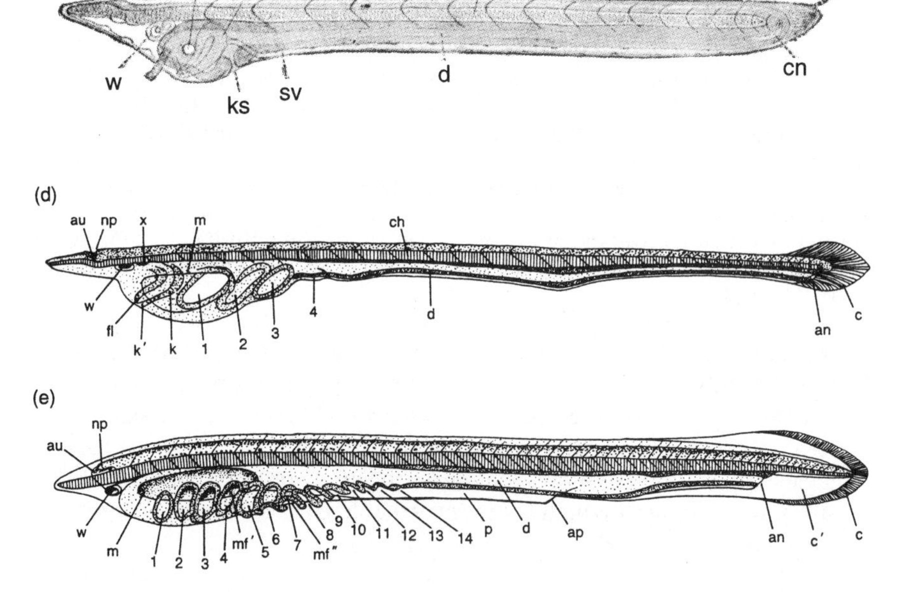

Figure 1.23 Five left-lateral views of amphioxus embryogeny (from Sedgwick, 1909, after various sources). (a) Gastrula; (b) early embryo with the rudiments of two somites, neurenteric canal and the beginnings of a medullary plate (mr); (c) 36-hour-old embryo with left-sided mouth (m), a further gill slit (ks), preoral pit (left head cavity) (w), nerve cord, notochord and myotomes; (d) larva with three gill slits and 36 myotomes; (e) with 14 gill slits and full complement of 61 myotomes.

> enough that in these differences the larva does not present the more archaic condition, but an adaptational arrangement. We do not at present know what are the conditions of life which render its excessive asymmetry advantageous to the larva. (p.461)

Eventually these asymmetries are largely ironed out, and the adult is buried in the sand, head up and tail down, with no lateral preference. As the atrium forms, the mouth moves ventrally and the gills become arranged in two series, left (or primary) and right (or secondary). Tertiary gill slits open on both sides, behind these primary and secondary gill slits. Asymmetries as regards distributions of internal organs do persist into adulthood, particularly in *Epigonichthys*.

The significance of these asymmetries has remained a mystery for more than a century. Bone (1958) examines two contrasting hypotheses[19].

First, that asymmetry is a consequence of adaptation to the larval lifestyle (Van Wijhe, 1913; Garstang, 1928): the fact that the asymmetries are smoothed out during metamorphosis is a testament to their necessity in larval, as opposed to adult life. They must, therefore, be connected with some feature of life in the plankton.

The second rests on the observation that the notochord in the amphioxus extends to the anterior tip, in complete contrast to that in tunicates and vertebrates. This essentially adult specialization (for burrowing) imposes certain functional constraints on the larva. Willey (1894), and later Medawar, supposed that the forward extension of the notochord displaced the mouth from a primitively anterior and dorsal position to the left side.

Bone (1958) adopts a slightly different view – not that the mouth is shoved aside by a violent notochord, but that it bows out gracefully before that confrontation occurs. Presumably, this torsion would occur whether the notochord were there or not, so there is some selective advantage for the mouth to be laterally placed. Thus, Bone subscribes to Garstang's view that a laterally placed mouth in a laterally compressed animal can be bigger than a dorsal or ventral mouth, or one at the apex. Asymmetry is an adaptation to present larval function, says Bone, not an adult adaptation or a holdover from an ancestor asymmetric for other reasons.

Somewhat earlier, Goodrich (1917) reviewed opinion on the function of Hatschek's Pit, a structure in the roof of the buccal cavity of the amphioxus, in which can be seen another asymmetry: 'the story of the origin of Hatschek's pit is one of the strangest episodes in the strange history of the development of Amphioxus', he writes (1917, p.541).

Essentially, Hatschek's Pit and some other structures develop from the left of two coelomic sacs that derive from the front end of the archenteron. The right-hand sac becomes the head coelom and is blind, but the left, through Hatschek's pit, opens to the outside as a ciliated canal.

Goodrich (1917) harks back to Bateson (1885), drawing a homology between this opening and the 'proboscis pores' of the enteropneust *Balanoglossus*, found on both sides of the animal in some species, but on the left side only in others. He speculates on further homologies; with the protocoel pores of *Cephalodiscus*, the echinoderm hydropore (likewise a left-handed structure), possibly the subneural gland of tunicates – and the hypophysial portion of the vertebrate pituitary.

Goodrich argued that the paired premandibular somites – the most anterior somites of the vertebrate body, each containing a coelomic cavity – bore, at least primitively, a duct each connecting the hypophysis with the external environment. Such ducts are certainly evident in fish and reptiles. With reference to sharks, he notes that 'there may be a right and a left tube, but – a significant fact – the left is usually better developed and persists longer than the right' (1917, p.548), the implication being, possibly, that this phenomenon echoes the asymmetry in amphioxus in which the right-hand tube fails to develop.

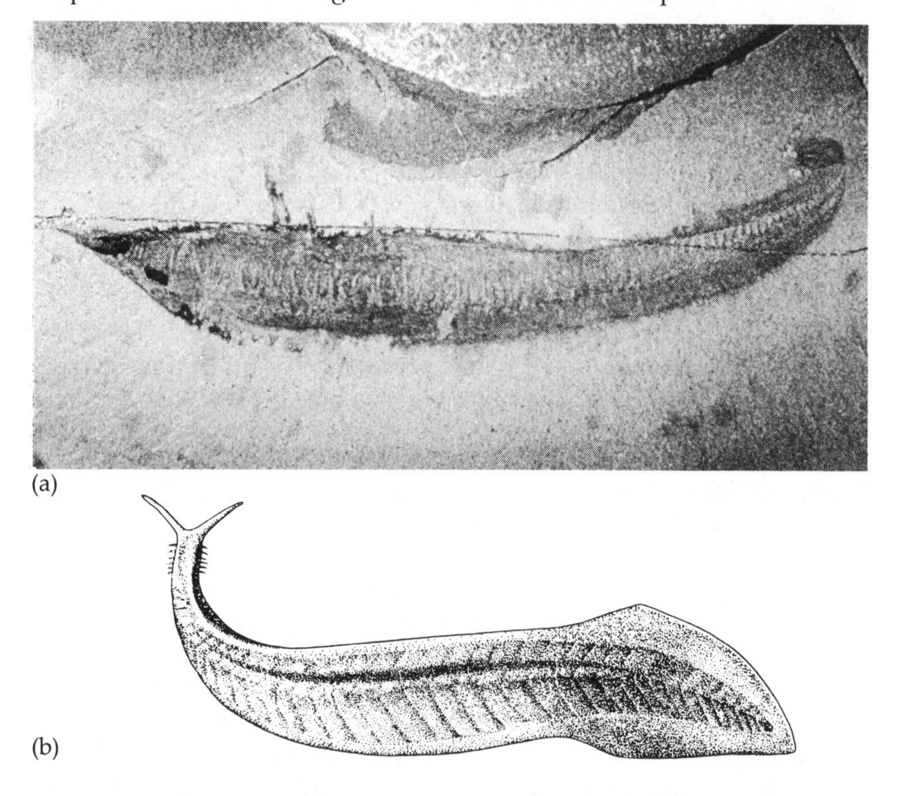

Figure 1.24 *Pikaia*, a possible cephalochordate from the mid-Cambrian Burgess Shales of British Columbia; (a) a specimen (courtesy Simon Conway Morris, University of Cambridge), and (b) a restoration (drawn by Marianne Collins, from Gould, 1989).

Turning to the fossil record of cephalochordates, the earliest representative is *Pikaia* from the Middle Cambrian Burgess Shales of Canada (Figure 1.24, and Gould, 1989). This animal had segmental, chevron-shaped muscle blocks, a bilobed 'head' and possibly fringing fin-folds. One wonders on the similarity between this creature and body-fossils of conodont animals (see below). More controversially, Jefferies (1973) claims that the Ordovician mitrate carpoid *Lagynocystis* is a stem cephalochordate: among many other features, it has a median atrium with tertiary gill slits in addition to left and right atria. More on this interesting fossil in Chapter 4. Chen *et al.* (1995) interpret the Cambrian Chinese fossil Yunnanozoon as a cephalochordate, though others prefer a hemichordate status (Shu *et al.*, 1996), or that it belongs to a previously unknown class of chordate (Dzik, 1995).

1.9.15 Hagfishes

These exclusively marine creatures are traditionally regarded as primitive vertebrates. However, they have neither scales nor paired fins, their external sense organs are reduced, and they lack jaws. Their long, snakelike bodies are supported by a notochord which, in contrast with other vertebrates, remains unadorned with the distinctive neural and haemal arches of vertebrae. It hardly seems appropriate, then, to class them as vertebrates. Nevertheless, they do have a cranium – so they may be called craniates (Janvier, 1981). Again, unlike all other vertebrates (but like echinoderms and other chordates) their body fluids are isotonic with sea water.

The fossil record of hagfishes is meagre. The earliest known comes from the Upper Carboniferous of Illinois (Bardack, 1991), although this should not be regarded as representative of the origin of the group.

1.9.16 Conodonts

Geologists have long had the benefit of tiny, tooth-like fossils called conodonts. These are so common in some Lower Palaeozoic sedimentary sequences that they can be used as zone fossils. But for many years, there was not the faintest idea of what conodont-bearing animals actually looked like. Some people even suggested that they were crystalline inclusions in plants.

One interesting diversion came with the description of the enigmatic creature *Typhloesus wellsi* from the remarkable Carboniferous Mazon Creek fauna (Conway Morris, 1990; Figure 1.25b). This animal contained conodont teeth in its gut – were they used like the pharyngeal teeth of some fish, or gizzard-stones of birds, to grind up prey swallowed whole?

Another was the suggestion, also by Conway Morris (1976; Figure 1.25a), that tooth-like structures on what might have been a lophophore in the flattened worm-like creature *Odontogriphus* from the Burgess

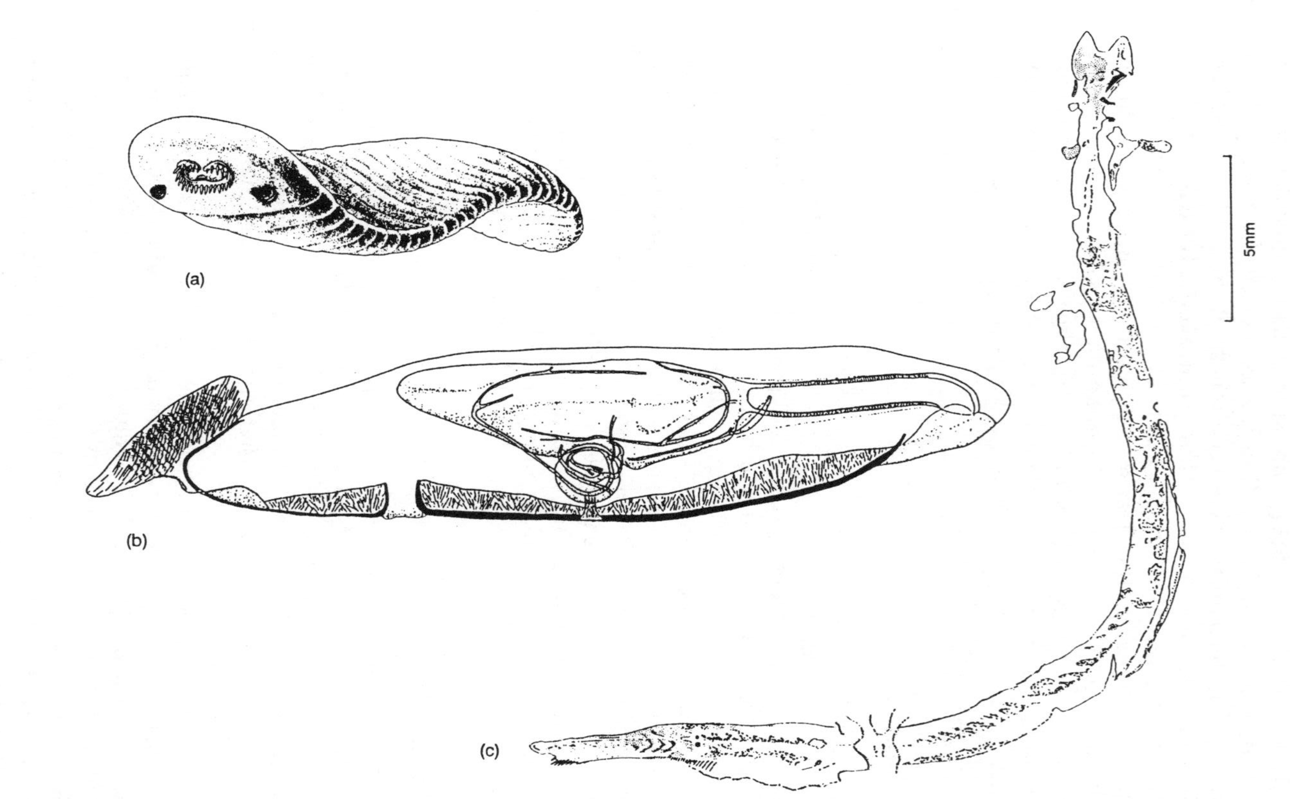

Figure 1.25 Conodont animals right and wrong: (a) *Odontogriphus* (drawn by Marianne Collins, from Gould, 1989), (b) *Typhloesus* (from Conway Morris, 1990) and (c) *Clydagnathus* (from Briggs *et al.*, 1983). Only the last is currently in contention for the palm as the genuine conodont animal.

Shales were similar to conodont elements – such that *Odontogriphus* could have been the conodont animal.

The mystery continued: it was a bizarre Borgesian plot made real, of future palaeontologists obliged to reconstruct human history based on several million sets of dentures, and nothing else.

The riddle seemed solved with the discovery of worm-like creatures in the Lower Carboniferous Granton Shrimp Bed of Scotland, each with conodont elements convincingly borne on one end (Briggs *et al.*, 1983; reviewed by Higgins, 1983; Figure 1.25c). The appearance of the animals was strongly suggestive of chordate, even vertebrate affinity, although other solutions could not be ruled out. As Briggs and colleagues pointed out, the animals could, for example, have been chaetognaths, or aberrant molluscs with highly specialized shells. As a result, Briggs *et al.* (1983) chose caution and the placement of conodonts in their own phylum.

The debate about the affiliation of conodonts continued (Rigby, 1983, Dzik, 1986; Aldridge and Briggs, 1986; Aldridge, 1987; Sweet, 1988; Conway Morris, 1989; Briggs, 1991; Peterson, 1994) during which time new material was discovered which added weight to the idea that conodonts were chordates (Aldridge *et al.*, 1986; reviewed by Benton, 1987).

In the 1990s, the case for vertebrate (rather than simply chordate) affinity was strengthened by chemical analyses and electron micrographs suggestive of two kinds of phosphatic tissue in conodont teeth, with histologies similar to vertebrate enamel and acellular bone (Sansom *et al.*, 1992) although this is controversial (Briggs, 1992; Forey and Janvier, 1993). More recent work suggests that the teeth of some conodont animals, at least, were arranged in jaw-like fashion and were capable of biting and occlusion, although from side-to-side, rather than up-and-down as in jawed vertebrates (Aldridge *et al.*, 1995; Purnell, 1995). Large rings of sclerotized material at the anterior end of some conodont animals, far anterior to the conodont apparatus, have been claimed to represent eyes (Purnell, in press), and fine soft-tissue preservation in the South African conodont *Promissum pulchrum* suggests the preservation of extrinsic eye musculature (Gabbott *et al.*, 1995).

All these features are strongly suggestive of vertebrate affinity, perhaps with one or other groups of the extinct jawless armoured fishes, the ostracoderms. However, as Janvier (1995) points out, the precise phylogenetic position of conodonts is clouded by uncertainty over the position of the eyes and teeth.

First, there is no trace of a branchial basket, despite the exquisite preservation in other respects. Second, the terminal position of the eyes, far anterior to the conodont apparatus, is somewhat unusual. It could be that a substantial rostral portion of the animal remains to be discovered, in which case parts of the conodont apparatus might represent a pharyngeal rather than a buccal structure – branchial supports with pharyngeal teeth, rather than 'jaws'.

On this evidence, at least, it seems clear that conodonts represent a very early radiation of 'jawed' vertebrates, possibly distinct from the radiation of more familiar jawed vertebrates, or 'gnathostomes'. In which case, they need not tell us very much about vertebrate origins as such. However, it should be noted that much of this work is the subject of vigorous dispute.

1.9.17 A surfeit of lampreys

Like hagfishes, lampreys are elongate, jawless animals that lack paired fins and scales, and are parasitic on fish. Nevertheless, they do have prominent paired sense organs, can tolerate variation in salinity, and have other structures suggestive of vertebrate affinity, notably weakly developed vertebrae that constrict the notochord. It is fair, then, to regard lampreys as true vertebrates. The earliest known fossil lampreys are of Carboniferous age (Bardack and Zangerl, 1968; Janvier and Lund, 1983)

Lampreys have traditionally been regarded as allied to hagfishes, in a group called the cyclostomes – referring to their circular, jawless mouths. It remains moot, though, how much this grouping reflects structures developed in common in support of a specialized predatory habit, rather than possessed as shared common heritage, and, also, how far the structures held in common are simply primitive for craniates. Lampreys and hagfishes cannot always have been specialized predators, but because they are so specialized, examination of the modern animals reveals little about what their ancestors were like.

Fortunately, the abundant fossil record of heavily armoured jawless vertebrates called ostracoderms, from rocks of between Ordovician[20] and Devonian age provides a clue: many seem to have been bottom-living filter-feeders (Forey and Janvier, 1993; Figures 1.26 and 1.27).

It is now thought that lampreys and hagfishes represent two distinct monophyletic groups, the hagfishes more distantly related than the lampreys to the jawed vertebrates (Janvier, 1981; Maisey, 1986). If so, one may speculate that the true heritage of lampreys lies among the armoured ostracoderms, and that lampreys lost their armour – but that the ancestry of hagfishes is devoid of armoured forms.

The development of lampreys is interesting in that it links vertebrates with protochordates. The adult lamprey is a parasitic predator, with fish-rasping teeth and a highly specialized muscular, sucking pharynx. The larva, or ammocoete, in complete contrast, is a peaceable filter-feeder. Just like the amphioxus, it pumps water through its pharyngeal gill slits, straining out food particles in a mucous filter secreted by an endostyle (Figure 1.28). Unlike the amphioxus, though, the water currents are maintained by muscular rather than ciliary action. Upon metamorphosis,

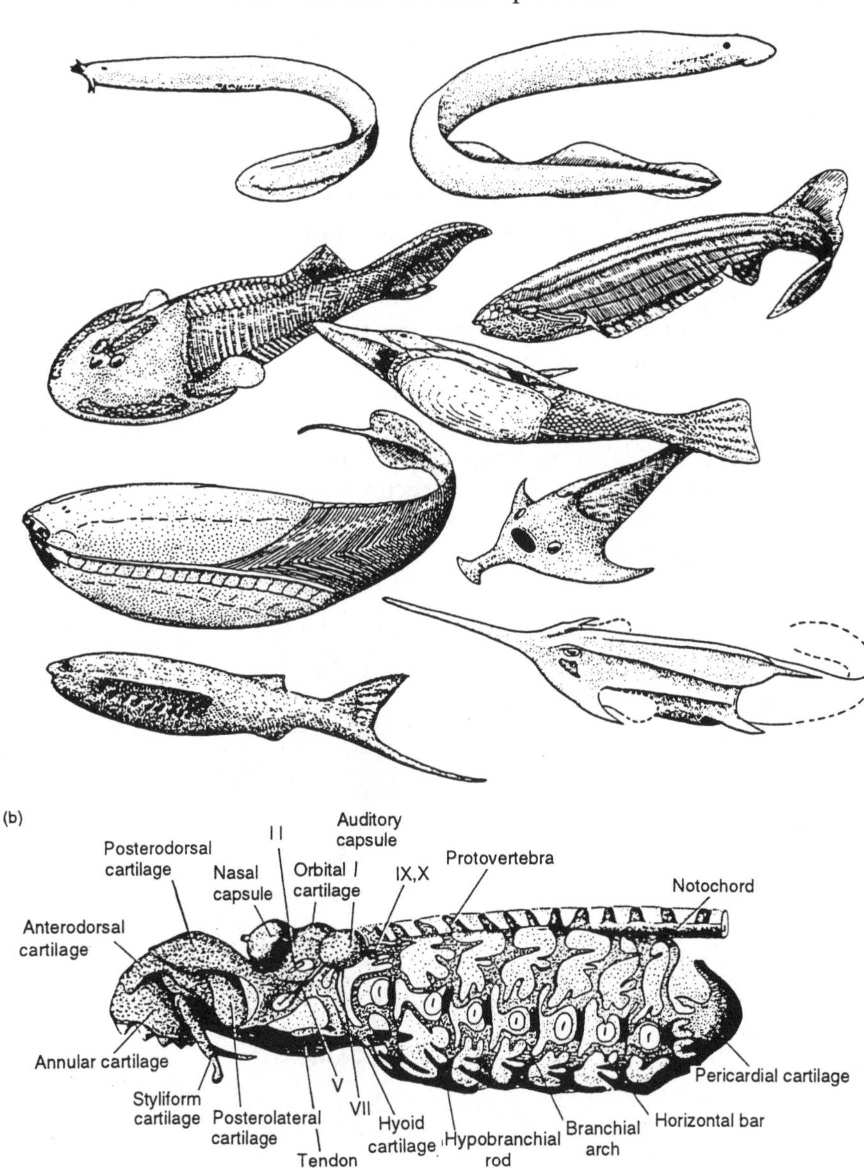

Figure 1.26 (a) Lampreys and hagfishes are two modern representatives of a once highly diverse and paraphyletic assemblage of agnathans more or less closely related to one another and to gnathostomes (jawed vertebrates). Clockwise from top left; a hagfish, a lamprey, an anaspid *Pharyngolepis*, a pteraspid heterostracan *Pteraspis*, a galeaspid *Sanchaspis*, a pituriaspid *Pituriaspis*, a thelodont *Thelodus*, an arandaspid *Sacambambaspis*, and a cephalaspid osteostracan *Cephalaspis*. All but the first two kinds are extinct (from Forey and Janvier, 1993). (b) Lateral view of skeleton of head and branchial arches of the lamprey *Petromyzon* (from Young, 1981, after Parker).

among other changes, the open endostyle curls up and closes in on itself to become the thyroid gland of the adult.

1.10 VERTEBRATE DEVELOPMENT IN PERSPECTIVE

What can be said, at this stage, about vertebrate development and its context within the relationships of vertebrates with other chordates and other deuterostomes?

The early course of ontogeny in vertebrates, including gastrulation, varies widely between species (for the reasons discussed above in the section on tunicates), although recent molecular work points up general similarities in gastrulation (De Robertis *et al.*, 1994). In all vertebrates, development converges on a remarkably uniform scheme (the so-called neurula-pharyngula) at the time of neural induction and the formation of axial structures, when segmental structures (including gill pouches) also become apparent. This corresponds to the phylotypic stage, during which time the homeobox genes which specify identity along the anterior–

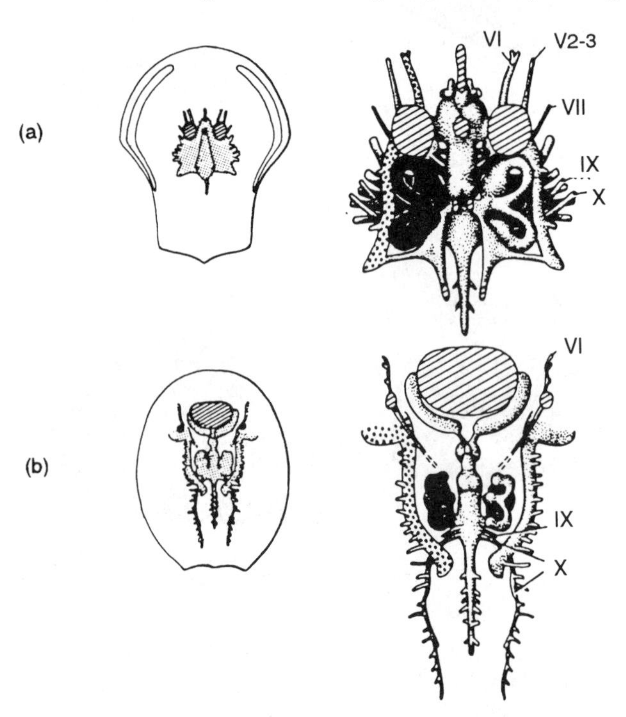

Figure 1.27 Reconstructed soft-part anatomies of agnathan heads. (a) An osteostracan *Norselaspis* (from Forey and Janvier, 1993, after Janvier); (b) a galeaspid *Duyunolepis* (from Forey and Janvier, 1993, after P'an and Wang). (c) Head shield of the osteostracan *Kieraspis* seen from below, with (d) reconstruction of brain and internal canals from an endocast (from Young, 1981, after Stensiö.)

(c)

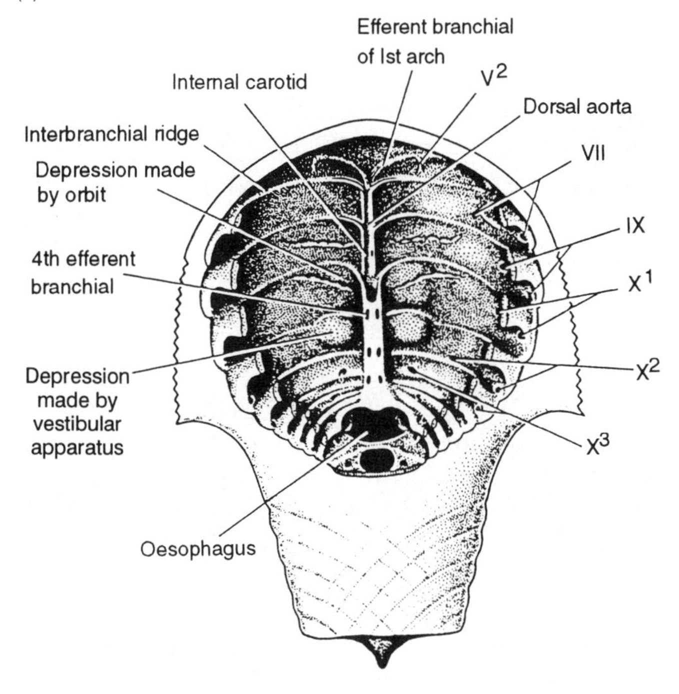

Efferent branchial of Ist arch
Internal carotid
V2
Dorsal aorta
Interbranchial ridge
VII
Depression made by orbit
IX
4th efferent branchial
X1
Depression made by vestibular apparatus
X2
X3
Oesophagus

(d)

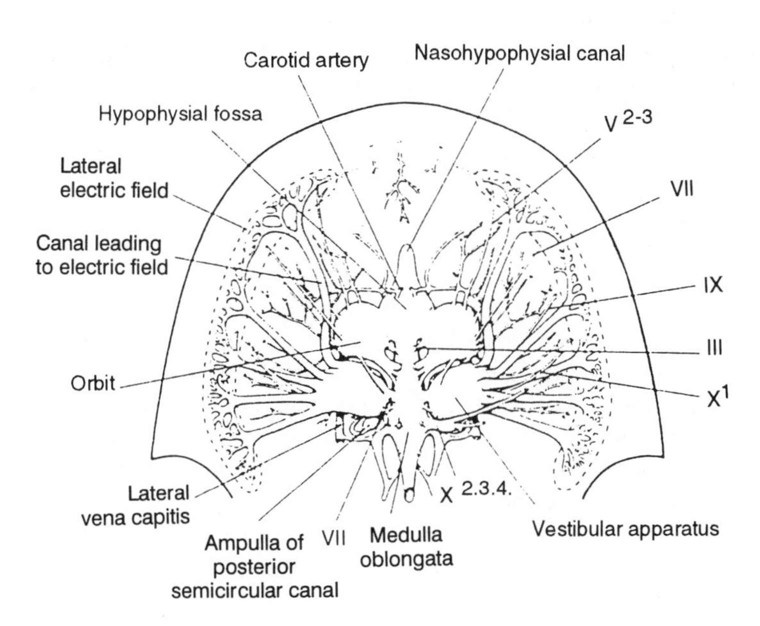

Carotid artery
Nasohypophysial canal
Hypophysial fossa
V 2-3
Lateral electric field
VII
Canal leading to electric field
IX
III
Orbit
X1
Lateral vena capitis
X 2.3.4.
Ampulla of posterior semicircular canal
VII
Medulla oblongata
Vestibular apparatus

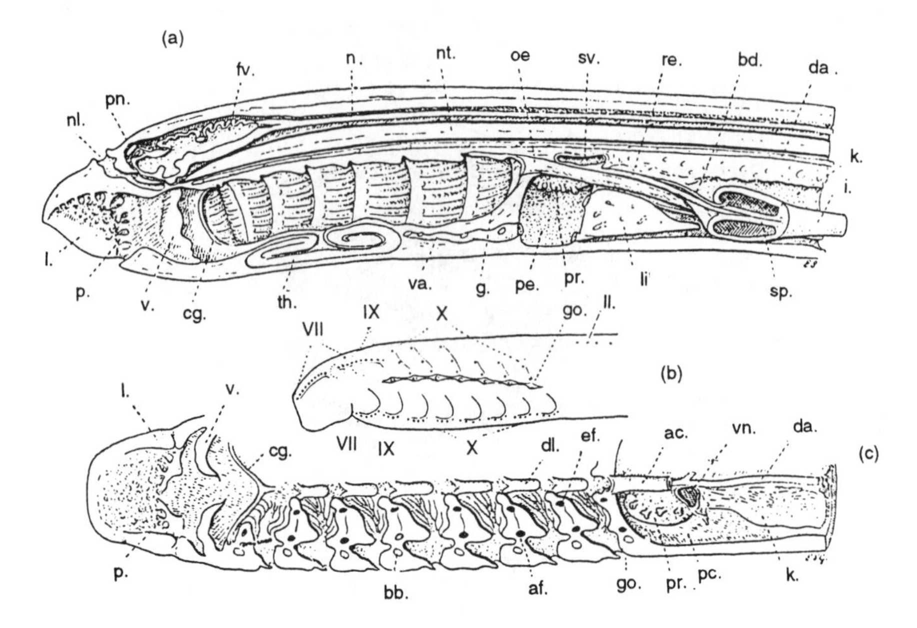

Figure 1.28 Ammocoete larva of *Petromyzon fluviatilis* in (a) sagittal section, (b) left lateral view and (c) horizontal longitudinal section passing through the gills (from Goodrich, 1930).

posterior axis are maximally expressed (Slack *et al.*, 1993; Duboule, 1994). Interestingly, the formation of the tail in vertebrates is not a separate process, but continuous with gastrulation (De Robertis *et al.*, 1994).

In his review of work on deuterostome ontogeny and phylogeny, Schaeffer (1987) notes that whatever the shape of early vertebrate embryos, there are three processes whereby the vertebrate body plan is formed.

The first is mesodermal induction in which the endoderm induces the cells of the ectoderm with which it is in contact to differentiate into mesoderm.

The second is dorsalization, in which the mesoderm is further regionated into notochordal (chordamesoderm), somitic, and lateral plate. Further interaction between somitic and lateral plate mesoderm results in the differentiation of pronephric mesoderm, which becomes the excretory system. The somites further differentiate into dermatome (the dermis), myotome (muscle blocks) and sclerotome (cartilaginous sheets on either side of the notochord).

The third is that to which I have already alluded, neural induction, in which the dorsal ectoderm and the endoderm of the archenteron roof interact to produce the neural plate, the germ of the central nervous system. The regionation of the central nervous system into spinal cord and a complex brain occurs at this time, a process very much more complicated in vertebrates than in other deuterostomes.

And, as we have seen, the development of vertebrates is typified by a special tissue called the neural crest, not seen in any other deuterostome. Neural crest, either of itself or through its interactions, gives rise to the organs of special sense, many of the cartilages and bones inside the head such as the trabeculae and nasal capsules, the dermal bones, the gill skeleton, the rudiments of the teeth, much of the mesodermal musculature of the head and face, the spinal ganglia and so on[21].

But neural crest or not, the essence of all three processes occurs in all chordate groups that have been investigated. For example, cephalochordates show some degree of mesodermal dorsalization, although they lack pronephric mesoderm (a vertebrate feature). Tunicate tadpoles have neither pronephros, lateral plate nor somites, but the notochord, mesodermal strands and neural plate are where one would expect to find them in other chordates.

These similarities, of course, should not disguise the many differences in detail. For example, the somites of vertebrates are formed after the chordamesoderm has clearly separated from the endoderm, not, as in cephalochordates (and, as it happens, in the first three somites in the lamprey *Petromyzon*), from direct evagination from the archenteron.

1.11 THE GAP BETWEEN HEMICHORDATES AND CHORDATES

For all that they represent a group of varied habit and appearance, a number of features appear time and time again in deuterostomes. These features echo one another: similar structures may be found in two very different organisms, such as gill slits in hemichordates and vertebrates.

Although the presence of these features enables one to recognize deuterostomes for what they are, at least in a broad sense, their diffuse distribution makes their arrangement into a phylogeny rather more difficult. It also confounds attempts to identify which, if any, of the extant deuterostomes represents the closest condition to ancestral vertebrates – and which, therefore, could shed light on vertebrate ancestry.

Hemichordates are very similar to echinoderms, as regards their tricoelomate organization. The somatocoels of echinoderms are probably homologues of the metacoels of hemichordates, except that as a result of torsion during echinoderm metamorphosis, the left metacoel is homologous with the oral somatocoel, and the right metacoel is homologous with the aboral somatocoel.

Further, the hydrocoels of echinoderms are probably homologues of the mesocoels of hemichordates. If so, then the (left) mesocoel of pterobranch hemichordates, including the space within the (left half of) the lophophore, is homologous with the water–vascular system of echinoderms.

The tornaria larva of enteropneusts is similar both in gross structure and detailed disposition of its ciliated bands to the auricularia larva of holothurians that some kind of common heritage is plausible, although the similarities have been questioned (Schaeffer, 1987). As we have seen, recent comparative work on 18S rRNA gene sequences suggests a close link between echinoderms and hemichordates (Wada and Satoh, 1994; Halanych, 1995).

It is therefore possible to surmise a common heritage for the coelomic organization in hemichordates and echinoderms. As discussed above, the lophophore in pterobranchs develops from both left and right mesocoels, but the water–vascular system in echinoderms develops from the left hydrocoel. The right hydrocoel atrophies, if it develops at all[22]. Again, the left somatocoel dominates over the right.

There seems to be no easy explanation for the suppression of structures on the right side. And yet this tendency is also found in chordates. In the amphioxus – as discussed above – the gills on the left side start to develop long before those on the right side appear. Jefferies (1986) uses this left-sidedness as evidence in a scheme in which chordates and echinoderms are more closely related to one another than either is to hemichordates or lophophorates, which do not display this asymmetry[23], even though a homologue of the lophophore or the water–vascular system does not exist in chordates, at least not in any living form.

This idea is controversial, and rather different from most of the others, which tend to suggest instead that hemichordates are more closely related to chordates than either is to echinoderms.

The argument centres on suggested changes in the ways the animals feed. Most deuterostomes feed by filtering particles from currents of water using some arrangement of external tentacles. With the water–vascular system, echinoderms used a modified version of the same system to the same ends, as the habit of stalked crinoids suggests. Later echinoderms modified this further, becoming more mobile and adopting other forms of feeding such as algal grazing (sea urchins) and carnivory (starfishes). The adoption of radial symmetry, with the various kinds of torsion that involves, including the dominance of left-hand structures, can be seen as a special feature of echinoderms.

Cephalodiscus, though, combines external collection (a lophophore) with a definite pharyngeal region behind the mouth, perforated by a pair of gill slits (Gilmour, 1978, 1979). Enteropneusts lack a lophophore, but retain a kind of detritus-feeding (with the ciliated proboscis) and have expanded the pharyngeal gill region.

But what of the link between hemichordates and chordates? Gill slits provide a clue, because they are present in both the hemichordates and the chordates. Enteropneust gill slits are similar to those of the amphioxus in many of their details, such as the subdivision of the primary slits by tongue bars. These similarities in structure are underpinned by a common developmental pathway: for the gill slits to form, the pharyngeal gill pouches, developing inside the pharynx, must make contact with the ectoderm of the body wall. This feature of embryonic induction may indicate common ancestry of hemichordates and chordates (Schaeffer, 1987), although not necessarily the monophyly of hemichordates (Jefferies, 1979).

The chordates, in contrast, share a very definite set of characteristic features not present in hemichordates. Besides the notochord and dorsal tubular nerve cord, many of these features are associated with, and integrated to form, a specialized kind of pharyngeal filter-feeding that is quite unlike that found in lophophorates or echinoderms. Even some larval vertebrates, notably the ammocoete larva, feed in ways essentially similar with that of non-vertebrate chordates. Although hemichordates have gill slits, external structures such as the pterobranch lophophore or the enteropneust proboscis are still responsible for food collection.

In the pharynx of a generalized chordate, tracts of cilia drive a current of water in through the mouth, diverting it laterally, through the gill slits. Mucus generated in a ventral gutter, the endostyle, is carried upwards along the gill bars by yet other ranks of cilia. This mucus, forming a filter-bag inside the pharynx, traps particles in the water current. The particle-laden mucus is then rolled up in further special structures in the dorsal part of the pharynx (such as the dorsal lamina of tunicates) to make a kind of rope, and is directed backwards, into the mouth.

This arrangement can only function by virtue of the interdependence of several specialized features – the endostyle, the ciliated tracts associated with it, the dorsal edge of the pharynx and the gill bars, the dorsal structures and so on – that it would be quite hard to imagine how any individual feature could exist on its own, except perhaps in another guise.

To bridge the gap between hemichordates and chordates, one must be able to answer the question of how the characteristic chordate feeding system came into being.

1.12 ECHINODERMS, CHORDATES AND HEMICHORDATES

In addition to the gap between hemichordates and chordates, the standard textbook story of the deuterostomes is complicated by a cladistic three-taxon problem. That is, given the known distribution of features in

Recent animals, do the hemichordates lie closer to the echinoderms, or to the chordates?

If one cannot decide between these two, might not one adopt the third solution, that chordates and echinoderms are sister groups, with hemichordates lying further away? A typical scenario, pointing up these problems, might run something as follows.

The deuterostomes originated (for the sake of argument) from sessile, protostome animals with an external, lophophore-like feeding structure, although whether it was truly homologous with the lophophore of brachiopods is a moot point (Gee, 1995a).

Echinoderms were early offshoots, acquiring calcite skeletons. Like the pterobranchs (and lophophorates), their feeding organs were derived from the mesocoel, although at some point only the left mesocoel was used and the result was the internal water–vascular system rather than the external lophophore. Perhaps the transition between one structure and the other was connected with the acquisition of the calcite skeleton. Early echinoderms had a biradial, triradial or irregular symmetry, with pentameral symmetry developing later.

The hemichordates may have been next to split away, but the presence of a lophophore in pterobranchs (and lophenteropneusts) suggests that these, at least, were earlier offshoots than echinoderms. This implies that the earliest echinoderms had gill slits, but then lost them: an implication that is likewise consistent with a close relationship between echinoderms and hemichordates, excluding chordates (Halanych, 1995).

On the other hand, the lack of gill slits in modern echinoderms, and their presence in hemichordates, might suggest that echinoderms never had them, in which case hemichordates branched off the deuterostome root-stock later than echinoderms.

On the face of it, there is little to tell between the two alternatives. But it could be that the assumption on which these alternatives are based is flawed and that some other branching order is possible. That assumption is that chordates are considered more 'advanced' than either echinoderms or hemichordates, such that the presence of gill slits in hemichordates is in some way indicative of a genetic link with chordates.

But are hemichordate gill slits really homologues of chordate gill slits? In terms of function, they seem quite different – but this is expected, as hemichordates lack the other features of the chordate pharynx.

The irregular Palaeozoic echinoderms known as cornutes had serial perforations which Gislén (1930), Jefferies (1967) and others have interpreted as gill slits. To Jefferies (1986), this suggested that cornutes are closer to chordates than (other) echinoderms, but this depends on a presumption of homology perhaps no more likely than that which links hemichordate and chordate gill slits. On this evidence it is just as likely that hemichordates are more closely related to echinoderms than *either* are to chordates. As Schaeffer (1987) notes (p.190) 'this hypothesis is not compromised by the presence of pharyngeal slits and their skeletons ... in the hemichordates, urochordates and cephalochordates'. For the present, though, it is hard to tell which of the two, echinoderms or hemichordates, are more closely related to chordates.

1.13 CHORDATES AND VERTEBRATES

The integrity of the chordate group seems more certain, although there is less consensus about interrelationships within the group – giving rise to another three-taxon problem. Perhaps the majority view is that cephalochordates are closer to vertebrates than are tunicates, by virtue of their possession of segmented muscle blocks and a notochord throughout life, and (as has been found more recently) a large number of similarities in detail (Gee, 1994).

Importantly, cephalochordates and vertebrates possess a particular kind of segmentation that seems to be lacking in tunicates and, it seems, other deuterostomes. Nevertheless, most commentators seem to have agreed that the vertebrate stock arose from animals with the chordate filter-feeding system, present in tunicates, cephalochordates and (in the larval lamprey) in vertebrates.

The presence of the notochord and dorsal nerve cord has aroused some debate, though. Filter-feeding as a habit, whether in protostomes or deuterostomes, is generally associated with a sessile life. But if ancestral craniates were filter-feeders, how is it that features connected with locomotion (notochord, elaborate central nervous system, segmented muscle blocks) are what make craniates distinctive? Put another way, at what point in deuterostome history did the sessile, filter-feeding lifestyle give way to an active, swimming and possibly predatory one?

If these structures originated at different times in deuterostome history, in which order did they appear, and at which point would one be entitled in calling the animal a craniate? Might the notochord and dorsal tubular nerve cord have been associated, originally, with some other activity besides locomotion, such as feeding?

As I show in Chapter 2, Garstang solved the problem by deriving the axial structures of vertebrates from modifications of the ciliated bands of an auricularia-like larva, to form something like the tunicate 'tadpole' larva – effectively bypassing the entire adult feeding apparatus. This scheme has many resemblances with the actual ontogenetic process of neural tube formation, and attracts approbation today for that reason (Lacalli *et al.*, 1994).

It is important to remember, though, that in almost all ways apart from the possession of a notochord, the tunicate tadpole larva is very different from cephalochordates or vertebrates, and could well represent a degenerate form of an organism, larval or adult, that had true metameric segmentation. Recent anatomical work, revealing the regular arrangement of cilia along the tail of the tadpole larva (Crowther and Whittaker, 1994), and genetic work which shows the segmental expression of a homeobox gene in the tadpole tail (Di Lauro, personal communication) tends to support this view. If so, then Garstang's idea, so good at explaining the functional transition between filter-feeding and active predation, is less compelling: it becomes harder to justify the sessile ancestry of the craniates.

One way to bridge the gap is to suppose that vertebrates made the transition through a form that, although in principle mobile, was essentially benthic and fed on detritus. Larval lampreys fill this gap, as do many of the first unambiguous vertebrates to appear in the fossil record, the ostracoderm agnathans.

However, this is simply a 'scenario', an *ad hoc* way to explain character distributions in the form of a story. As such, it cannot be used as evidence, particularly as other stories are possible (pelagic lophophorates or tunicates, armoured predatory hemichordates and so on, enteropneusts as degenerate echinoderms, as fanciful as one could wish).

As we have seen, many of the characters used to define groups such as the deuterostomes are based not on the form of the adult, but on the shape and development of larvae and embryos. Nowhere is this more true than with the ideas of Garstang, which I explore in the next chapter.

1.14 A DEUTEROSTOME PHYLOGENY

Schaeffer (1987) draws character-state information from comparative deuterostome embryology to produce a cladistic phylogeny for the group (Figure 1.29). Given that it is both very clear, and explicitly based on similarities of developmental process rather than adult form, it can be seen as a kind of ground-state for the phylogenies I discuss later on in this book, some of which concern embryonic form. However, it does not account for much recent work, particularly molecular phylogeny, which challenges this accepted wisdom. The cladogram is as in Figure 1.29.

The character-state information used to define the groups at each node are worth mention, and I paraphrase them here[24]. Note that the protostomes are held to be a monophyletic-sister-group of the deuterostomes, which may be at variance with current thinking that the protostomes are, as a group, probably paraphyletic. Note also that not all the characters (such as the first) define nested sets, suggesting either guesswork in places

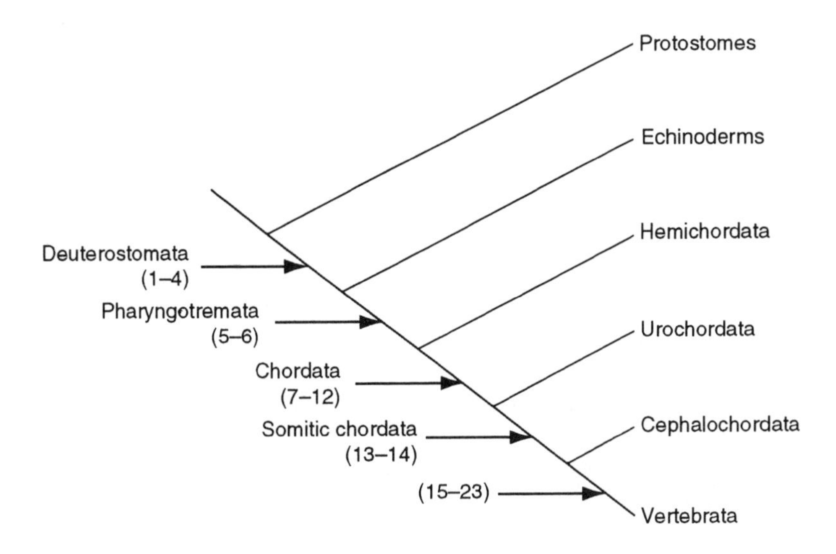

Figure 1.29 Phylogeny of the deuterostomes (adapted from Schaeffer, 1987).

where information is insufficient, or the possibility of convergence minimized here by the application of the principle of parsimony.

Deuterostomata

1. Induction of basic body plan along animal–vegetal axis of the early blastula[25].
2. Enterocoely of coelom and mesoderm.
3. Blastopore becomes anus of larva (deuterostomy).
4. Free-swimming larva with subdivided digestive tract and ciliated bands.

Pharyngotremata[26]

5. Ciliated pharyngeal slits from ectoderm–endoderm fusion.
6. Distinctive gill skeleton of tongue bars and synapticulae.

Chordata

7. 'Fate Map' of blastula has an equatorial region with the potential to become mesoderm, notochord and neurectoderm (central nervous system).
8. Mesodermal induction.
9. Dorsalization, indicated by presence of notochord and lateral muscle bands.
10. Neural induction prompted by archenteron roof.
11. Endostyle.
12. Free-swimming, bilateral, elongated larva with locomotory tail.

Somitic Chordates[27]

13. Further dorsalization, indicated by presence of differentiated somites (myotome, sclerotome) and lateral plate mesoderm.
14. Dorsal, hollow nerve cord with dorsal nerve roots that carry sensory and motor fibres.

Vertebrates[28]

15. Further dorsalization of mesoderm (nephrotome).
16. Somitic coeloms develop independently of archenteron, except the first three in lampreys.
17. Presence of partially differentiated cranial somites called somitomeres (Jacobson, 1987, 1988), which form branchial and extrinsic eye muscles and contribute to cranium and visceral skeleton.
18. Neural crest, including axial elaboration of brain mediated through archenteron roof.
19. Epidermal placodes from ectoderm–mesoderm interactions, give rise to organs of special sense.
20. Gills (as opposed to gill slits).
21. Differentiated endoerm organs such as pancreas and liver.
22. Oxygenated blood pigment.
23. Complex endocrine system.

Several points emerge from this analysis. The first is that hemichordates are linked with chordates (as 'pharyngotremates') solely on the basis of similarities in the structure and development of the gill skeleton. Schaeffer (1987, p.222) concedes that hemichordates could, alternatively, constitute a sister-group of the echinoderms, because they possess the equivalent of an echinoderm hydropore, something that chordates lack.

The hydropore is, primarily, the left opening of the protocoel (axocoel). In echinoderms, it is termed the madreporite and connects with the mesocoel (or hydrocoel) via the stone canal. The equivalent in hemichordates is the left member of the paired protocoel pores. The decision between the two phylogenies – echinoderm (hemichordate, chordate) and (echinoderm, hemichordate) chordate – depends on the weight that one attaches to either character; the shared features of the gill skeleton in hemichordates and chordates, or the presence of a hydropore in echinoderms and hemichordates. It is a classic cladistic dilemma: one of the characters must be parallel – but which?

Second, it is unclear how segmentation arose in 'somitic' chordates. Segmentation in the amphioxus is considered primitive, in that the lateral plate is segmented as well as the somites, and the somites extend to the anterior end. In vertebrates, the lateral plate smooths out to form a continuous sheet, and the somitomeres – the somites in the head, which Schaeffer asserts are broadly homologous with the anterior somites in the amphioxus – are less clearly defined (Jacobson, 1988). This implies that the trend has been to *reduce* the degree of segmentation from what must have been an even more highly segmented condition. In which case, one could speculate that tunicates, rather than being primitively unsegmented, might have lost segmentation to an even greater degree (for example, Crowther and Whittaker, 1994). This sharpens the dilemma about the relative placement of tunicates and cephalochordates with respect to craniates, as well as the origin of segmentation in general – something that Schaeffer does not resolve.

NOTES

1 For detailed discussion on invertebrate structure and relationships, consult Hyman (1959); Barnes (1968, 1980) and the more recent textbooks by Barnes *et al.* (1993) and Willmer (1990). For a palaeontological perspective, see Conway Morris, 1993 and for a new view based on comparative embryology, see Davidson *et al.*, 1995.

2 Hall (1992) goes into the history behind the development of the 'body plan' concept that underlies systematic zoology, and Gould (1977) into the relationships between ontogeny and phylogeny.

3 Every roundworm or nematode of the species *Caenorhabditis elegans*, for example, has precisely 959 cells, excluding sex cells, and the complete pattern of connections of the 302 nerve cells has been worked out (reviewed by Sulston *et al.*, 1992). The ability to plot cell fate makes this animal a favourite with scientists interested in how cells cooperate to create multicellular structures in development.

4 Some have used comparisons of genetic sequences to suggest that cnidarians represent a separate radiation from unicellular protists, so that the Metazoa reflects a polyphyletic assemblage of creatures with different ancestries that look the same by convergence (Field *et al.*, 1988, 1989). But for all that this analysis might be flawed (Lake, 1990), molecular sequence data do suggest that diploblasts and triploblasts diverged at a very early date, certainly before the 'Cambrian Explosion' during which time most triploblasts diverged from a flatworm-like ancestor (Philippe *et al.*, 1994). See Willmer, 1990, for further discussion on metazoan monophyly.

5 Echinoderms are radially symmetrical, like jellyfish, but this is secondary. Echinoderm larvae are bilaterally symmetrical and become radial during metamorphosis.

6 However, molecular evidence (Turbeville *et al.*, 1992) suggests that the nemertean worms, conventionally thought to be aceolomate, may actually be closely related to higher protostomes. In other words, the coelom has been secondarily lost.

7 Willmer (1990) describes a metazoan phylogeny in which the deuterostomes and various protostome groups derive separately from a pool of flatworm ancestors. If so, then flatworms are paraphyletic, and once again the protostomes cease to exist as a group.

8 Although the terms 'stem' and 'crown' make for convenient metaphors, their use in phylogenetic reconstruction does, strictly, imply a difference in

the treatment of living and fossil forms. More on this in Chapter 4: for now, a 'crown' group is strictly a monophyletic group which includes all extant members of that group together with the latest common ancestor of that group. This ancestor is, in turn, the latest of a 'stem' lineage, all of whose members are extinct. Side-branches from the stem group represent extinct groups, by definition – unless they include the latest common ancestor of a different 'crown' group.

9 Berrill (1955) uses the terms 'monophyletic' and 'polyphyletic' in the senses in which modern systematists would understand them. Indeed, they are found and defined in *Chambers' 20th Century Dictionary*, New Edition, 1983. 'Paraphyletic', though, occurs in neither.

10 Except in the filter-feeding ammocoete larva of lampreys.

11 Nichols (1967b) gives a round-up of living and fossil echinoderms.

12 See Jefferies (1981a).

13 Gilmour (1979) suggests that the gills of *Cephalodiscus* may have originated by the partial fusion of paired oral lamellae as part of a mechanism to increase the efficiency of ciliary currents during feeding. Similar currents (though no gills) are seen in the pterobranch *Rhabdopleura* and other lophophorate phyla (Gilmour, 1978), a fact which may argue for a close relationship between pterobranchs and other lophophorate groups.

14 A comprehensive review of the physiology of sessile, adult ascidians can be found in Goodbody (1974), who considers the animals very much in their own right, rather than as the adjuncts to discussions of chordate phylogeny that they so often find themselves. Much of Berrill's *The Origin of Vertebrates* (1955) concerns the natural history of tunicates and makes interesting reading on that account alone.

15 Alldredge (1976, 1977) reviews the home life of the larvacean tunicates.

16 The scarcity of resources in the deep sea has forced other filter- or suspension-feeders into macrophagy, including at least one species of sponge (Vacelet and Boury-Esnault, 1995).

17 The key sentence reads:

Nach allen diesen Gründen glaube ich mit vollem Rechte den Achsencylinder des Schwanzes der Ascidien mit der *Chorda dorsalis* des Amphioxus sowohl functionel, als auch genetisch vergleichen zu können.

This translates as 'On all these grounds, I believe that, with full justification, I can compare the axial cylinder of the ascidian tail with the notochord of amphioxus, both as to their function and on their mode of genesis'.

Not that the entire world immediately acquiesced to this view. In the sequel to the 1866 study, Kowalevsky (1871, p.102) reports that whereas Kupffer generally stood in agreement, Metschnikoff (the same that pointed out the similarities between enteropneusts and echinoderms) saw more resemblance between the development of tunicates and that of arthropods and leeches than with vertebrates (Kowalevsky notes of Metchnikoff that 'Er schloss aus seinem Studien, dass die Ascidien eine grössere Aenlichkeit in ihrer Entwicklung mit den Arthropoden und Hirudineen als den Wirbelthieren besässen').

Kowalevsky may have had the last laugh, however, in that recent genetic work supports his contention that the notochords of tunicate tadpoles, vertebrates and the amphioxus share a common genetic origin. A gene called *brachyury* which, in vertebrates, is closely concerned with notochord formation, is believed to have a similar function in the amphioxus (Ortner *et al.*, submitted). In tunicates, though, *brachyury* has the explicit function of specifying the cells which develop into the notochord (Yasuo and Satoh, 1993, 1994).18 Perhaps the definitive descriptive work on the development of amphioxus is that of Conklin (1932). This classic paper is a detailed review with no fewer than 200 figures, as well as comparisons with the early embryogenesis of ascidians. One can tell that this paper is of a certain age when one reads the prefatory remark 'Some excuse, or at least some explanation, seems to be needed for the publication of another paper on such a well-worked subject as the embryology of Amphioxus.' (1932, p.69).

19 A third possibility, not discussed by Bone, is that asymmetry is recapitulatory, in the Haeckelian sense that it provides for evidence of an asymmetric adult stage in the ancestry of the amphioxus.

20 Or, more tentatively, the Cambrian (Repetski, 1978).

21 For reviews on neural crest, and related topics such as the origins and development of hard tissue in vertebrates, see for example Northcutt and Gans (1983), Gans and Northcutt (1985), Gans (1987), Halstead (1969, 1987), Le Douarin (1983), Lumsden (1987), Hall (1987, 1988, 1992), Thomson (1987a, 1991), and Hanken and Thorogood (1993).

22 The heart vesicle of enteropneusts could be homologous with the embryonic dorsal sac of echinoids, derived from the right protocoel (Hyman, 1959, cited in Schaeffer, 1987).

23 Except, perhaps, for the presence of a left protocoel pore only, without a corresponding right, in some enteropneusts.

24 I do not discuss them all, so my numbering system differs from that found in Schaeffer (1987).

25 Not known in hemichordates, and awaits verification in tunicates.

26 To denote those deuterostomes with pharynges perforated by slits.

27 That is, those chordates with metameric segmentation.

28 Schaeffer uses the term 'craniata' for vertebrates, throughout, as the latter arguably excludes hagfishes (Janvier, 1981).

2

The origins of vertebrates

She swore in faith 'twas strange, 'twas passing strange; 'twas pitiful, 'twas wondrous pitiful. She wished she had not heard it ...

Shakespeare, *Othello*

2.1 VERTEBRATE ORIGINS BEFORE GARSTANG

Many groups of animals have been proposed as the closest relatives of, or ancestral to, the vertebrates. Some hypotheses make strange reading today, but one should remember that this strangeness could, in part, be a reflection of our familiarity with ideas of a different kind, rather than any inherent crankiness or error in these others[29].

The reason why these fail where (say) Garstang succeeds lies not in their seeming implausibility, but in their failure to accommodate all the available evidence in a single scheme. For example, ideas that espouse close links between vertebrates and arthropods ignore embryological evidence that links vertebrates with echinoderms, tunicates and other deuterostomes: those that link vertebrates with annelids or nemerteans offer a solution only to put greater problems in place.

2.1.1 Kovalevsky and the tadpole larva

As discussed above, Kowalevsky (1866a,b) showed that tunicates were allied to vertebrates, by virtue of the presence of the notochord and other axial structures in the tadpole larva. This counts as one of the most important discoveries ever made in zoology – and one of the most systematically misinterpreted.

One might expect that this link would be seen as evidence for the primitiveness of tunicates with respect to the vertebrates. Far from it – the loss of axial structures in tunicate metamorphosis was quickly seen neither as a symptom of primitiveness, nor as a peculiarity of tunicates, but of degeneracy.

The immediate impact of Kowalevsky's work was a general perception of tunicates as highly modified vertebrates – 'vertebrates fallen from their high estate' as Gregory puts it (1946, pp.352–353) – which had lost all traces of segmentation in both larva and adult, and, but for their brief appearance in the larva, such distinctive vertebrate features as the notochord[30].

Hubrecht (1883) presented an illuminating survey of how things stood in the 1880s. 'In 1868 the solution appeared to have been found' he writes,

> when Kowalevsky's splendid researches concerning the development both of Amphioxus and of the Ascidians could be compared side by side. The Tunicate larva was for the time being proclaimed to be the missing link, to be of all Invertebrates the closest approach to the much-looked-for parent form.
>
> Since then the aspect of things has changed and later investigations, more especially those of Dohrn and of Ray Lankester, have rendered it nearly certain that the Tunicata must, on the contrary, be looked upon as degenerate Vertebrates which can be hardly of much use in helping us to the failing clue (1883, p.349).

Some, such as Haeckel, saw the metamorphosis of larval tunicates into a sessile, unsegmented state as a stage in development lacking in (other) vertebrates, which were motile and segmented. In other words, tunicates were in a sense more 'evolved' than vertebrates, recapitulating their vertebrate heritage as a larval stage. Not surprisingly, the ontogeny of tunicates was regarded as good evidence for the Haeckelian notion of recapitulation.

2.1.2 Annelids and other segmented animals

With tunicates relegated to the status of decadent specialists, and the realization that primitive vertebrates were segmented and motile, it became obvious that the vertebrates shared their ancestry with other segmented, motile creatures. It was in this guise that annelids and arthropods presented themselves.

Annelids, in particular, resemble vertebrates and the amphioxus in that they are motile and segmented, with the axial repetition of structures such as muscles and nerves.

Never mind that this resemblance is superficial: apart from the mere fact of segmentation, the style of segmentation in annelids is quite different from that of vertebrates. In annelids, segmentation is complete – the whole body is divided into rings. But in vertebrates (and the amphioxus), only the dorsal mesoderm is rigidly segmented. Given that the arrangement of the organs (muscles and so on) of the dorsal mesoderm is connected with locomotion (as is the entire hydrostatically supported body of an annelid) it may be more likely that segmentation in these two groups is a convergence, a common solution to the problem of forward motion.

But perhaps the most blatant obstacle before this alliance is that annelids lack a notochord, and their primary nerve cord is solid and ventral, rather than hollow and dorsal.

Yet even with all these difficulties, the transformation from annelid to vertebrate by the apparently simple expedient of turning the annelid (or arthropod) body upside down (Figure 2.1) is an idea that has had broad appeal. It goes right back to Cuvier's old opponent Geoffroy St Hilaire[31] (who also noted similarities between vertebrates and cephalopod molluscs as did Lamarck, 1809), and has been discussed by others notably Dohrn, Beard and Minot (reviewed by Cunningham, 1886; Gregory, 1946) and Hatschek.

An inverted annelid has a nerve cord lying above the gut, just like a vertebrate. Put another way, the gut and nerve cord swap places, so that the hollow gut of the annelid/arthropod becomes the hollow nerve cord of the vertebrate, with the loss of 'old' annelid/arthropod mouth to create a blind-ended sac, the brain. The connection between nerve cord and gut is not quite as fanciful as it sounds, given the close embryonic connection between the embryonic gut and nerve cord in vertebrates, through the neurenteric canal which forms during gastrulation and neurulation (De Robertis *et al.*, 1994). A new gut develops from an infolding of the new ventral (old dorsal) surface.

Several elaborate schemes (notably by Patten, 1890, 1912: see below) have been proposed to link the hypophysis (an organ at the base of the brain which, in primitive vertebrates such as lampreys and hagfishes

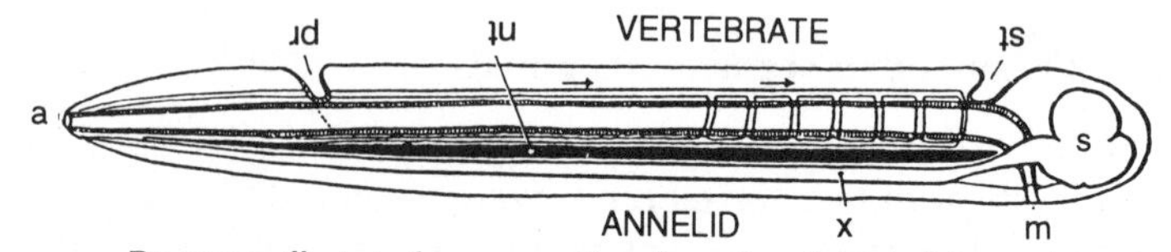

Diagram to illustrate the supposed transformation of an annelid worm into a vertebrate. In normal position this represents the annelid with a "brain" (s) at the front end and a nerve cord (x) running along the underside of the body. The mouth (m) is on the underside of the animal, the anus (a) at the end of the tail; the blood stream (indicated by arrows) flows forward on the upper side of the body, back on the underside. Turn the book upside down

and now we have the vertebrate, with nerve cord and blood streams reversed. But it is necessary to build a new mouth (st) and anus (pr) and close the old one; the worm really had no notochord (nt); and the supposed change is not as simple as it seems. (From Wilder, History of the Human Body, by permission of Henry Holt & Co., publishers.)

Figure 2.1 The worm that turned: how to turn an annelid into a vertebrate by inversion. This figure (with its ingenious caption) is from Romer (1970).

connects the ventricles of the brain with the outside by a duct) with the old arthropod/annelid mouth and buccal cavity.

Now, it is easy to imagine an animal suddenly turning upside-down and assuming this inverted position as its regular habit. Examples abound, from cnidarian polyps (upside-down medusae), water-fleas and water-beetles to 'upside-down' catfish, barnacles and sloths. But it is harder to envisage how such a sudden inversion of one body form might, in real life, be accompanied by the attendant profound internal reorganization necessary to re-orient the organs of the inverted animal into a semblance of being, once again, the right way up. For example, it is difficult to understand how a right-way-up annelid could invert to become a right-way-up vertebrate without great chaos and tumult engendered during the inversion process.

Again, the unsegmented state of the notochord in chordates is hard to reconcile with the lack of any such structure in annelids. A contemporary objection to the annelid hypothesis, owing something to Haeckelian recapitulation, notes that derivation of chordates and annelids would imply the appearance of the notochord later in ontogeny than segmentation, in which case the notochord would show signs of segmentation. Instead of which, the unsegmented notochord appears early in ontogeny – a circumstance suggestive of the origin of chordates from an unsegmented ancestor, rather than a segmented worm.

Third, embryological evidence suggests that annelids and arthropods belong to an altogether different animal group from the chordates. In particular, their embryos exhibit spiral cleavage, and the mesoderm develops from one particular blastomere (Chapter 1).

Cunningham's (1886) analysis of Dohrn's views is light-hearted, if not arch:

> A great anatomist once said that if he wished to read romances he knew better specimens than histories of creation wherewith to amuse himself, à propos of which Dohrn points out that if phylogenies are to be compared with romances it is as well to remember that the most sensational are not always the best works of art (1886, pp.283–284).

Recently – and remarkably – discussion on the annelid/arthropod link has enjoyed something of a renaissance. Nübler-Jung and Arendt (1994) review the evidence for the inversion first suggested by St Hilaire, and Arendt and Nübler-Jung (1994) cite genetic evidence (discussed further in Chapter 3) in support of the contention that annelids and arthropods share a common (if distant) ancestry with chordates, and that the embryonic dorsoventral axis became inverted in the early evolution of chordates (for a review, see De Robertis and Sasai, 1996).

If these correspondences are real, then the nerve cord of insects is homologous with that of chordates; only the polarity has changed. This inversion is posited to have happened at the gastrulation stage, and may have been connected with the origin of deuterostomes from protostomes, rather than the origin of chordates or vertebrates directly from annelid-like or arthropod-like ancestors.

If Arendt and Nübler-Jung (1994) are correct, and the ventral nerve cords of arthropods are homologues of the chordate dorsal nerve cord, then we should not expect to see both types of cord in the same animal. Peterson (1995b) challenges this by pointing out how this statement is invalidated by the condition in enteropneusts, which have both a dorsal, hollow nerve cord as well as ventral cords. The presence of both in the same animal questions the assertion that one is the homologue of the other (the homology between chordate and hemichordate nerve cords has been questioned: references in Peterson, 1995b).

Arendt and Nübler-Jung (1994) have also been challenged by Lacalli (1995), but this discussion properly belongs with matters relating to Garstang's auricularia theory, of which more below.

2.1.3 Nemerteans

But to return to the 1880s, long before molecular genetics had been invented: the ventral nerve cord of annelids and arthropods can be made homologous with the dorsal cord of chordates only if every other structure is turned upside down – and even this radical proposal cannot account for the presence of the notochord in vertebrates. These problems can be made to disappear thanks to (of all things) nemertean worms which, according to Hubrecht (1883), could have been ancestral to both annelids/arthropods and vertebrates.

Rather than a single dorsal or ventral nerve cord, nemerteans have a pair of lateral cords. Ventral coalescence of these could have produced an annelid nerve cord – dorsal fusion, on the other hand, would have made a vertebrate nerve cord.

In addition, nemerteans have, running down their length, a peculiar cavity known as the rhynchocoel, more or less dorsally placed, that contains an eversible proboscis that can be projected through an anterior opening. Hubrecht interprets this structure as the homologue of the notochord:

> According to my opinion the proboscis of the Nemerteans, which arises as an invaginable structure (entirely derived, both phylo- and ontogenetically, from the epiblast [ectoderm]), and which passes through a part of the cerebral ganglion, is homologous with the rudimentary organ which is found in the whole series of verte-

> brates without exception – the hypophysis cerebri. The proboscidean sheath of the Nemerteans is comparable in situation (and development?) with the chorda dorsalis [notochord] of Vertebrates. (1883, p. 351)

This idea has the following advantages: first, it is more credible to derive vertebrates, arthropods and annelids from the relatively simple nemertean condition than from one another; second, the presence of the rhynchocoel as a ready precursor of the notochord; and third, it is no longer necessary to turn anything upside down.

It has the disadvantages nonetheless that, first, the ectodermal derivation of the rhynchocoel is at variance with the origin of the notochord, budded off from the archenteron. Second, although nemerteans are relatively unspecialized, the hypothesis hangs on relations between the notochord and the rhynchocoel, both of which are specialized structures peculiar to their groups. This criticism, at least, Hubrecht deflects with aplomb, insisting that 'I do not advocate any direct relation between existing Nemertines and existing Vertebrates' (p. 362), only proposing the idea as one way to get on up from the flatworm grade. But, like flatworms, nemerteans lack a coelom: Hubrecht tentatively links nemertean gut diverticula with the embryonic sacs of amphioxus.

Bateson (1886) is sympathetic to this thesis, although he notes the lack of evidence in the development of the chordate nerve cord for the coalescence of twin lateral structures. The union of lateral folds to produce the hollow nerve cord of vertebrates cannot be advanced in homology, as the 'medullary plate of Amphioxus is distinctly single' (Bateson, 1886; p.555) even if the development of the dorsal nerve trunk of *Balanoglossus* – by delamination from the dorsal ectoderm – did not rule it out altogether.

The nemertean idea was revived by Jensen (1960, 1963) and extended by Willmer (1974, 1975). Jensen (1963) sees in nemerteans direct ancestors of primitive vertebrates of the hagfish grade, with the subsequent appearance of gnathostomes and lampreys ('petromyzonts'). Not only cephalochordates and tunicates, but enteropneusts, pterobranchs and even echinoderms are viewed as having degenerated from primitive lamprey-like animals.

Willmer is less extreme than Jensen, suggesting that vertebrates and the amphioxus evolved in parallel, from different nemertean stocks. Enteropneusts could also have evolved from nemerteans, the proboscis of the former having arisen from the permanently everted proboscis of the latter.

The nemertean hypothesis has always been some way from the zoological mainstream, which tends to place nemerteans as acoelomate animals akin to flatworms. However, it could be that the circulatory system and rhynchocoel cavity of nemerteans are of coelomic origin, a suggestion strongly supported by the placement of a partial 18S rRNA gene sequence of the nemertean *Cerebratulus lacteus* among the coelomate protostomes (Turbeville *et al.*, 1992).

2.1.4 Arachnids

Arthropods, like annelids, are segmented animals, and have repeatedly been advanced as candidate vertebrate ancestors. But if the morphological similarities between annelids and vertebrates are superficial, how can arachnids fare any better? In seeking to answer this question, a paper by William Patten (1890) makes intriguing reading.

Patten finds many close correspondences between arachnids (in particular, the scorpions) and vertebrates, mainly in the organization of the central nervous system[32]. In addition, and according to Gregory (1946), the striking correspondences Patten identified between the shapes of the head-shields of early ostracoderm fishes on the one hand, and arthropods such as the extant *Limulus* and extinct eurypterids on the other made 'even unbelievers ... waver in opposition to his theory'[33].

His thesis, though, starts as a rebuttal of the annelid hypothesis. The nerve cords of arthropods are as ventral and as solid as those of annelids, but the segmentation which, in annelids, continues right to the anterior end is not apparent in advanced arthropods such as spiders and scorpions – the arachnids – which have a well-developed complex of head and thorax called a cephalothorax. Of course, any similarity between annelids and the amphioxus in this regard is, however, of no importance, given the prevailing view that the amphioxus represents a degeneration, rather than a harbinger, of the vertebrate condition.

In its internal organization – particularly in the arrangement of the brain and its associated nerves – the cephalothorax looks somewhat like that of the head and trunk of vertebrates. Patten (1890) records structures in the nerve cord that correspond with the notochord; features in the arrangement of thoracic legs in scorpions that mirror the structure of the gill arches[34]; similarities between the histologies and dispositions of various nerves and sense organs, and so on.

For all this detail, the idea suffers from the same kinds of problems that beset the annelid idea. There, locomotor segmentation is as likely to be convergent on the vertebrate body plan as akin to it, both having derived independently from simpler, separate forms. With arachnids, the convergences seem to be the overall similarities between the cephalothorax and the head and trunk of vertebrates, in particular the fusion and concentration of nerve roots and ganglia, which may both be consequences of cephalization in general rather than suggestive of shared common ancestry. Annelids, such as the earthworm, are not as cephalized as either arachnids or vertebrates.

Two aspects of Patten's ideas are diverting. First is the undoubted superficial similarity between the arachnid exoskeleton (especially that of the extinct eurypterid 'water-scorpions') and that of primitive, extinct armoured fishes, in particular the placoderms *Pterichthyodes* and *Bothriolepis*, which Patten explores in some detail.

The second concerns the style and nature of the segmentation of the vertebrate head. Patten criticizes the view, then-widely-held, that vertebrates were primitively segmented nose-to-tail, in the manner of annelids. Unfortunately, it is very hard to discern segmentation in the heads of modern, adult vertebrates. In criticizing the annelid idea, Patten notes that:

> In failing to add materially to what the anatomy and embryology of the Vertebrates themselves can demonstrate, the Annelid theory not only is sterile, but is likely to remain so; because unspecialised segments being characteristic of Annelids, it cannot hope to elucidate that profound specialisation of the Vertebrate head which it is the goal of Vertebrate morphology to expound. Moreover, since Vertebrate morphology itself reflects as an ancestral image only the dim outlines of a segmented animal – but still not less a Vertebrate than any now living, – it is clear that the problem must be solved, if at all, by the discovery of some form in which the specialisation of the Vertebrate head is already foreshadowed. Since of all the Invertebrates, concentration and specialisation of head segments is greatest in the Arachnids, it is in these, on a priori grounds, that we should expect to find traces of the characteristic features of the Vertebrate head. (1890, p. 318)

If vertebrates were once wholly segmented, it should be possible to envisage the cranium as a series of modified vertebrae. This is hard to make out in modern crania, which are more or less simple boxes, with little or no trace of segmental origin. Patten pointed out that, in scorpions, there is an unsegmented cartilaginous plate, the sternum, and suggests that this structure is a homologue of the cranium.

> ... the absence of segmentation, and the relation of the primordial cranium to the vertebrae are obscure points which the anatomy and embryology of the Vertebrate head failed to elucidate, but which the Arachnid theory resolves into the comparatively simple question as to the origin of the cartilaginous sternum. Vertebrate embryology, it seems, told all there was to tell: the fault was not in the answer, but in the interpretation; or rather, in the conviction that if ontogeny did not show what was expected, it was due to the imperfections of the ontogenetic record, not of the expectations. (Patten, 1890; p. 355)

As Patten sees it, there is clearly a way of dividing the vertebrate body into regions that transcend the segments. Although nothing else whatsoever connects them, this feature of Patten's idea seems to prefigure the 'somatico-visceral animal' idea of Romer (1972), discussed in Chapter 3. There, Romer sees segmentation as a vertebrate feature superimposed on an earlier bipartite division.

(a) Lobster quadrille

Immediately following Patten's paper in the *Quarterly Journal of Microscopical Science* (1890) is a paper by the physiologist W. H. Gaskell. In 1890, Gaskell was in the middle of a series of papers purporting to show that the tubular central nervous system of vertebrates, exemplified by the ammocoete larva of the lamprey *Petromyzon* (on account of its primitiveness, wherein such ancestral links could be most clearly seen) could be derived from the tubular gut of a 'crustacean-like' ancestor (similar to the horseshoe-crab *Limulus*), accompanied by the development of a 'new' gut of a type peculiar to vertebrates.

This transformation would not require an inversion – the arthropod gut becomes the vertebrate nervous system, and everything else follows from that.

Gaskell notes that the vertebrate central nervous system is disposed as a hollow tube bounded on the interior by epithelial, supporting tissue, in the same way that the crustacean nervous system is disposed in a cylindrical manner about the gut epithelia. He thus makes the link, pursued in detail, that the vertebrate nervous system evolved from the gut of the crustacean-like ancestor.

Simply stated, the vertebrate nervous system is a tube, distended anteriorly into a sac-like brain, bounded dorsally by the thin, vascular choroid plexus. The hypophysis is effectively a continuation of this brain forwards and downwards, as a tube. Gaskell suggested that the hypophysis is what remains of the mouth and foregut of a crustacean-like ancestor, leading into an arthropod cephalic stomach (the brain) and into a simple intestine (the dorsal nerve tube).

Bereft of a gut, the ancestral vertebrate evolved a new one: Gaskell suggests the extension of a segmented respiratory chamber to include the gill-bearing legs of the arthropod ancestor – as gill bars. This explains the otherwise extraordinary physical link between the pharynx, an organ which in vertebrates is designed for respiration, and the alimentary tract[35].

In the wake of the doubts cast on the position of tunicates and the amphioxus in vertebrate ancestry, Gaskell's work filled a void. His work was widely discussed, and he pursued his theme from its inception in the 1880s for more than twenty years, finally summarizing it in book form (Gaskell, 1908).

(b) The Darwin Jubilee debate

The next year, 1909, was the Golden Jubilee of the publication of Darwin's *Origin*. The Darwin–Wallace idea of natural selection had, of course, been first aired in public in the rooms of the Linnean Society of London, which duly celebrated the 50th anniversary with special meetings and lectures. A discussion on the origins of vertebrates took up the best part of two meetings of the Society, on 20 January and 3 February 1910, and was devoted to the exposition and discussion of Gaskell's thesis.

The published record of those meetings (cited here as Gaskell *et al.*, 1910) makes interesting reading. It is a regular Lobster Quadrille, with bravura performances by Gaskell and his (mostly physiologist) supporters parried by well-oiled set pieces from prominent morphologists of the day. It is a grand opportunity to read the views of the leading players of the time, in the form of brief, frank transcripts from a meeting, rather than long and sometimes obscure papers and monographs.

Today, it seems strange that Darwin should be commemorated by a discussion on the origins of vertebrates, let alone a two-day meeting devoted to a concept now regarded as eccentric, and deserving of a footnote at best. Indeed, the published proceedings have never been prominently cited (a suspicion confirmed by N. D. Holland, personal communication).

The extended commentary that follows is designed, in part, to mitigate this omission, but mainly to show that such ideas were once taken so seriously that they could command two whole sessions of a senior biological society.

There is another reason, which is that Gaskell's thesis represented the apotheosis of the protochordates-as-degenerates strand of biological thought. The Darwin Jubilee Debate offers an unparalleled opportunity to watch this ornate, very Victorian, terribly Neo-Gothic edifice in the act of crumbling.

Gaskell himself opened the batting on 20 January, 1910, with the conventional assertion that, as an organ system, it is the nervous system that dictates the shape and form of an animal, and occupies a central place in the evolution of that shape and form. He qualifies these statements by putting the alimentary tract very much in second place, in clear opposition to those who use the gut as a central reference point – such that the transition from segmented worm or arthropod to vertebrate would have necessitated an inversion. His view can also be seen as aspersions cast on creatures such as enteropneusts, tunicates and the amphioxus – all animals big on feeding mechanisms but small on brains.

He then documents the long and illustrious pedigree of the view that vertebrates had arisen from a segmented form. Ever since Geoffroy St Hilaire, advocates of this view, far from being eccentric outsiders, had been big players in zoological thought, 'all of whom based their views on the presence of the infundibulum in the Vertebrate in exactly the same position in the brain as the oesophagus in the Invertebrate group' (pp.9–10). However:

> So powerful was the fetish[36] of the inviolability of the alimentary canal, that no one of these observers ever noticed that if the infundibulum is the old oesophagus, it leads directly into the great cavity of the ventricles of the brain, which again lead into the straight narrow canal of the spinal cord and so through the neurenteric canal to the anus; that in fact if the infundibulum is the oesophagus, the rest of the lining-walls of the cavity of the central nervous system corresponds word for word with the rest of the Invertebrate alimentary canal. On the contrary, they considered the homology could only hold good by turning the animal topsy-turvy and making the back of the Invertebrate correspond to the ventral surface of the Vertebrate. Such a method was doomed to failure and is now universally discredited. (p.10)

Given the view then prevalent that tunicates and the amphioxus represented degenerate vertebrates, Gaskell asks his critics to produce any unsegmented invertebrate animal in which the alimentary canal is disposed ventral to the nerve cord, and to explain why the vertebrate nerve cord is uniquely tubular, before dismissing his ideas as fanciful.

Nevertheless, Gaskell's evolutionary rationale for his ideas must have seemed strange, even in 1910. It is an odd scheme in which one can hear distinct echoes of predarwinian thought, and a negation of the multiple phyletic branching and independent evolution from primitive common stock of the kind Darwin envisaged. Gaskell's rationale stood equally at odds with that of Geoffroy and the other 'transcendental morphologists', who pointed out structural affinities between vertebrates and other animals in a 'philosophical' sense, with no thought of the evolutionary transformation of one from another (see Richards, 1992 for a discussion of transcendental morphology and its place in predarwinian biology).

Given the 'paramount importance' of the nervous system, Gaskell suggests that each new major animal group must have arisen from the most highly evolved animals (that is, those with the most elaborate nervous systems) of the time. Thus, Man (with his large brain) came from mammals, 'the highest race in Tertiary times', which came from the reptiles (the same, in the Mesozoic), which came from Amphibians, 'the lords of the Carboniferous epoch', and so down to fishes. By extension, fishes should have evolved from the 'highest race' then present in the sea, the arthropods.

This explains, at root, Gaskell's comparison between the larval lamprey and the horseshoe crab *Limulus*, then seen as a 'living fossil', a

Palaeozoic arthropod incarnate. Of course, many early fishes were heavily armoured, and the superficial similarity of ostracoderms and arthropods such as eurypterids has been drawn several times.

Gaskell's developmental rationale is similarly uneasy. He first finds support for the 'Law of Recapitulation' in his ideas (see the discussion on Garstang, below), only then casting doubt on one of its instances, the germ-layer theory. For example, he finds comfort with the anterior opening of the neural canal at the neuropore in the amphioxus embryo, the simple sac-like shape of the embryonic vertebrate brain and so on – finding the arthropod heritage of vertebrates even more strongly expressed in the embryo, as one would expect according to recapitulation:

> The close comparison which it is possible to make between the eye-muscles of the Vertebrate and the recti muscles of the Scorpion group on the one hand, and between the pituitary and coxal glands on the other, are based upon, or at all events are strikingly confirmed by, the study of the coelomic cavities and the origin of these muscles in the two groups. In fact the embryological evidence of the double segmentation in the head and the whole nature of the cranial segments, is one of the main foundation stones on which the whole of my theory rests. (p.14)

How amazed Gaskell would have been, had he lived to learn of the close correspondences between arthropods and vertebrates as regards the embryological specification of head and trunk segmentation (see Chapter 3).

But Gaskell is picking those bits of Haeckelian theory he likes and dismissing those he likes less, notably the 'germ layer' idea. He objects to what he sees as the conclusion, made *a posteriori*, that as the nervous system in modern vertebrates is demonstrably epiblastic (that is, ectodermal) in origin, it cannot have developed from hypoblast (endoderm), a different germ layer, as one would expect were it derived from the gut, as Gaskell's theory demands.

However, he asserts that the hypoblast is more of a physiological conception than a morphological reality: rather than being a presumptive tissue that forms the gut, the hypoblast is defined after the fact as that tissue from which the gut is formed. So, were a tissue (such as nervous tissue) hypoblastic in origin but not concerned with digestion, it would not be recognized as hypoblastic by those who subscribe to the germ-layer theory.

Sceptical from the start, E. W. MacBride disputes Gaskell's reasoning as, literally, unprecedented. Although one can see how (say) the hermit crab is descended from a more typical crustacean[37], Gaskell's movement between established phyla requires him to 'reconstruct the entire animal, leaving only the central nervous system standing', with the origination of a new alimentary canal by invagination of the ventral surface. That this idea is unprecedented has good reason, and is in no way a credit to Gaskell: for no mechanism can be identified whereby, in real life, such a transformation could take place.

Throughout the debate, Gaskell's views on the development of the eye receive particular attention. The vertebrate eye develops as an outpocketing of the brain that grows out to meet the skin. In arthropods, though, the eye develops as an epidermal thickening. How can these views be reconciled were the vertebrate eye equivalent to a gut diverticulum? As MacBride puts it:

> Are the lateral eyes of the two groups homologous or are they not? If they are homologous, how is their different origin explained? ... If the eyes in the two cases are not homologous, why did the Arachnid ancestor of Vertebrates give up its external eyes and develop a new pair from its old alimentary canal? To say that there is no precedent for such a change is to put it mildly. (p.16)

Then MacBride touches a nerve, so to speak;

> Dr Gaskell indulges in a polemic against the germ-layer theory, whilst maintaining strongly the theory that the development of the embryo recapitulates the history of the race. He seems to be unaware that the germ-layer theory is only a special instance of the recapitulation theory. (p.17)

and passes over criticisms of minor implausibilities in the theory with a cutting play on words:

> ... for the mind which accepts the main ideas of the theory will be capable of digesting such trifles also. (p.17)

Stirring stuff. He then demolishes the similarities between ostracoderms and arthropods as even more superficial than those between whales and fishes, and notes that ostracoderms could, in turn, have been descended from less well-armoured vertebrate forms. 'I have no doubt at all', he says, that whilst '*Cephalaspis, Pterichthys* [armoured fishes], and their congeners were practising this sluggish mode of life [i.e. crawling on the sea floor], the real ancestors of the dominant Vertebrates of the sea were ranging like flashes of living light through the waters above' (p.18). And then, significantly:

> It is customary to speak of *Amphioxus* as a degraded creature, but no one who has ever seen it swim will fail to realize the immeasurable superiority of the Vertebrate motor system over that of the Arachnid. The comparison of the one to the screw of a steamer and of the other to an eight-oared boat gives some idea of the difference. (p.18)

The proposed uniqueness of the tubular nervous system in vertebrates, and the special 'enduring' character of the nervous system get similar short shrift. Tubular neuronal structures are met with in the amphioxus, and if that fails to convince, in echinoderms, where exposed neuronal plates are folded over and closed by flaps of non-neuronal tissue.

And the very structure of the nervous system gives the lie to its supposedly eternal nature. The arthropod system is highly ganglionated to best serve a regionated system of locomotion; the vertebrate system is highly centralized to govern a system of movement that depends on central coordination. In each case, the nervous system of a particular phylum can be traced back to a 'mist' of simple nervous ectoderm. There is no need (or sense) in translating one highly developed system into another which, when looked at critically, is quite different in both form and function.

Instead, MacBride looks to echinoderms, and (through the tornaria larva), enteropneusts, tunicates and amphioxus as 'degenerate off-shoots' of the 'free-swimming group of pelagic animals' from which the vertebrates arose directly.

Lastly, MacBride condemns Gaskell's own condemnation of the parallel evolution of similar forms (such as nervous systems) as a negation of the *Origin* itself:

> Dr Gaskell states that his theory strikes at the root of the conception of parallel development. In this case I venture to predict that the root will prove to be more resistant than the axe with which it is struck. (p.20)

Strong words, indeed. But not strong enough to deter E. H. Starling from sticking up for his fellow physiologist. Starling asserts that physiology plays at least as great a part as morphology in the Darwinian struggle for existence. This may be fair comment. But Starling has plainly not been listening to MacBride's demolition of Gaskell's non-darwinian ideas on parallel development.

First, Starling suggests that the evolution of 'dominant' forms (such as Man) 'must have been continuous and progressive'. This misreading of Darwin favours the chain-of-being model of evolution rather than a fan-like adaptive radiation, in that it requires new 'higher' forms to evolve from 'lower' creatures less well-adapted to environments already occupied by 'higher' forms of a different kind. Thus he cannot accept the origin of vertebrates from lowly unsegmented forms once arthropods had already appeared. Whatever one thinks about the anatomical arguments, vertebrates must have evolved from arthropods or they would not have evolved at all.

Again, the primacy of the nervous system is stressed. For evolution to be progressive and to culminate with Man, the pace of evolution must have been set – not followed – by changes in the central nervous system.

Starling then descends into a curious quasi-biblical diatribe against the morphologist, who 'while professing a lip service to the doctrine of Evolution, has really forsaken the teachings of Darwin and gone back to the worship of his old idol, the study of form for itself'. Terms like 'impious' and even 'Author of all evil' crop up in what amounts to an accusation of suppression of Gaskell's views as a threat to conventional orthodoxy. But theories destined for success never achieve universal approbation by calling names. He concludes:

> It is a happy augury for the revival of freedom of thought in English biology that the Linnean Society should, in this jubilee year of Darwin, have devoted an evening to the discussion of a theory, which, I believe, will prove to be the most important contribution to the history of our race since the publication of the 'Descent of Man'. (p.24)

Such are hostages to fortune[38]. Propriety is regained with the contribution by E. S. Goodrich who, as a student of the late Francis Maitland Balfour, was a leading light in what came to be known as the 'segmentationist' school. The Linnean debate provides an interesting opportunity to see how he came to adopt a set of ideas that carry weight and influence to this day.

In what is almost a cladistic way, he takes the features that extant vertebrate forms hold in common and works backwards to discern the essential features of a vertebrate ancestor. For example, the first jawed vertebrate would have been shark-like, with a superficial covering of denticles, cartilaginous endoskeleton, open branchial slits and so on.

The cyclostomes (lampreys and hagfish), he feels, have a simplicity of structure that, to him, seems primitive rather than degenerate. There are, for example, absolutely no vestiges of paired limbs, dermal skeleton or true teeth. Segmentation is more complete, more uniform, and more evident in the head of the lamprey than in that of the jawed vertebrates – all of which is evidence for the formation of the head from a number of segments, a process that has gone much further in the jawed vertebrates than cyclostomes. For example, in cyclostomes, cranial nerves IX (glossopharyngeal) and X (vagus) emerge behind the skull and are associated with muscle-blocks, whereas in jawed vertebrates these nerves are incorporated into the head and the associated muscle blocks are suppressed.

Going further back, the amphioxus-like condition is reached, in which the clean ranks of segments are all but undisturbed by a distinct skeleton, cranium, organs of special sense or cranial nerves. To suppose that the amphioxus reached this almost ideal state by degeneration, leaving no sign of the organs and tissues which, in vertebrates, have so profoundly modified the basic structure, is 'ridiculous': no such change could occur and cover its tracks so completely.

The open endostyle and the simple nephridia link the amphioxus with other invertebrate coelomates. But which? The enteropneusts, although distant, 'point to a remote common ancestor in which the supporting notochord was not yet formed, the nervous system was superficial and more diffuse, and the segmentation less perfect'. (p.26)[39]

By the same means, the complex forms of arthropods and molluscs can be traced to simpler and more generalized ones, along lines of descent 'fundamentally different' from that of vertebrates. It may be true that the nervous system has played an important part in the evolution of these three groups (or represents the organ-system wherein the effects of increasing complexity are best summarized): but it is not surprising that the nervous system in each case has adopted similar solutions to this complexity. After all, sensory stimuli such as light and sound present the same problems for animals irrespective of ancestry. Importantly, organs of special sense are present in highly evolved molluscs, arthropods and chordates, but are absent in the more primitive members of these phyla, suggestive of the independent origins of these and other structures. Thus Goodrich rests his case, his argument sound enough without rhetorical adornment.

H. Gadow then pleads sympathetic treatment of Gaskell's case, urging that it has survived despite the endurance of fourteen years of 'pitiful contempt and ridicule'. He goes on to propose a scheme of vertebrate ancestry that depends not on what he sees as the segmentationist 'Elasmobranch worship' of Balfour, Gegenbaur and their school (a pointed reference to Goodrich), but on a fundamental difference between the primitive 'head' and metameric 'tail':

> With a mouth not terminal but ventral: their [the ancestors of vertebrates] bulk consisting of a large anterior complex and a short, tapering tail, both segmented and metameric. Condensation and fusion produced a head which was so large because it contained all the principal organic systems, as nervous, digestive, respiratory, vascular, and possibly excretory and generative.
>
> Metamerism in this anterior complex, the incipient head, was doomed, but in the posterior portion it underwent renewed activity. Not only were more segments formed by interstitial budding, but metamerism ran wild, culminating, besides other features, in vertebralization.

> The latter proceeded from the tail end forwards, and it is idle to seek for vertebrae in the primitive head, excepting in the part from the vagus backwards, which in the early creature we are dealing with, was a very recent formation.
>
> Meanwhile, the posterior or tail portion becoming larger, part of it, from before backwards, was converted into a trunk, as this was receiving most of those organs which were crowded out from the consolidating head, and also no doubt owing to the repetitional budding backwards of some of these organs. (p.27)

The final result was a tadpole-like vertebrate. I quote this discussion at length because, despite its clear allegiances to Patten's ideas (and thus to those of Gaskell), it resonates at certain points with the ideas of Romer (the 'somatico-visceral animal') of 1972; certain ideas relating to homeotic gene interaction such as 'posterior prevalence' (see Chapter 3), as well as Jefferies' ideas of the transformation of a mitrate carpoid into a vertebrate, what he calls 'head–tail intergrowth' (see Chapter 4).

Importantly, Gadow sees the spinal cord and notochord as rearward growths from the head end that arise relatively late in ontogeny as a consequence of that 'metamerism run wild', the pronounced rearward interstitial budding that gives rise to a new section, the trunk. This is reminiscent of Berrill's (1955) description of the development of the notochord in tunicate tadpole larvae (see later in this chapter).

Gadow is quite happy that the cartilage of the notochord is endodermal in origin, consistent with it having budded from the new gut of a protovertebrate, its old gut having been made into a nervous system in Gaskellian fashion. He is also swayed by Gaskell's 'detailed, almost too minute' comparisons between arthropod and vertebrate, so much to say that such correspondences cannot have arisen by chance.

He also plays down the germ-layer theory as an inflexible dogma that fails to recognize the importance of present function over past heritage. There exist gills of both ectodermal and endodermal origin, but gills for all that: conversely, amphibians breathe through their (ectodermal) skin as well as, and in some cases instead of (mesodermal) lungs. Thus 'it is false dogma that the gut must be *the* organ which is homologous in all gut- possessing animals'.

Finally he defends Gaskell's hypothesis on the reasonable grounds that although it may be mistaken here and there, it attempts to encompass a great deal of evidence within a single theoretical framework. It must therefore be studied as a serious contribution, for all that it seems odd to begin with.

Arthur Smith-Woodward, noted palaeoichthyologist[40], started part two of the meeting on 3 February 1910, with an immediate counter to Gaskell's idea that evolution is always progressive, in the sense that the first representatives of a dominant group evolve from the highest members of the immediately preceding dominant type. In contrast, new groups seem to spring from the more primitive, less highly evolved members of the group preceding, so that if vertebrates evolved from arthropods, it was from a generalized and probably soft-bodied form beyond the reach of fossilization.

Smith-Woodward feels that ostracoderms – and their modern relatives, the lampreys and hagfishes – are more primitive than modern 'true' fishes, and thus finds Gaskell's work on the ammocoete larva 'striking and interesting'. However, comparisons between ancient armoured ostracoderms and contemporary arthropods must be confined to the superficial. It is faulty observation or misinterpretation that converts ostracoderm scales into the paired arthropod-like appendages observed by some (including Patten). Such are the dangers of reconstructing extinct forms from fossils.

Arthur Dendy, the Secretary of the Linnean (and, like Goodrich, an advocate of the view that the amphioxus is a primitive rather than a degenerate form) takes Gaskell to task on his comparisons between vertebrate and arachnid lateral and median (i.e. pineal) eyes, finding them wanting, and even sloppy. However, he cannot comment on the essential difference between lateral eyes – the modes of origin of arthropod and vertebrate eyes, one from a thickening of the epidermis, the other an outpocketing of the brain, except to suppose Gaskell's reconciliation (a double origin of the retina from the epidermis and the former gut wall) 'far-fetched', and the attempt to compare vertebrate and arthropod eyes – both highly specialized – as 'foredoomed to failure'.

Dendy can, though, refute Gaskellian ideas on the pineal eye directly from his own researches, in which he shows that the epiblast plays no part in the ontogeny of the well-developed pineal eye of the tuatara *Sphenodon*.

Generally, Dendy accuses Gaskell of having failed to distinguish analogy and homology. Structures may look the same despite having different origins, phyletically and ontogenetically. There are good, although different reasons why the arthropod gut and the vertebrate nervous system are tubular. The cephalopod eye is far more similar to that of the vertebrate than that of the arthropod, despite equal remoteness of ancestry, 'and yet no one, so far as I am aware, has ventured to include the Octopus in the ancestral portrait gallery of the Vertebrata'. This point makes a nonsense of the views of Gaskell's physiologist supporters that present function, not past morphology, is the key to explaining ancestry.

Indeed, Dendy makes a fine case against the Gaskellian view that changes in the nervous system, by virtue of the perceived importance of this system, offer the most informative clues about evolution: a view which

> ... is entirely contrary to the usually accepted views of systematic zoologists, who find in structures which are apparently of the least use to their possessors the best guides to genetic affinity. Organs which are of great use must be subject to adaptive modification in accordance with the changing needs of the organism. (p.37)

It is because of the very importance of the nervous system that such similarities as there are between the systems in arthropods and vertebrates – distinct on the phylum level – are more likely to reflect functional convergence than vestiges of shared common ancestry.

Sir Ray Lankester took responsibility for planting, some years before, the seed in Gaskell's mind that blossomed forth so abundantly. But he found little to commend the result: in contrast to the characters that link tunicates and vertebrates (notochord, dorsal tubular nerve cord, pharyngeal slits, cerebral eye), Lankester can find no shared derived features that group arthropods with vertebrates. He affirmed the views of many critics, that each major phylum traced its own unique descent to primitive ancestors, there being little to connect one phylum with another, to the exclusion of other groups.

As for himself, Lankester regards both tunicates and amphioxus as too specialized and degenerate to shed much light on the condition of the vertebrate ancestor. He, as did others, sees some hope in *Balanoglossus*, and draws attention to 'the remarkably complex brain and cerebral respiratory pits' of nemertean worms as offering a clue.

Lankester's account is given as reported speech rather than direct quotation, although such a device dulls not his barbs. He all but asks whether Gaskell might not be light of brain:

> Sir Ray Lankester held and he desired to state it without any offence, that in searching by long and strenuous enquiry for evidence in favour of such a hypothesis as that adopted by Dr. Gaskell, the mind is liable to a kind of 'suggestion', and that of those who too readily admit all sorts of coincidence as evidence that Bacon wrote the plays of Shakespeare. The heroic nature of the task which it is sought to accomplish undoubtedly in many enterprising and devoted investigators has re-acted unfavourably on the judgment. All are liable to it and it may be that something of the kind is here at work. (pp.39–40)

Given that Gaskell set such store by the importance of the nervous system, P. Chalmers Mitchell reviews the general form of the central nervous system in the major animal groups, and finds nothing special about the chordate version that links it with that of the arthropod in particular. Intriguingly, he refers to (then) recent work by the American physiologist Herrick[41], who compared the two systems without reference to questions of origin, finding between them fundamental differences underlying 'all superficial resemblances'.

Further apologia come from Stanley Gardiner, who complained that 'of the many speakers only Dr. Gaskell has put forward a connected theory which the rest have merely attempted to destroy'. The problem, he rightly notes, is that there are very few facts on which to construct any hypothesis regarding vertebrate origins. Thus free to gambol in an empty field, Gardiner re-asserts the primacy of the nervous system as a guide – as if the contributions of the previous speakers on the subject were as a tale told by an idiot, full of sound and fury, signifying nothing.

However, from his perspective, Gaskell's idea seems more rounded than the sketchy arguments of MacBride, Goodrich and so on that the amphioxus should be considered as a primitive representative of the vertebrate stock. Mischievously, though, Gardiner misrepresents his opponents by loudly proclaiming that the amphioxus cannot but be somewhat degenerate and off the main line of vertebrate ancestry – as if MacBride (in particular) said anything else. Gardiner conveniently omits Lankester's direct (if reported) statement on the position of the amphioxus. Gardiner then dismisses the constraints of the germ-layer hypothesis, by demonstrating the interchangeability of layers in the development of certain creatures:

> Again, in budding there are difficulties with this [germ-layer] theory, the gut of some budded-off Polyzoa being formed from mesoderm, while of Tunicates, supposed relations of the Vertebrates, *Clavellina* buds from the endoderm and *Botryllus* from the ectoderm, giving ectoderm and endoderm respectively; and do not some Sponges turn inside out to give the adult? (p.44)

Gardiner's wider error rests on an over-literal reading of recapitulation (which is not his fault, but no reason to throw out recapitulation altogether); and, second, the affirmation of the validity of one hypothesis by the presentation of evidence that pertains to another (which is). That is, even if tunicates *do* play games with their germ layers during their individual development, is that any argument for linking vertebrates with arthropods in a broader, phyletic way?

The position of T. R. R. Stebbing as a cleric allows him licence to make comments of the 'plague-on-both-your-houses' variety.

> When we return home and our friends gleefully enquire, 'What then has been decided as to the Origin of Vertebrates?', so far we seem to have no reply ready, except that the disputants agreed on one single point, namely, that their opponents were all in the wrong. (p.45)

For his part, Stebbing advances the jointed exoskeleton and the notochord as two independent responses to the need for an actively 'wriggling' animal to achieve support, whether or not the ancient soft-bodied wrigglers came from a common stock.

Gaskell was called upon to have the final say. He started by poking fun at MacBride, whose assertion of 'diabolical ingenuity' must be on the part of Nature's facts rather than Gaskell's own interpretation – given, says Gaskell, the close correspondences between arthropod gut and vertebrate nervous system. Again, he affects the air of an Olympian pretending to be wronged by impudent mortals. Where, he asks, are the 'violent' changes of function that MacBride asserts as central to the Gaskellian view? Gaskell counters this with a welter of detail, avoiding the point that, at some time in evolution, a gut was transformed into a structure of radically different function, and a new gut was produced. Instead, he misconstrues MacBride by deflecting the error, suggesting that the arthropod idea is seen as outlandish only through an inappropriate adherence to the germ-layer theory.

The injured godhead thus hurls his bolts at Goodrich's championing of the amphioxus as representative of vertebrate ancestry.

> Surely this [Goodrich's view] is a unique position! All other morphologists look upon *Amphioxus* as a degenerate animal ... What conception has Goodrich of the evolutionary process, of the struggle for existence, of the survival of the fittest? (p.48)

But Gaskell's rationale for these insults rests on several misconceptions about evolution – that the vertebrate lineage must, presumably because of its culmination with Man, be somehow progressive and 'special'; that evolution in general is always 'progressive'; that evolution demands of necessity a single arena in which all creatures compete with one another; and, perhaps most importantly, that change proceeds purely through adaptation without limits imposed by historical contingency:

> Just consider it: here is a wretched animal without brains, without eyes, without a nose, victorious in the struggle for existence over the whole of the Invertebrate world. What is the driving force; how could it have taken place? Only, it seems to me, by some beneficient power taking special charge of him and assisting him in the growth of brain and eyes and nose. (p.48)

His remaining comments are similarly dismissive, except where he answers some of Dendy's more technical criticisms.

If there is one take-home message from this discussion, it is this: Gaskell's error lies in his overall conception of evolution as progressive, directed and panselective. His treatment of recapitulation is also flawed:

he cannot subscribe to it as it suits him, and omit a central tenet of Haeckelian thought, namely the 'germ-layer' theory. Subsequent history and experiment have modified the germ-layer theory substantially, but vertebrate origins in Gaskellian mode has gained no credence thereby.

The dismissal of Gaskell's views must be grounded in these wider issues of theory, not on the grounds that an alliance between vertebrates and arthropods seems somehow outlandish. After all, the vertebrates had to come from somewhere, so why not a *Limulus*-like arthropod?

Instead, one should ask oneself whether Gaskell's ideas would stand in higher regard today were he to cast tunicates or echinoderms in place of arthropods onto his misconceived evolutionary stage. I venture that they would not, given the manifest theoretical failures of his scheme that transcend such particulars.

2.1.5 Enteropneusts

Long before Goodrich's defence of the amphioxus in the Darwin Jubilee Debate, Bateson (1886) had dismissed the shared segmentation of annelids/arthropods and vertebrates as functional convergence, and had built a theory round the idea that protochordates were primitive, and that they had something to contribute to work on vertebrate origins.

Bateson proposed a close link between enteropneusts and chordates, particularly cephalochordates, and put tunicates some distance away. This was based on the idea that tunicates primitively had but a single pair of gill slits, and that the elaboration of the ascidian pharynx was a secondary specialization. The single pair of slits in larvaceans such as *Oikopleura* was held to indicate the primitive status of this motile form with respect to other tunicates, which was the general view of these creatures prior to Garstang's work on paedomorphosis (Garstang, 1922).

However, enteropneusts quite clearly had a well-developed series of gill slits similar to those of the amphioxus even though the animals are, in general, not segmented: they also had the rudiments of a notochord (the stomochord) and a chordate-like nervous system, neither of which are accorded much validity today (but see Peterson, 1995b).

Bateson found the presence of gill slits – and the absence of segmentation – in enteropneusts compelling evidence that they were close relatives of vertebrates, and that vertebrates had acquired segmentation later on, independently of annelids and arthropods. The poor development of gill slits in larvaceans seemed, to Bateson, to indicate that tunicates in general had branched from the chordate stock somewhere before enteropneusts.

As shown in Chapter 1, the current consensus overturns this state of affairs: enteropneusts are nowadays generally thought to be related to pterobranchs, but their place among deuterostomes in general is uncertain – except to say that they are less closely related to vertebrates than either tunicates or amphioxus.

Not that enteropneusts do not have their advocates today. Peterson (1995b) argues (citing recent evidence) for a revival of the idea that the enteropneust stomochord is a homologue of the notochord, and that the dorsal nerve cord of enteropneusts is a homologue of the cognate structure in chordates. This adds to the resemblances of structure between amphioxus and enteropneust gill slits.

2.2 GARSTANG

The early theoretical work of the marine biologist Walter Garstang (1868–1949), positing connections between echinoderms, tunicates and vertebrates, would have been somewhat unfashionable in 1910, so it is not surprising that it is not mentioned at the Linnean meetings (quite apart from the fact of the obscurity and brevity of its publication). At the time, Garstang was more concerned with the conservation of plaice stocks in the North Sea (Baker and Bayliss, 1984), and would not return to the subject of vertebrate origins for another twelve years.

But like Goodrich and MacBride's wretched creature as parodied by Gaskell, sans brain, sans eyes, sans nose, sans almost everything, Garstang's ideas have inherited the Earth, and represent arguably the most influential body of work on the origins of vertebrates.

Garstang was one of the first to work at the Plymouth Marine Laboratory, but moved on via Oxford until coming to rest as the first Professor of Zoology at the University of Leeds. He was fascinated by the larval forms of marine animals, and was preoccupied by one particular question: why was it that planktonic larvae have such different shapes and lifestyles from the adults into which they metamorphose?

This question runs through all his work, but is most apparent, surprisingly, not in a weighty treatise but a collection of light verse, *Larval Forms* (1951), largely written for the amusement and edification of students. This posthumous compilation benefits from a lucid introduction by A. C. Hardy, which along with the biographical memoir by Baker and Bayliss (1984) and his obituary (Eastham, 1949) forms a good summary of Garstang's life and thought.

To return to the question, then: Garstang reasoned that larvae look different from adults by virtue of the different selective pressures that they must endure. Indeed, the selective pressures on planktonic larvae are at least as great as those on adults. But the constraints of selection on a population of large, possibly sessile, benthic adults will differ from those on a population of small, possibly motile, planktonic larvae.

Larvae are torn by pressures in opposition. On the one hand, they must spend as much time as possible in the larval state in order that they might be dispersed as widely as possible – which is, of course, one of the primary functions of planktonic larvae. On the other hand is strong selection to shorten larval life (driven by the risk of predation) such that larvae must become sexually mature and reproduce as soon as they are able.

These ideas were radical in Garstang's time, given the importance attached to Haeckelian recapitulation, a concept that had been partly inspired by the discovery (Kowalevsky, 1866a, 1871) of the chordate nature of tunicates (reviewed by Schaeffer, 1987). According to this idea, the stages in development of an animal could be interpreted as a record of its evolutionary history – to use a phrase that has since become something of a *cliché*, 'ontogeny recapitulates phylogeny'. In the course of an animal's evolutionary history, old-fashioned adult forms were pushed backwards in ontogeny, that is, relegated to larval stages, as each new adult form evolved.

To put it the other way round, the evolution of complex forms is a result of new, adult stages being tacked onto the end of ontogeny, such that previously adult stages become larval. In any given ontogenetic sequence, the result is a breathless record of evolution, compressed to form and encapsulated within a single embryogeny.

For example, a human being develops from a single cell, like the protozoan condition; through a blastula and gastrula, reminiscent of coelenterates; through yet further stages that nod respectfully to fishes (presumptive gill slits that remain imperforate)[42].

Although one might sympathize with this idea, it plays down the importance of larvae as functional entities in their own right. In terms of natural selection, there is no point in having a distinct larval stage, if its only function is to act as a passive repository of phylogenetic history, there for the convenience of anatomists. Garstang, who studied larval forms at first hand rather than as theoretical constructs, soon found many reasons to disagree with the idea of recapitulation.

Instead, he proposed an alternative view of ontogeny that was governed by selection. One solution to the conflict between dispersal and maturation in larvae is for them to forego metamorphosis and become sexually mature while still in the larval state. This phenomenon is called neoteny and is seen in some present-day animals, such as the axolotl (*Ambystoma mexicanum*), a species of salamander.

Neoteny in these cases can usually be ascribed to special circumstances such as physiological deficiency of trace nutrients, but Garstang went further, to suggest that through the action of natural selection, ontogeny could be distorted such that neoteny could become a process of evolutionary change. Garstang called this process paedomorphosis. Animals that shed their adult state, maturing while still larvae, could take a completely new evolutionary course. Instead of larvae preserving the relics of long-lost adults, as Haeckel had proposed, they may herald the adults of the future. Although not in its original formulation, paedomorphosis came to occupy a central place in Garstang's ideas about vertebrateevolution.

In Garstang's time, it was also common to suppose that chordates evolved from animals that could swim freely, unlike modern ascidian tunicates, which are fixed in the same spot throughout their adult lives. The ancestor could have been fully pelagic or a semi-sessile mud-burrower, provided that it was unattached (Bateson, 1886; a student of enteropneusts) or visualized as something between a tadpole larva and an amphioxus (Willey, 1894; a student of the amphioxus).

The discovery of the chordate affinities of the tunicate tadpole larva was, in part, responsible for this view. In this scheme, the free-living condition of the enteropneusts and the amphioxus, for example, as well as that of the tunicate tadpole larva and the free-living adult phase of larvacean tunicates such as *Oikopleura*, were held to be primitive, and the sessile state of ascidian tunicates was a secondary development. That Garstang overturned this idea, invoking a sessile rather than a motile ancestry for chordates, was a consequence of paedomorphosis.

Nonetheless, Garstang worked in a milieu that had come to recognize the rudiments of all that made a chordate, in particular structures such as the endostyle, pharyngeal gill slits and notochord.

2.2.1 Garstang before paedomorphosis

Garstang's own published ideas on chordate phylogeny were at first very modest. In particular, they were motivated by an effort to bridge the gap between hemichordate and chordate that I discussed in Chapter 1.

These early ideas appeared in only his third scientific paper (the first two were about the marine fauna around Plymouth), which was published in 1894, soon after he had left his post at Plymouth for Lincoln College, Oxford (although the paper is dated 'Plymouth, Febr. 2nd 1894'.)

The 1894 paper had been inspired by a brief report by Ritter (1894) on a tornaria larva of the enteropneust *Balanoglossus*, collected off southern California.

Ritter records a band of thickened epithelium on the floor of the oesophagus of the tornaria larva which he suggests might be 'functionally, at least, an endostyle, meaning by this that it performs the same office as the similarly situated and similarly named organ in Tunicates', but 'whether it be homologous with this organ in the Chordata is quite another matter' (1894, p.29).

One should not be deceived by this note of caution: Ritter records several other instances of structures in tornariae which others had compared with the endostyle, and Ritter is clearly in sympathy with the idea of a close relationship between enteropneusts and chordates.

In response, Garstang published a preliminary note in the same journal on his ideas about chordate ancestry, results which:

> have proved so harmonious with one another that I have been able to synthesize them into a consistent theory of chordate phylogeny of which I briefly sketch the outlines. The complete evidence for this view will shortly be published in the Quarterly Journal of Microscopical Science. (1894, p.123)

And so it did, but 'shortly' turned out to be much later – in the 1920s, after he had sharpened his ideas on paedomorphosis (Garstang, 1922), which he believed necessary to make his ideas on vertebrate origins really work.

Back to the endostyle, though. Although this structure is found in all chordates (even if, in vertebrates, it is transmuted into the thyroid gland) it was (and is) not known to occur in hemichordates. As I showed above, it is in some respects easier to link hemichordates with echinoderms rather than with chordates. Obviously, the discovery of a hemichordate endostyle that would, through hemichordates, have linked echinoderms firmly with chordates would have found in Garstang a most welcoming recipient.

Garstang started by emphasizing the close resemblances between the tornaria and the auricularia larva of holothurians. Simply described, the auricularia larva is bilaterally symmetrical and has a mouth, an anus (blastopore) and two rings or tracts of ciliated cells. One, the aboral, snakes around the middle of the animal like the seam around a tennis ball, separating the oral field from the anal field, and is underlain by a nerve ring. The other, the adoral band, fringes the mouth closely and all the way round, except for a loop that snakes ventrally into the mouth, flooring the oesophagus for a short distance. The tornaria larvae of enteropneusts look a little different, but the topological relations of the ciliated tracts are broadly similar. That is, one band separates the larva into oral and anal halves, and another band ornaments the mouth.

Radially symmetrical echinoderm adults are elaborated from this simple, bilaterally symmetrical condition largely, as I described above, by the pre-eminence of the left hydrocoel. Garstang hypothesized that chordates not only retained the primitive bilateral symmetry, but emphasized it with a radical change in the disposition of the ciliated tracts.

The aboral tract, far from pursuing its sinuous course round the body, was swept up dorsally on both sides to form a pair of ridges. Still climbing, the ridges met to enclose a ciliated canal into which opened the blastopore: a course of events very similar to the formation of the neural tube in vertebrate embryogenesis, and presumably homologous with it (Figure 2.2, and see Lacalli *et al.*, 1994). The associated nerve ring, forced to realign itself along this axis, became the central nervous system.

Garstang suggested – and Ritter's work on the tornaria larva appeared at first to strengthen that suggestion – that the extension of the other ciliated band, the adoral band, along the foregut floor in echinoderms was a homologue, through the hemichordates, of the endostyle in tunicates, the amphioxus and larval lampreys. It is easy to see why – the endostyle is largely formed from a hairpin loop of ciliated epithelium that has much the same relations as the adoral band in auriculariae and, it seemed, tornariae.

It is interesting to note that Garstang mentions, in passing, the degeneration of the right anterior coelom with the elaboration of the left as a feature of the amphioxus and by inference ancestral chordates, drawing an analogy with the same situation in echinoderms – although the comparison remains an analogy and nothing more.

The pattern of enteropneust development, though, is different. Although the aboral ciliated band gathers itself up into a ridge as in the hypothetical ancestral chordate, it only manages a full fusion in the central portion of its length. Garstang linked this with the 'collar' region of the adult enteropneust.

Unfortunately, the premise on which the 1894 paper had been based turned out to be false. Ritter's report did not describe an unambiguous endostyle, but what might have been a ventral ciliated tract in the

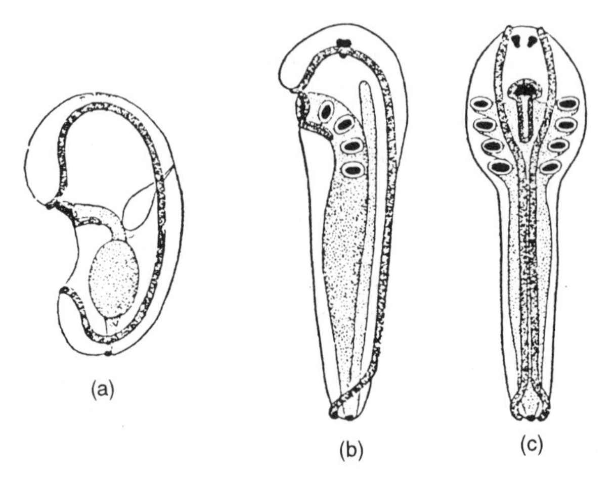

Figure 2.2 Garstang's own figure (from Garstang, 1928) to illustrate the 'auricularia' theory. (a) shows an auricularia, whereas (b) and (c) show respectively the side and top views of the hypothetical early protochordate larva.

oesophagus of a single specimen. Indeed, an endostyle has never yet been found in a hemichordate, either adult or larval.

Despite its being built on a misapprehension, Garstang had sketched a tentative hypothesis about vertebrate origins, in which a re-arrangement of the aboral ciliated band in an auricularia-like larva could form a dorsal, tubular nerve cord, initially connected to the gut (by the neurenteric canal), just as it is in early chordate embryos.

Implied in this idea is that the chordate 'ancestor' formed by this process was motile, at least as a larva. Long before Garstang had proposed paedomorphosis, of course, nothing is said in the 1894 paper concerning the evolutionary implications of alterations in life history. We are not told whether – or even if – this hypothetical chordate ancestor metamorphosed into an adult or, if so, what this adult would have been like, sessile or motile, enteropneust, echinoderm or tunicate.

Again, it was in response to work by another that Garstang was led to comment on chordate phylogeny for the second time, in a paper read to the Oxford University Junior Scientific Club on 6 November 1896 (Garstang, 1897). The work in question was an address to the Physiological Section of the British Association meeting in Liverpool earlier that year, by none other than W. H. Gaskell (Gaskell, 1896).

When Gaskell claimed to have found homologies between the thyroid gland of the ammocoete and parts of the sex organs of scorpions, Garstang felt driven to air his opposition in print.

Garstang noted, once again, the similarities between the filter-feeding arrangements found in tunicates, the amphioxus and ammocoete, and the importance of the endostyle in their function. He then re-emphasized that of all the structures found in invertebrates, the adoral band of the auricularia repays investigation more fruitfully than does the genital duct of scorpions, at least in respect of comparisons with the endostyle.

As noted above, the 1890s version of the auricularia idea differs from that found in modern textbooks, in that Garstang assumed, as did most others at the time, that chordates evolved from free-swimming ancestors that in principle could have evolved by muscular elaborations on the theme of ciliated larvae. The later and last version of the theory, expounded more fully in the late 1920s (Garstang, 1928; Garstang and Garstang, 1926), is different, in that the primitive sessility of chordate ancestors (as adults) is stressed.

The difference lay not in the facts themselves but in their interpretation. As I discuss below, the early form of the theory was implicitly rooted in recapitulation, the later explicitly renouncing it. The great divide came in 1922, with his 'critical restatement' of recapitulation as a biogenetic 'law', and his coinage of the term 'paedomorphosis'.

2.2.2 Paedomorphosis

Ernst Haeckel was Darwin's greatest ally in the German-speaking world. In 1866, not long after the *Origin*, Haeckel formalized what many had seen as a general tendency – that creatures echo their evolutionary past during their individual ontogenies. Garstang (1922) summarizes this recapitulation in two statements, one of fact, the other about causation.

First, that ontogeny is the recapitulation of phylogeny; second, that phylogeny is the mechanical cause of ontogeny. In other words, ontogeny recapitulates phylogeny *because* ontogeny is a direct result of phylogenetic process. 'In these now familiar terms the new conception of evolution was wedded, fifty years ago, to current ideas of ancestry, heredity and development:' Garstang writes (1922, p.82) 'Ancestors created, heredity transmitted, and development repeated the order of creation.'

Whether it had been Haeckel's intention or not, recapitulation became seen as a strictly regulated phenomenon, in which evolution progressed by the addition of new stages at the end of the life cycle, the preceding stages being pushed back in ontogeny to create a series of linear but ever more intricate ontogenies as evolution progressed.

The essential flaw in this argument was obvious to one so interested in larval forms: adults do not spawn new adults, but gametes and zygotes which must undergo the complicated and dangerous processes of differentiation and development before they, in turn, become adults. Ontogeny is not fixed by fate. Instead, anything can happen, any slip, between the zygote and the new adult, diverting ontogeny into new and unexpected courses. Thus, ontogeny is a vitally important arena in which the drama of natural selection can be staged – so is evolution shaped. In Garstang's words, 'Ontogeny does not recapitulate Phylogeny: it creates it' (1922, p.82).

Nevertheless, it can hardly be denied that ontogenies nod to phylogenies in a very general way (it was such an observation that Haeckel sought to formalize and express in the dynamic way demanded by the new concept of darwinian evolution). This, said Garstang, was inevitable given the linked but parallel progress of adult forms and their ontogenies.

However, it is plain that life cycles do not evolve by the simple terminal addition of stages. Rather, that the entire ontogeny changes in all sorts of ways, large and small, to accommodate a structure that is more elaborate overall. This process may compress earlier stages until they are barely recognizable ciphers, or extinguish them altogether. Some ontogenetically 'primitive' structures, though, may be necessary for the elaboration of further ones. For example, that vertebrates retain a notochord could be seen as a necessity for the formation of a vertebral column, not as a mere relic of a former adult condition on which the vertebral column has been appended as a terminal stage.

> Ontogeny ... reproduces those successive grades, not because successive adult types have been included in it, but because each ontogeny is a modification, within limits, of its predecessor; and by those predessors the phyletic chain of adults was organised and equipped. (1922, p.85)

Garstang's own metaphor is particularly vivid. 'A house is not a cottage with an extra storey on top' he wrote (1922, p.84), but an altogether higher order of residence. Only the presence of such formalisms as four walls and a roof betray a relationship. But the transformation from cottage to house comes at the earliest stage, when the architect is drawing the plans, not when the cottage is complete. Indeed, 'it is impossible to overlook the fact that some of the most pregnant changes in the characteristics of the higher Vertebrates', says Garstang (1922, p.91) 'are directly or indirectly traceable to changes in the earliest stages of the ontogeny'. The notochord, the dorsal tubular nerve cord and the neural crest, for example, are not terminal additions, but develop very early in the ontogeny of vertebrates.

This observation does not on its own invalidate recapitulation, because recapitulation makes no prediction about the time that must elapse between fertilization (say) and the appearance of this or that structure. But recapitulation *does* imply the appearance of structures in a particular *order*: and the development of many other fundamentals of vertebrate organization depends on the proper construction of notochord, nerve cord, neural crest and so on. In short, features that are distinctively vertebrate and chordate appear in ontogeny not only very early, but simultaneously or earlier than features that one might expect would be phylogenetically more primitive, and may exert influence on these features.

In a sense, the recognition of the primacy of ontogeny over adulthood turns the whole recapitulation argument upside down. The tadpole of a frog, for example, does not represent the fishy adult ancestor of amphibian, but a modification of the *larva* of that ancestor, *irrespective* of the adult appearance of that ancestor. This phenomenon – in which larval stages of different creatures look more alike than the corresponding adults – had been recognized by the anatomist von Baer long before Haeckel turned his mind to the problem. (Indeed, von Baer's 'principle of paedogenesis' dates back to 1828.) Garstang contends that prominent Haeckelians of his day, notably MacBride, had conflated this (real) tendency with an (imaginary) recapitulation.

For example (one of several given by Garstang), MacBride sees recapitulation in the asymmetrical hermit crab, the young of which have the symmetrical abdomens and swimming habits of adult shrimps and prawns. The implication is that the shelled habit of the hermit is a terminal addition to a shrimp-like life cycle. But, says Garstang, *young* shrimps and prawns of all kinds have symmetrical abdomens and a pelagic habit, just like adults. In that case, the life cycle of the hermit crab does not display recapitulation – rather than building on a demonstrably *adult* stage, it shows that similar larvae have the potential to alter their ontogenies at different times and in different ways.

> Nowhere does he [i.e. MacBride] show, or claim to show, that the young stages of any of these animals resemble the *adult more closely than the young* stage of typical members of their respective orders. He does not show this because he cannot. (p.89, original emphasis)

MacBride had fallen into the same trap with the normal-looking, symmetrical larvae of fish of the family Pleuronectidae, which metamorphose to become asymmetrical flatfish:

> It follows that, for all he [MacBride] has shown to the contrary, the 'typical' or 'normal' larvae, which the Pleuronectid larva resembles, might have grown into Cod, Mackerel, or any other type of Teleost, and that the *adult* ancestors of Pleuronectids, so far from being 'normal', may have carried themselves upside down like a *Remora*, or stood on their tails like Pipefishes. (p.90, original emphasis)

In short, with ontogeny as such a rich source of potential change, it does not necessarily follow that ontogenies can be used as a record of phylogenetic change – particularly in respect of adult morphology.

To go further, many larval forms exhibit morphological peculiarities and adaptations not seen in the adults which undoubtedly bear on their survival, and thus on evolution and subsequent phylogeny. But because Haeckelian recapitulation only 'sees' adults, these adaptations are usually thought of as trifling, rather than ontogenetic interpolations of potentially great importance.

The proof of the recapitulationist pudding should be seen in palaeontology. 'It is, however, the palaeontologists who are the real defenders of the Biogenetic stronghold', writes Garstang (p.92). 'With them the Law is a faith that inspires to deeds, while to the embryologist it is merely a text for disputation'. Were recapitulation true, it should be possible to match the ontogenies of living species with the phylogenies of their ancestors as deduced from the fossil record. But Garstang examines two case histories (one involving the ontogeny and phylogeny of the crinoid *Antedon*, as reconstructed by Bather, echinoderm specialist and staunch recapitulationist) and finds no match to speak of.

In one short paper, then, Garstang demolishes fifty years of recapitulation. Incidentally, the word 'paedomorphosis' appears just once – it is the last word of the article before the references[43].

Four years later, in what was effectively a preview of the long 1928 paper on tunicates (Garstang and Garstang, 1926), he wrote equally frankly about why he thought recapitulation in error, and self-evidently so. This brief note is illuminating in that it gives us a glimpse of what Garstang thought about his own ideas, and how they had changed since the 1890s. He starts by complaining that the earliest papers had been almost completely ignored, 'probably because they left the question of the adult differences between the various groups almost as perplexing as before' (Garstang and Garstang, 1926, p. 81). That recapitulation was popular in the 1890s might have been another reason for the difficulties his ideas encountered when he first presented them, for at the time

> ... they were interpreted by the author as indications of the common origin of Echinoderms, Enteropneusta, and Chordata from an *Auricularia*-like adult ancestor. A detailed account of the theory along these lines was written, and the MS. is still in our possession, but the difficulties entailed by this idea increased on reflection, and the full paper was never submitted for publication. (1926, pp.81–82: emphasis mine)

So, at a time when the idea of recapitulation received wide subscription, Garstang framed his earliest ideas along recapitulationist lines. His views on recapitulation in 1926, encapsulate the thoughts expressed in the 1922 paper.

> There is as yet no crucial fact to prove, or even to render probable, the view that gill-slits or external gills were ever limited to the adult condition of a fish-like creature. In strictness, therefore, the characteristics of a tadpole recall the larval, not the adult, characteristics of its immediate aquatic ancestors. For all we know from embryology alone the adult stages of those ancestors may have been creatures glued to the rock by suckers and lacking both gill-slits and tail. It is comparative anatomy, not embryology, that renders such an idea improbable. If these points are admitted, it follows that no safe conclusion can be drawn from embryonic or larval stages alone as to the characteristics of adult ancestry – in other words, there is no 'law' of adult recapitulation. If there were such a law the Perennibranchiate Amphibia[44], which are admittedly derived from the terrestrial Urodeles[45], ought to pass through such a stage in their ontogeny. The fact that they do not is sufficient in itself to establish our conclusion. (1926, pp.82–83)

Two years later, in 1928, Garstang is, if anything, even more direct. 'The theory of adult recapitulation is dead', he writes, 'and need no longer limit and warp us in the study of phylogeny' (1928, p.62).

But paedomorphosis as the evolutionary analogue of neoteny is only one facet of a larger concept, that the timing of development may be flexible, provided that stages in one ontogeny tend to preserve the characters of equivalent stages in another ontogeny (presumably so they could be recognized as so equivalent) and:

> ... instead of new characters tending to arise only towards the end of the ontogeny, they may arise at any stage of the ontogenetic sequence; and ... instead of new characters always tending to push their way backwards in ontogeny, they may extend in adjoining stages in either direction, either backwards from the adult towards the larva and the embryo (tachygenesis) or forwards to the adult from the embryo (paedomorphosis). (1928, p.62)

2.2.3 Garstang after paedomorphosis

In the 1928 paper, Garstang combines paedomorphosis with an emphatic rebuttal of the idea that chordates were primitively free-living. He does this largely by analogy with the free-living larvacean tunicates. These were then thought to represent either degenerate vertebrates or very primitive tunicates, exhibiting supposedly primitive features such as a small number of gill slits, the presence of a tail and so on.

In a sustained argument, Garstang suggests that they are far from the throwbacks they appear to be. Instead they are paedomorphic offshoots of a group of free-living tunicates called doliolids, themselves explicable as derivatives of sessile stock. It is the sessile ascidians, not the motile appendicularians, that retain the primitive state. And as with tunicates, so as with chordates as a whole – the ancestor was not motile but sessile, transformed by paedomorphosis into a retention of its larval motility.

The earliest chordates, then, developed from sessile lophophorates that gradually expanded their capacity for pharyngeal filtration through gills at the expense of the external tentacles, which were then lost. (Incidentally, Garstang pokes fun at the notion of fast-moving pelagic ancestors which would, by phylogenetic necessity, have to have borne delicate lophophores at the front.)

Although Garstang's inversion of the prevailing view of larvacean ancestry makes for a good story, it applies *only* to the larvacea[46]. It cannot explain why vertebrates and the amphioxus are motile – and not sessile, like ascidians. For this, Garstang introduces the muscular tadpole larva as an interpolation in the tunicate life cycle between ciliated, planktonic auricularia-type larva and sessile adult.

The development of muscles, endostyle and a nervous system in a ciliated larva does not demand a leap of faith of the vigour that one might at first imagine, given that the planktonic larvae of various aquatic forms develop muscular extensions and buoyancy aids of various sorts once they outgrow the small size that cilia alone can comfortably propel. Large echinoderm larvae, for example, have muscular 'wings' that flap up and down, to propel them through the water.

Reshaping of ontogeny would have forced the tadpole larva backwards in ontogeny, so that it hatched directly from the egg, eliminating the ciliated larval stage altogether. And in the final step, paedomorphosis would have led to a tadpole larva that achieved sexual maturity before fixation: something that, inasmuch as it is pelagic, is the same thing as an amphioxus or a vertebrate. Here is the process in Garstang's own words, in 1926:

> The story of Chordate evolution accordingly begins with an unprogressive sessile or tubiculous stock, obscurely related both to Pelmatozoa, Phoronis and Cephalodiscus, provided with an external ciliary apparatus for the capture of micro-plankton, and having a free-swimming Dipleurula larva. The larva then underwent continuous advances in locomotive and plankton-catching powers until it was able to carry the larval method of endopharyngeal feeding into the adult sessile stage (Tunicata), and finally to dispense with fixation and metamorphosis (Amphioxus). From that condition the next step was the pursuit of individual and larger prey, involving the development of jaws from branchial bars, and of paired limbs, and a transformation of superseded larval organs to endocrine glands (*e.g.* hypophysis, thyroid, parathyroid), and other organs subservient to greater growth, power and locomotive activity (Craniata). It is perhaps the most striking example of what one of us elsewhere has called 'paedomorphosis' – evolution from the larval, not the adult half of the life-history. The old sessile adult organisation has been first rejuvenated (Tunicata), then superseded altogether (Amphioxus–Craniata) by an organization primarily acquired by the larva in response to the conditions of an active pelagic life. (1926, pp.85–86)

There are, of course, problems with this scheme. As Berrill (1955) was to emphasize, the modern tunicate tadpole-larva does not feed, and is concerned solely with seeking a suitable anchorage for the sedentary adult it will become. It does this as quickly as possible, within days or even hours, and it metamorphoses immediately on settlement. In ascidian species that live in a uniform, muddy benthos where one prospective settlement site is much like another, the tadpole stage may be suppressed altogether. As a putative vertebrate ancestor, then, the ascidian tadpole larva makes a poor contender.

But what we are asked to examine is not the present-day ascidian tadpole, but an animal with a general, morphological resemblance to it. After all, and as Garstang no doubt realized, the function of the modern tadpole is very specialized – to locate an attachment site for the adult it will become – and it is designed to fulfil that particular purpose and no other. As with larvae in general, it does not exist wholly (or even partly) as a memento of things past.

According to Garstang, the existence of the larvacea suggests that this sort of transformation can happen. But, as I suggested above, Garstang uses them in much the same way that Darwin used pigeon-breeding in the *Origin of Species*: as a metaphor for a grander process that might be at work in the wider world. For Darwin it was natural rather than artificial selection. For Garstang it was paedomorphosis in the protochordate ancestors of vertebrates, not just the sessile, ascidian-like ancestors of the larvacea[47].

Apart from this, the 1928 paper is incomplete, in that it does not discuss the development of the chordate and vertebrate central nervous system and associated structures, such as the hypophysis (matters that preoccupied those who advocated vertebrate kinship with arthropods). A discussion of these issues was promised but never materialized.

Also omitted was the problem of the origin of segmentation in cephalochordates and chordates, all the more pointed in view of its lack in tunicates and their tadpole larvae. Again, Garstang's views on segmentation would have been valuable given the preoccupation with this subject in the minds of people with views as different as Goodrich and Gaskell.

For all its lacunae, it is easy to see why Garstang's ideas continue to cast a long shadow. The thesis combines a sensitive yet level-headed approach to the needs of living animals, with sound knowledge of their integrated functional anatomies. This aspect of Garstang's work shows up the garage-mechanic ideas of Gaskell and others for what they are, static and sterile, despite loud claims of adherence to darwinian precepts. Garstang's ideas retain their appeal even today, probably for the same reasons, so much so that its merit is generally seen to outweigh its otherwise speculative content. For example, Young writes:

> The theory may seem at first sight fantastic. It is necessarily speculative, but has certain strong marks of inherent probability. It violates no established morphological principles and certainly enables us to see how a ciliated auricularia-like larva could be converted by progressive stages into a fish-like creature with muscular locomo-

> tion, while the adults, at first sedentary, substituted gill slits and endostyle for the original lophophore. (1981, p.71)

It was not always so. In his introduction to Garstang's *Larval Forms* (Garstang, 1951), Hardy feels that the speculative aspects of Garstang's work were seen as a significant demerit, at least during Garstang's lifetime:

> Why did so few of his contemporaries grasp the full significance of what he was driving at? Why, even now, among the younger generation, are there not more who acknowledge the importance of the contribution he had made to the theory of evolution? There are, I think, three principal reasons. For two of them, it must be admitted, he was himself largely responsible. Firstly he was apt to let his passion for theoretical speculation run away with him; along with his more fully worked out ideas he would put forward, as if of equal value, others which many felt to be based on much more slender evidence, and these consequently tended to prevent a more serious consideration of the former. (in Garstang, 1951; pp.1–2)[48]

Many other hypotheses have long since fallen from favour, perhaps because they were built on a small number of possibly coincidental apparent resemblances without full consideration of the ontogenetic, embryological and physiological implications for the animals concerned.

Garstang's ideas, in contrast, were based on argument built on detailed and wide-ranging knowledge of the anatomy, physiology and in particular the embryology of extant species. Perhaps most importantly, Garstang worked to find a coherent pattern in all the evidence available to him, whether from anatomy, physiology or embryology, rather than selecting the aspects he liked and ignoring the rest. An hypothesis that accommodates more evidence than its neighbour usually makes for a more robust and longer-lasting theory.

2.2.4 Variations on Garstang's ideas

The embrace of natural selection as the driving force for evolution may also explain the lasting appeal of Garstang's ideas. Although Haeckel was an enthusiastic Darwinian, Haeckelian recapitulation is, in a sense, more of a morphological abstraction rather than a mechanism of change. Like punctuated equilibrium (Gould and Eldredge, 1993), its importance lies more in a recognition of present pattern than a proposed mechanism for the production of that pattern, or any other. But in the real-life struggle for existence, every feature of an animal, larval or adult, is subject to selective pressure. No animal can afford to carry too many souvenirs of its past unless each affords survival value in the here-and-now. Natural selection was certainly the guiding star for Garstang's followers.

Of less concern was the caveat that natural selection cannot create something out of nothing – it must work with what is available, within the anatomical and developmental constraints of the organisms concerned. This leaves the door on recapitulation slightly ajar, in that there is still room for structures that owe their existence more to the accidents of history than omniscient selection (Gould, 1989).

In the same year that Garstang died (1949), Romer's textbook *The Vertebrate Body* appeared. Romer's view, like Garstang's, was that chordates shared with echinoderms a simple, bilaterally symmetrical and soft-bodied marine ancestor (presumably similar to what Garstang would have called a 'Dipleurula'), and that the hemichordates, tunicates and the amphioxus represented specialized, somewhat degenerate offshoots from the vertebrate line. (However, the amphioxus could represent an instance of what Romer calls 'paedogenesis', and be the 'permanent larva' of some early vertebrate type.)

The ancestral chordate would have had an elaborate filter-feeding pharynx, and would have retained the larval tail when the vertebrate 'line' migrated from the seas towards fresh water – where it would have come in handy for swimming against the current.

This retention of the larval habit led to the evolution of free-swimming vertebrates, at first in fresh water: this theme was an essential feature of Berrill's ideas about vertebrate origins.

(a) Berrill's The Origin of Vertebrates

Perhaps the most influential of Garstang's followers was N. J. Berrill, an authority on tunicates, who summarized his view on the subject in a short and readable book, *The Origin of Vertebrates* (1955). In the book, Berrill bangs the drum for the tunicates, asserting that vertebrates and the amphioxus evolved from ascidian-like stock, through the mechanism of paedomorphosis.

However, his ideas differ from Garstang's in that they are even heavier on the natural selection and lighter on the old-style morphological argument. Although one learns a great deal about tunicate comparative embryology, Berrill resorts to natural selection as an explanation for most differences.

He is also inclined to dismiss competing ideas simply because they do not appeal to him. This propensity towards unsupported statement (together with its unavoidably pre-Hennigian treatment of evolutionary pattern and process) makes *The Origin of Vertebrates* an unsatisfying read today, although one can quite understand its popularity in the 1950s and 1960s.

So confident is he that the truth of his ideas is self-evident, that Berrill begins the book by demolishing everyone else's views as casually as putting out the trash. He starts with Delsman (1922) on annelids, and moves on to Patten (1912) and Gaskell on arthropods, and the supposed similarities between arthropods and ostracoderms.

As I discussed above, these ideas do not work because they depend on the importance attached to certain arbitrarily chosen key characters, the rest of the body form being cut to fit around them. They fail, again, through fundamental misapprehensions of darwinian natural selection. Berrill's reasons for dismissing them echo those of the disputants in the Darwin Jubilee Debate, particularly MacBride. The effort to derive chordates from annelids or arthropods, Berrill notes:

> ... shows the fallacy in attempting to derive any one highly differentiated animal type from another of comparable complexity. Similarities in such cases are almost inevitably the result of parallel evolution. (1955, p.4)

He then goes on to dismiss Gislén's (1930) hypothesized link between chordates and carpoid echinoderms (of which more later) in the same terms:

> I feel that connecting the heavily armoured, bottom-crawling, fully differentiated ostracoderms with the complexly organized, heavily armoured, and more or less coextant echinoderms is on a par with the attempt to equate a vertebrate with an annelid. (1955, p.5)

Although one can sympathize with Berrill's unease at the derivation of new body plans from others fully formed, there is no sense of the incongruity of this statement in a book that goes on to discuss the links between the 'more or less coextant' vertebrates and tunicates: or the fact that tunicate larvae are no less complex, in their own way, for being larvae. He continues:

> To find our sources we need to go much farther back than the Ordovician and even the Cambrian periods, to evolutionary times that have left us no fossil evidence of any kind; our clues can come for the most part only from the study of living organisms, the manner of development, and the nature of their adaptations. (1955, p.5)

At a time, just after Garstang's death, when his ideas were less keenly appreciated than they are today[49], Berrill's aim in *Origin* is to put tunicates centre-stage. He sets out his general argument in a brief three-page chapter (1955, pp.11–13), the rest of the book is devoted to a fuller exposition of each of these in turn.

He starts by making two assumptions. First, that hemichordates have little direct bearing on chordate evolution, their gill clefts and dorsally placed nerve centre (then thought to reflect chordate allegiance) having been independently acquired. Second, that the amphioxus is a degenerate rather than a primitive form, one that has lost its head rather than failed to acquire one. Answers will not come from that quarter:

> On the other hand, ascidians have never been taken at their face value. This I propose to do, not so much as an evangelical protagonist of their ancestral role as with the belief that it is the one approach to the problem that has yet to be made whole-heartedly, and that it is in essentials a simpler and more direct approach than any other. (1955, p.11)

His thesis goes like this.

1. Ascidians are primitively sessile animals from which evolved pelagic forms, on one or more occasions.
2. The tunicate tadpole larva has evolved within the group to meet specific ascidian needs, notably site-selection, and has not been inherited from any other source.
3. Neotenous tapole larvae became sexually mature, allowing them to exploit the resources of the plankton. Doliolids (thaliacians) and larvaceans are modified descendants of the original neotenous form.
4. Soon after neoteny but before the acquisition of specialized thaliacian features, one neotenous stock appeared that took advantage of detritus flowing into the sea from rivers. This stock began to find its way upstream through elaboration of sensory and locomotor equipment, in particular the acquisition of organs of special sense and segmentation:

    > Segmentation was called forth by the need to maintain or improve position in the face of down-flowing freshwater currents. The derivative problem is the origin of segmentation in terms of developmental mechanics. (1955, p.12)

5. The ascent of estuaries and rivers was attempted, first, only by adults, who returned to the sea to breed. Their eggs were numerous and small.
6. The amphioxus is a relic of this phase that lost its fondness for freshwater and returned to the sea: losing, too, its brain and organs of special sense.
7. The first true vertebrates, the armoured ostracoderms, evolved in freshwater. Along with their armour (proposed as a defence against osmotic oedema as well as a store for excreted phosphate) they acquired the relatively large freshwater vertebrate eggs as still laid by lampreys, lungfishes, the coelacanth and many amphibians.

8. With the arrival of this 'vertebrate prototype', Berrill's tale comes to an end. 'At this point', he says (p.13) 'the conception of vertebrate evolution becomes that of current palaeontological thinking and extends beyond the scope of this discussion'.

(b) Berrill in depth

Let us look at Berrill's hypothesis, point by point.

1. *Ascidians are primitively sessile animals from which evolved pelagic forms, on one or more occasions.*

Finding the usual morphological comparisons between echinoderm, hemichordate and chordate development inconclusive (1955, pp.110–118), Berrill deduces primitive sessility from a detailed examination of extant tunicate form, and in particular the form of the tadpole and its relationship to the life habits of the particular species under consideration. I shall discuss this further, under point 2.

However, he suggests (albeit tentatively) that the whole group of what we would now term deuterostomes comprised primitively free-living animals. On settling to assume a sessile filter-feeding life, some – the pterobranchs – elaborated the lophophore, whereas others – the primitive ascidians – developed the pre-existing internal gills.

His hesitancy comes from an unwillingness to accept the usual links between the various deuterostome groups, couched as they are in terms of development. This is because tunicates do not fit into the usual scheme of things, defined by indeterminate cleavage, a blastula with a large blastocoel, a strongly invaginative process of gastrulation, a subdivided coelom arising from the archenteron and so on.

The reason, says Berrill, is that tunicate embryos have, in general, far fewer cells than other chordates at equivalent stages of development. This arises because tunicate tissues tend to differentiate rather sooner than amphioxus or vertebrate tissues, for a given number of embryonic cell divisions. This small change has a dramatic effect, in that it makes development by large-scale movements of cells very difficult, and enforces, effectively, a kind of determinate cell fate on the animal[50].

But just because the initial difference is small – a shift in time by just one or two cell divisions – it should not prejudice the position of ascidians as chordates. By the same token, it should not be used to bar ascidians as vertebrate ancestors: a small delay in the timing of tissue differentiation, and the embryo destined to become a tunicate could become something else entirely. Berrill's complaints are similar to those of Willmer (1990) and Willmer and Holland (1991), who warn of undue phylogenetic importance being attached to such seemingly fundamental characters as style of cleavage, formation of the coelom and so on.

The mechanism whereby pelagic forms evolve from this sessile ancestor is where Berrill meets Garstang. Berrill devotes a whole chapter to Garstang's ideas (1955, pp.37–43). Perhaps surprisingly, the similarities between the two are rather few.

Garstang's initial insight came in 1894, in relation to the similarities between the tornariae of enteropneusts and certain planktonic larvae, at a time when some Haeckelian weight was attached to larval morphology.

But by Berrill's time, it was clear that the forms of planktonic echinoderm larvae reflected particular specializations concerning locomotion, and were relatively lean phylogenetic pickings. It is no surprise, therefore, that similar larval forms – with similar specializations – should be found among enteropneusts (Berrill, 1955, p.41). Echinoderm and hemichordate larvae may give structural pointers, but say nothing about ancestry.

But Berrill is more concerned with Garstang's 1928 paper, written, as it were, after paedomorphosis, in which the tadpole larva was proposed as an interpolation in the life cycle of an ascidian-like creature. Given the rejection of homology between the body parts of specialized echinoderm and hemichordate larvae, Berrill is unhappy about Garstang's contention that the tadpole larva, or the primitive chordate, came into being when ciliated bands in an echinoderm-like larva gathered to form a neural tube. But:

> On the other hand, the manner in which the neural tube of chordates actually is formed during individual development is almost exactly the same as the way in which, according to Garstang, it evolved in the course of evolution[51]. If the course of development in general, and that of the nervous system in particular, is recapitulatory, then the evidence of vertebrate embryology strongly supports the concept of neural-tube origin phylogenetically as well as ontogenetically.
>
> In the final analysis it seems to me that Garstang's thesis rests too heavily upon the validity of the general theory of recapitulation, although it was Garstang himself who first emphasized the possibility that new evolutionary ventures may take larval stages, as distinct from adult, as points for new departures – what has since been termed clandestine evolution. (1955, p.42)

Berrill doubts the plausibility of Garstang's scheme, but this leaves him in a hole. If tadpole larvae did not originate in Garstang's recapitulatory way, how *did* they originate? The answer, says Berrill, is 'suddenly' – a developmental innovation (what we would call a 'hopeful monster') that had immediate adaptive value. Aware of the improbability of this, Berrill spends the next 33 pages justifying it with an exhaustive comparative account of tunicate development.

2. The tunicate tadpole larva has evolved within the group to meet specific ascidian needs, notably site-selection, and has not been inherited from any other source.

Berrill's hypothesis depends on the tunicate tadpole larva as a unique specialization of ascidians that was not inherited from some earlier kind of animal. If not unique, notes Berrill (1955, p.13) then ascidians will be no more important in chordate phylogeny than any other group with this attribute. Therefore, one of the aims of the book is to substantiate the claim that the tadpole larva is an ascidian innovation.

But as Berrill does not hold with Garstang's auricularia theory (the 1894 version as incorporated into the 1928 paper), he must find some other way to derive the tadpole larva. His method is ingenious, and involves arguments both from natural selection and development.

Garstang supposed that the prolongation of larval life in the plankton was the selective force behind the development of the tadpole larva. Berrill demonstrates the exact opposite – tadpole larvae have a lifetime measurable in days, if not hours: many ciliated larvae last far longer. If longevity were the key, tunicate larvae would have stayed ciliated. Instead, the function of the tadpole larva is to find a suitable site for metamorphosis and attachment as quickly as possible.

Given that many ascidians prefer to anchor on the dark undersurfaces of rocks, tadpoles need to be able to detect and respond to light and gravity, and to swim towards the chosen target. Hence the two main attributes of the tadpole as distinct from a ciliated larva: the sensory capsule containing a simple eye (ocellus) and gravity detecting organ (otolith), and the notochord-supported muscular tail.

Hatched ascidian tadpoles at first swim towards the light and against gravity, upwards, towards the surface of the sea. As they do not feed, this response is, presumably, to facilitate dispersal. Suddenly, their mood changes as they seek out the depths and the dark. The tail provides the necessary torque for the rapid, sideways movements necessary to find vertical or overhanging surfaces of rock on which to anchor.

The function of the tadpole purely as a short-range dispersal phase with the ability for site selection is proven in the breach. Many ascidians of the family Styelidae live on fairly featureless sandy or muddy bottoms, where one site is as good or poor as another: their tadpoles lack ocelli and are blind. Molgulid ascidians live in similar situations and the larvae may lack tails as well as eyes.

The natural history of the ascidians shows that certain features of the tadpole larvae are selective necessities, dropped as soon as they are no longer required. But what of the developmental basis for these organs? And how could they have arisen so suddenly? Berrill's exploration of these topics is the subject of *Origin*'s longest and perhaps most valuable chapter, Chapter 7 (pp.44–77).

The key, says Berrill (pp.44–45), is that the ascidian egg is a 'dual' system, which presumably contains the germ of both larva and adult – each developing from largely different structures. But many adult tunicates have the capacity to produce new adults by budding, by-passing the egg and tadpole stages. Subtracting the second process from the first, as it were, should shed light on the origins of the distinctive tadpole stage within the egg.

The surprise is that tunicates derived by budding are as complete as those that develop through the tadpole stage. This begs the question of the necessity for having a tadpole stage at all. It is maintained by selective pressure as a dispersal mechanism, but the question is how it originates in the first place. Berrill thus goes back to the egg.

As noted above, ascidian eggs are remarkable in that tissue differentiation happens extremely early in embryogenesis. Indeed, presumptive tissues can be followed back to regions in the undivided egg. At the 64-cell stage (i.e. after six cleavages):

> ... the so-called animal or presumptive ventral half of the cleaving egg consists of 26 ectoderm and 6 neural plate cells; the vegetal or presumptive dorsal half consists of 4 neural plate, 4 chordal, 10 mesenchyme, 4 muscle, and 10 endoderm cells. (1955, p.52)

This is as far as the embryo gets before gastrulation, which happens between the sixth and seventh cleavages. Gastrulation brings the widely separated blocks of chordal cells into a single dorsal mass, bounded by muscle cells. These chordal cells divide no more than two or three times more, to give a notochord of 42 cells or so, irrespective of the original size or yolk content of the egg (see above).

In general, the number of accompanying muscle cells is similarly restricted. These are the essential tissues of locomotion, absent in the adult but necessary for the larva, and their origins are traceable right back to the undivided egg. If the tunicate tadpole started out as an interpolation in development, this interpolation can be traced to the very earliest phases of development.

Nevertheless, these chordal and muscle cells are yet in a single mass, and have not extended to form the distinctive tadpole tail – the essentially chordate feature of the tadpole, and with it tunicates as a whole.

Extension happens after gastrulation and the last division of the chordal cells, and results simply from the rearrangement of the mass of chordal cells into a single file 42 cells long. During this process the cells become vacuolated and swell to several times their original size. In some species, the notochord becomes syncytial, the vacuole extending from one end to the other. As the notochord swells and extends, the muscle cells get dragged along, each muscle band extending and differentiating as it goes.

Once again, the point is proved in the breach, with the tailless larvae of some molgulids. Just like in other ascidians, the chordal cells form and aggregate into a dorsal mass, the neural plate tucks itself in to form a tube, the groom is nervous, the best man has remembered the rings, the congregation is ready and waiting. But at the last minute the bride decides that her heart belongs to Daddy:

> ... a stage is reached which is virtually identical with that which in other ascidians immediately precedes the outgrowth of the tail. But the chordal cells do not thereafter increase in volume, either individually or in aggregate, so that there is a negligible tendency to extend posteriorly, to obliterate the neurenteric canal or allow neural folds to extend and fuse above the residual blastopore. It is at this stage, in fact, that tadpole development fails. The chordal cells are present, but they do not mature histologically; no posteriorly extending notochord forms, and no tail grows out ... Accordingly, the absence of a tadpole stage in the development of these ascidians appears to be the result of the failure of the chordal cells to undergo their typical vacuolation and consequent swelling. An adequate number of chordal cells are present at the right time in the right location, but they remain inert and no tail grows out. The cells that might have formed the two bands of tail-muscle tissue are also present, but they, too, remain inert at the critical period ... (1955, pp.66–67)

Thus, the entire formation of the tail depends on a simple cellular process whereby the presumptive chordal cells rearrange themselves and are vacuolated. No vacuolation, no tail. It is that simple. So simple, that Berrill argues that a single mutation would be all that would be needed to produce a tail from a primitively tailless ancestor (presuming of course that the ancestor had the capacity, at the egg stage, to determine presumptive chordal and tail-muscle tissue).

Biochemical and genetic work appears to bear out Berrill's predictions. Reviewing work on tunicate development that indicates how taillessness in some molgulid larvae seems to be the result of a relatively small number of genetic mutations, Ruddle *et al.* (1994) note that the tailless *Molgula occulta* apparently lacks a homeobox gene, the homologue of which is expressed in the hinder parts of vertebrates. The related *Molgula oculatu*, though, has both this homeobox gene and a tail. The two species are related closely enough to interbreed and produce hybrids with short-tailed larvae.

However, this simple story, which is based on somewhat negative evidence, conceals a system of developmental complexity that bears comparison with that of vertebrates. Jeffery and Swalla (1992) report the presence of a protein in the tailed *M. oculata* but not in the tailless *M. occulta*. They speculate that this protein is connected with the cytoskeleton, the network of tubules that braces all cells, and could be involved in the kind of cell–cell interaction which, fundamentally, governs all embryogeny.

Swalla *et al.* (1993) describe the isolation of two proteins, *Uro-2* and *Uro-11*, found in both molgulid species, but the genes encoding them are expressed more strongly in the tailed *M. oculata*. Messenger-RNA transcripts of both genes are found in the oocyte – they are first expressed in maternal tissues – and mostly finish up in presumptive ectoderm. However, *Uro-11* continues to be transcribed in the zygote, but only in *M. oculata*: Swalla *et al.* (1993) predict on the basis of the transcript sequence that it is a DNA-binding protein.

In addition, Yasuo and Satoh (1993, 1994) report the expression of the gene *brachyury* in tunicate cells destined to become notochord. This gene, in vertebrates and the amphioxus, is involved in the specification of the notochord, among other things. The relationship between tunicate *brachyury* and other factors governing notochord tail development such as *Uro-11* is not clear. But whatever it turns out to be, these studies indicate that the 'decision' underlying the formation of notochord depends on a number of other events having already happened, an inductive 'cascade', in which the nerve tube forms in response to a 'signal' (mediated by a diffusible molecule and/or cell-to-cell interaction) generated by the notochord, which in turn forms in response to a signal generated by presumptive endoderm (Jeffery, 1994). This kind of development, by inductive interactions, is typical of animals with indeterminate cleavage, such as vertebrates, but in tunicates it is masked by the constraints of limited cell number, which enforces more of a deterministic style of development.

The other essential chordate feature is the dorsally situated neural tube. This forms from the closure of the neural plate before the extension of the tail, with an anterior neuropore and, for a time, a posterior neurenteric canal. The system is tripartite: a region at the front develops into the vesicle that houses the larval ocellus and otolith; the neural tube in the tail; and an undifferentiated region in the posterior part of the vesicle from which the neural ganglion of the adult develops.

In Chapter 7, however, Berrill is really concerned with the second part – the neural tube. He regards the infolding of the neural plate to form a neural tube to be a simple, mechanistic process: simply, the best and most logical way of forming an internal tube from an external sheet. As with the deterministic character of tunicate development in general, it need not have any evolutionary significance either in general or in relation to the tadpole larva: neural structures develop in a similarly simple way in adults budded from the stolon of thaliaceans:

> The point I wish to make here is that neural-tube formation occurs in fundamentally the same manner in the egg and in a stolonic structure that has no possible connection with any hypothetical ancestral larval form. (1955, p.63)

Thus, although the notochord and its musculature are features peculiar to tadpoles that can be traced back to the egg, the capacity to form neural structures by invagination is not. This is a feature imposed on the tadpole because that is the way ascidians in general make nerve cords, whether they go through a tadpole stage or not.

But for all the structural similarities between tadpole and vertebrate nerve cords, the tadpole nerve cord is not terribly nervous. Its constituent cells are small; there are no neurons, no ganglia, and no neural connections between the nerve cord and the muscle bands (which seem to be innervated at their anterior ends only, from a ganglion-like mass in the trunk of the tadpole). In fact, the nerve cord serves no function at all.

This is another poser for Berrill, for either the nerve cord is vestigial (which would endanger the entire contention that the tadpole larva is an ascidian invention, and, with it, the place of ascidians as chordate ancestors), or it has no selective value. Berrill plumps for the latter but only as the lesser of two evils[52]. Nevertheless, the conclusion is satisfying in that one can assume that the posterior extension of the nerve cord is as much a consequence of notochordal extension and differentiation of the muscle cells, as of serving some neural function.

Chapter 7 concludes thus:

> I have not proven that the tadpole larva evolved from a type of ascidian egg which originally had no such larval stage, or that the tadpole larva is actually an ascidian creation. Yet I believe that the fact that it has been possible to show on the one hand that the tadpole larva is of vital importance to ascidians in their present way of life, and on the other that its origin can be conceived from the point of view of developmental tissue mechanics as a sudden effective acquisition goes far to establish the possibility and perhaps the probability that such has been the actual course of events. It is admittedly no more than speculative thinking, but it is speculation tied closely to the known facts of tadpole ecology and tadpole and ascidian development, and to a great extent is self-contained and self-supporting. (1955, pp.76–77)

3. Neotenous tadpole larvae became sexually mature, allowing them to exploit the resources of the plankton. Doliolids (thaliacians) and larvaceans are modified descendants of the original neotenous form.

For all his doubts about Garstang's auricularia theory, Berrill is quite happy with the origin of the chordate stock from paedomorphic tadpole larvae, and demonstrates the possibility with a 20-page discussion of the secondarily pelagic thaliacean and larvacean tunicates. The theme is the same as in Garstang's scheme – larvacean tunicates are derived, not primitive, and their ancestry can be traced through doliolids back to an ascidian root[53].

Nevertheless, I have the impression that whereas Garstang used the evolution of pelagic tunicates as an analogy for vertebrate evolution – as a kind of morphological type – Berrill conceived of the evolution of thaliaceans, larvaceans and vertebrates all together, as a real event.

Points 1–3 cover the core of Berrill's hypothesis. In these, he replaces Garstang's somewhat abstruse auricularia theory with a scheme which – although simple storytelling by today's standards – is based on observable embryological and ecological results. He has no disagreement with Garstang on the central point of paedomorphosis, and his arguments are more or less the same, based on analogy with thaliacean and larvacean evolution. The *Origin* breaks down over points 4–7, as I shall show below.

4. The paedomorphic prechordate stock took advantage of detritus flowing into the sea from rivers. This stock began to find its way upstream through the acquisition of organs of special sense and segmentation.

Couched as it is in terms of natural selection, Berrill needs some excuse for vertebrates to have acquired their special features, in particular the segmentation of the dorsal region and organs of special sense. Otherwise they would presumably have turned into something like the larvaceans. This explains the discourse on freshwater migration, and a rationale for segmentation that depends on its acquisition as a response to selection for improved locomotion against a current. How it actually arises is another matter, and Berrill (as did Bateson in 1886) identifies the analysis of development and growth as the probable source of an answer (1955, p.171).

He starts by identifying metamerism as essentially the same phenomenon whether applied to arthropods or vertebrates. After that he says little more than that segmentation is a natural consequence of a cyclic 'pulsing character' somehow inherent in mesodermal growth, and which can be manifested quite suddenly even if invisible beforehand, by attaining an amplitude over some threshold value (pp.172–173, and again on pp.183–184). Such is the modest content of a 17-page chapter headed 'Amphioxus And The Segmentation Problem'.

Viewed with hindsight and knowledge of neural crest dynamics and homeobox genes, Berrill was thinking along the right lines. But without this information, the acquisition of segmentation in Berrill's world could only but look like Lamarckian *besoin* among creatures with a deeply felt need to swim upstream to seek their fortunes.

5. The ascent of estuaries and rivers was attempted, first, only by adults, who returned to the sea to breed. Their eggs were numerous and small.

This proposal is supported by just one piece of evidence, that tunicate eggs are small, and vertebrate eggs, especially those of primitive extant vertebrates living in freshwater, are big.

6. The amphioxus is a relic of this phase that lost its fondness for freshwater and returned to the sea: losing, too, its brain and organs of special sense.

The suggestion that the amphioxus is a relic organism that was a 'freshwater reject' depends on one piece of evidence – that the nephridial excretory system of the amphioxus is the leftover relic of an estuarine phase in its history: it never made the grade of the prevertebrates that made it upstream. But if Berrill is so fond of adaptation, is he to be allowed this piece of evolutionary baggage, just when it suits him?

7. The first true vertebrates, the armoured ostracoderms, evolved in freshwater. Along with their armour (proposed as a defence against osmotic oedema as well as a store for excreted phosphate) they acquired the relatively large freshwater vertebrate eggs as still laid by lampreys, lungfishes, the coelacanth and many amphibians.

The first sentence is the key here – Berrill's process of tadpole-to-vertebrate transmogrification depends on the idea that the first true vertebrates evolved in freshwater. Should it be found that the first vertebrates appeared in the sea, Berrill's idea will be seriously weakened. Of course, unequivocal proof is probably impossible to come by in principle, either for or against, but the general consensus today is that vertebrates first appeared in the sea.

The origin of vertebrates is tied up with the origin of bone, a unique vertebrate specialization in turn connected with the evolution of the neural crest. The reason *why* bone evolved is another question entirely – Romer (1933) suggests that it might have evolved in primitive freshwater fishes as a defence against eurypterids, the large scorpion-like predatory arthropods that Patten had linked more closely with vertebrate ancestry.

Romer followed the then conventional line that vertebrates had originated in fresh water, a conclusion reaffirmed in subsequent reviews of Ordovician and Silurian deposits (Romer and Grove, 1935; Romer 1955), a view echoed as late as 1972 (Romer, 1972), although he had by then conceded that the palaeontological evidence was equivocal.

Homer Smith (1939, cited by Gregory, 1946) suggested that bone evolved as a response to the osmotic stress felt by an animal moving upstream from the sea. Bone would be a useful physical barrier against the flux of fresh water into an animal which, presumably, had body fluids isotonic with seawater. However, as Gregory notes, the earliest vertebrates (for all their bony shielding) had internal anatomies similar to those of ammocoete larvae, begging the question of whether the vertebrate state might have been achieved in the sea, prior to migration upstream.

The hagfish, though primitive, is far closer to the vertebrate estate than a tunicate or amphioxus[54], and yet, like tunicates – and unlike lampreys – its body fluids are isotonic with sea water (Robertson, 1954, 1959). Actually, the blood plasma of the hagfish *Myxine glutinosa* is hypertonic to sea water, and the urine hypotonic to the blood, and the glomerular kidneys are probably used to regulate ionic balance (Morris, 1965). Of course, the presence of kidneys in the hagfish – for any purpose – is argument against the contention that kidneys evolved as an adaptation to the osmotic challenge of fresh water.

Turning to the fossil evidence, Denison (1956), in direct response to Romer and Grove's (1935) study, suggested that the earliest vertebrates appeared in nearshore, marine habitats. Robertson (1957) recorded the Ordovician deposits of the Harding Sandstone of Colorado, then as now the oldest well-established vertebrate locality, as 'undoubtedly' marine, containing as it did trilobites, brachiopods and cephalopods, as well as conodonts (but see Graffin, 1992). The freshwater habitat was colonized at least twice during the Silurian by microphagous, detritus-feeding ostracoderms – already fully formed vertebrates – followed soon after by predatory jawed vertebrates, the placoderms and acanthodian fishes (Halstead, 1985).

Matters come full circle with the work of Moya Smith and her colleagues (Sansom *et al.*, 1996) who have been working on vertebrate and conodont material from the Harding Sandstone with particular emphasis on its histology. In the light of the histological affinities between conodont and vertebrate teeth (Sansom *et al.*, 1992, 1994), they suggest that distinctive vertebrate hard tissues such as dentine evolved in tooth-like structures (denticles, conodont elements) before the appearance of bone in the sense that we would recognize it today.

Although strictly defined as belonging to the pharyngeal region of vertebrates, teeth can be shown from genetic and embryological evidence connected with neural crest to be homologous with structures growing elsewhere, for example the armour of dermal denticles carried by some ostracoderms (Forey and Janvier, 1993), notably thelodonts.

In conclusion, then, there is compelling evidence to suggest that vertebrates evolved into substantially their modern state while still in the marine environment. If true, this invalidates Berrill's contention, namely that vertebrates acquired many of their distinctive features by virtue of a migration from salt water to fresh water.

(c) Berrill's contribution

Berrill took Garstang's ideas further by replacing the auricularia theory with some plausible Modern Synthesis argument. The idea of paedomorphosis remains the same, although Berrill, like Garstang, does not say very much about how paedomorphosis changes a non-feeding, short-term dispersal phase into a fully functional larvacean, thaliacean or chordate. It is rather like proposing that a whole new group of animals evolved from the paedomorphic transformation of the highly specialized cercaria larvae of liver flukes. The migration-to-fresh water idea is similarly shaky.

Not that *Origin* received universal plaudits even when it was published. The essence of G. S. Carter's criticism (Carter, 1957) is that the entire scheme (the saltational appearance of the tadpole larva, the paedomorphosis, the necessity for a sessile ancestor) is over-elaborate, given that one could turn the entire edifice upside-down and come up with a simpler solution – that the vertebrate ancestor could have been motile, and that the tadpole larva (or something like it) could have been a plausible ancestor. Where, demands Carter, are the traces of sessile ancestry that must exist in Berrill's tadpoles? Where are the traces of stalked habit, tiny nervous system, forwardly directed anus?

These are there, to be sure, in the adult tunicate – but the tadpole has a relatively elaborate nervous system, a rudimentary (if not absent) gut and no trace of a stalk. Berrill's explanation that the tadpole evolved as an adaptive response is over-welcoming both of natural selection and paedomorphosis. Never once does he seriously consider the simpler alternative, that these vestiges of sessile creation are absent because they were never there to start with, and that the tadpole is the relic of an ancestral adult. Perhaps Garstang's refutation of recapitulation still held Berrill in thrall.

Carter prefers to trade paedomorphosis for the retention of primitive traits: so the tadpole is an echo of past adult mobility, with sessility a novel habit in certain ascidians. And nowhere does Berrill discuss the important question of the echinoderm relationship to chordates, except in the vaguest terms. The amphioxus, of course, is drawn into his fairy tale as the bad penny that returns to the sea to wallow in what might have been.

Whitear (1957) is similarly unconvinced, though less scathing. Berrill's scheme depends on the transition to fresh water – Whitear fails to see how it is even necessary. The increase in size that would have favoured segmentation is as advantageous to a marine as a freshwater animal, so there is no need to postulate a freshwater phase in the ancestry of amphioxus to explain its segmentation: the existence of the hagfish is testament enough to that. Indeed, given Berrill's weak account of how tunicate-like vertebrate ancestors would have coped with the osmotic challenge of fresh water, one might propose instead that vertebrates had evolved a kidney in the sea for some other purpose (excretion), exploiting it later on for osmoregulation. The existence of kidneys in hagfish suggests that this is indeed the case.

Again, Whitear suggests that the tadpole larva need not have been a unique ascidian invention, for this begs the question of how the vertebrates could have originated from what seems to be a specialized dispersal phase, leaving Berrill no option but to construct the freshwater-migration scenario. Whitear's ideas are discussed further below.

Bone (1960) is critical of any theories that derive chordate adults from tunicate larvae (both Berrill and Whitear draw his fire) because they fail to account for the functional difficulties imposed by the presumption of neoteny. He does not go quite as far as Carter (1957), who felt that shared common ancestry – not neoteny – is enough to explain morphological similarities between the larval and adult stages of related organisms: rather that the invocation of neoteny had been too liberal and inclined to be misplaced.

For my part, I think Berrill's fault lies ultimately in an over-literal adherence to natural selection, in which the retention of non-adaptive characters was *a priori* forbidden. It is important to remember that although Garstang valued adaptation by natural selection very highly, his refutation of recapitulation was a denial of the strict Haeckelian vision of that idea only. As the 1922 paper shows, he is willing to entertain recapitulation in the general sense of von Baer, that close relationship may be judged in ontogeny as well as adult form.

(d) Vertebrate origins after Garstang and Berrill

Berrill, of course, was not the only researcher to use Garstang's work as a starting point for a scheme of vertebrate origins. As discussed above, Whitear (1957) took her cue from Berrill. She felt that Berrill had painted himself into a corner with the assertion that the tadpole larva is a unique feature of ascidians. Instead, one might imagine that the free-swimming tadpole-like phase was once much more common among chordates than it is now, and that vertebrates could owe their origins to one strain in which the neotenous tadpole was disposed to large size and muscularity. Whitear sees the origin of the amphioxus as similarly neotenous, but from a separate lineage of pre-ascidian tunicate-like creatures, phylogenetically closer to modern tunicates than to vertebrates. The amphioxus having diverged, doliolids, appendicularians and thaliaceans evolved from ascidians in much the way that Garstang supposed.

To follow this line of thought further, it is evident that the invention of the tadpole larva *precedes* the evolution of ascidians, and was presumably the issue of some remote chordate ancestor that may have resembled a pterobranch: although at that point chordate ancestry gets lost amid the Cambrian swirl of echinoderms and graptolites.

Bone (1960) is less welcoming of neoteny as a mechanism whereby chordates may have originated from a tunicate tadpole larva. In his scenario, the paedomorphosis of tadpole larvae is a gradual process, in which planktonic larvae delay metamorphosis into a sessile adult until after the maturation of their own gonads. In which case, there must be some selective advantage in the prolongation of planktonic life, presumably that of access to abundant food.

Such considerations make the neotenous origin of chordates from ascidians unlikely, because (as Berrill himself demonstrates) ascidian larvae are highly specialized as short-range, short-lived dispersal organisms that do not feed. Instead, Bone envisages the origin of vertebrates through the gradual neoteny of a kind of larva that would have spent an extended period in the plankton. Ascidians are ruled out, but many other deuterostome larvae adopt this habit, including those of echinoderms, hemichordates and the amphioxus. Many of these creatures are not sessile as adults, so the final resting place is less critical than is the case for ascidians.

Neoteny from planktonic larvae can be observed today, claims Bone, in the 'amphioxides' larvae of certain cephalochordates (Bone, 1957), in which the gonads may develop while the animal is still resident in the plankton.

Ascidians are ruled out again in that it is hard to see how their larvae could be, or have been, anything other than short-range site-selection devices, and how they might have been transmogrified into long-range, long-lived plankton feeders. Bone agrees with Whitear (1957) that it is probably simpler to suggest that the tadpole larva was invented by some ancestral chordate prior to the appearance of ascidians, or even tunicates. The chordate features of tadpole larvae reflect their generalized chordate heritage, rather than a connection with vertebrates in particular. This view was endorsed in a useful review by Denison (1971).

On the other hand, Bone disagrees with Carter's (1957) view that vertebrates could have evolved, without neoteny, from an ancestor that was free-swimming to start with. Although many kinds of deuterostome are motile in principle, only the vertebrates can be counted as genuinely free-swimming. Therefore, many deuterostomes require a planktonic dispersal phase of some kind, even if not as acutely as do the ascidians. This leaves open the possibility of the evolution of vertebrates by neoteny from the planktonic larva of some sedentary (if not sessile) ancestor. The question, though, is which? Whitear (1957) suggests hemichordates, but only in rather vague terms. Bone (1960) is more specific, though not much.

By 'hemichordates', Bone is really referring to enteropneusts, which lack a tadpole larva. As discussed above, those that do not develop directly have a ciliated planktonic larva called the tornaria, which bears a striking resemblance to certain larvae of echinoderms.

As the way in which some ancestral hemichordate produced a tadpole larva, Bone adopts Garstang's (1894) scheme for the evolution for the chordate central nervous system by the modification of ciliated bands. The appearance of the notochord and myotomes presents Bone with particular difficulties, and he is forced to concede that the appearance of such things may have been sudden – despite earlier strong words against 'hopeful monsters'.

Urochordates, chordates and cephalochordates would, each in their own way, have evolved from the hypothetical 'protohemichordate' ancestor, which need not have looked like the modern variety:

> The specializations of the adult form of these organisms, led to the Urochordates on the one hand (by adoption of a completely sessile habit, and the consequent reduction of the larval phase); on the other, to the chordates by a process of gradual neoteny, and the omission of the adult type; lastly, to the modern Hemichorda [*sic*] by the paedomorphosis of the original adult form (this perhaps accounting for the re-appearance of the ancestral larval form, and the absence of any trace of the tadpole larva). (p.262)

Cephalochordates are regarded as somewhat aberrant stock, having (like vertebrates) evolved a free-swimming habit from planktonic larvae; secondarily assuming a semi-sessile habit as adults, only then reacquiring a planktonic dispersal phase which – in some forms such as the amphioxides larva – seems to be recapitulating Bone's scheme of chordate evolution.

What might these protohemichordates have been like? Of course, it is impossible to say, as Bone is well aware. However, he offers the Pogonophora as possible protohemichordate relatives. As discussed in Chapter 1, though, the possible affinities of pogonophores with deuterostomes have since been definitively discounted. All that need follow from Bone's idea is that protohemichordates were animals of a sufficiently sedentary nature to require a planktonic dispersal phase.

In the discussion following the presentation of Bone's paper, as in his critique of Berrill's book, G. S. Carter professed himself adamant against the application of neoteny as an evolutionary mechanism, especially in cases in which it was unnecessary. For chordates, neoteny becomes irrelevant were one to suppose that free-swimming chordates evolved from free-swimming ancestors that developed directly, without the complications of metamorphosis or a larval stage. In Carter's scheme, the vertebrate ancestor can be pictured as a larger, more elaborate version of the tadpole larva.

Whitear objected to Bone's scheme on the grounds that he could neither account for the presence of a notochord in the tadpole larva – or the lack of one in a hemichordate. But her disagreement with Carter about neoteny seems almost flippant: 'neoteny is such a useful concept, as a means of escaping from specialization, that it would be a waste not to use it' (in Bone, 1960, p.268). That evolution should have been as tidy-minded as the well-ordered Whitear!

In their inability to produce some plausible ancestor for the vertebrates, Bone (1960) and Whitear (1957) fail at the last hurdle. Such was noted by Tarlo (the late L. B. Halstead) in the discussion following Bone's 1960 paper – that there was no real difference between Whitear's proto-Urochordate and Bone's proto-Hemichordate, in that they both converged on something like a pterobranch, lost in the mists of prehistory.

Jollie (1973) summarizes the litany of problems afflicting the usual ideas about vertebrate origins, preferring a migration to fresh water (*sensu* Berrill) accompanied by the adoption of macrophagous predation, which would have spurred the development of the locomotory and sensory adaptations characteristic of vertebrates.

Harvey (1961) summarizes the work of Garstang, Berrill, Carter, Whitear and Bone in a brief and useful review that concludes with a hopeful note about the potential of molecular information for resolving questions of phylogeny:

> ... it is only a matter of time before biochemical evidence of this nature may enable us to determine accurately the interrelationships of major groups of the animal kingdom. (p.513)

Molecular evidence is close to resolving long-standing questions about segmentation and the origin of the head, neither of which were adequately addressed by Garstang or Berrill. They bolster Berrill's idea for the simple genetic basis for the origin of the tunicate tadpole tail. But as we have seen, molecular evidence has also been used to support some very ancient ideas: most notably, the revival of St Hilaire's idea that chordates are really inverted arthropods, by Arendt and Nübler-Jung (1994) and Nübler-Jung and Arendt (1994).

In his challenge to this idea, Lacalli (1995) demonstrates an abiding interest in Garstang's auricularia theory in its original and unadulterated form, free from concerns about paedomorphosis[55]. Garstang's auricularia is seen as a bean-shaped object, the surface of which is divided by the aboral band into an oral and an aboral field. According to Garstang, the dorsal tubular nerve cord was formed by the upward migration of the aboral band on each side until it met the dorsal edge, whereupon it became internalized as a tubular structure (Figure 2.2). This, Lacalli points out, resulted in the extension of the oral field (a unique deuterostome structure) at the expense of the aboral field, which on neurulation became the neural tube, internalized and inverted with respected to its original orientation – literally turned inside out.

This turn of events might explain why genes expressed over the ectoderm in insects such as *Drosophila* find their homologues in chordates (at least primitively) restricted, internalized, to the neural tube or its derivatives. It might also explain the apparent inversion of expression because, as the neural tube rolls up, the dorsal midline becomes the tube's ventral floor plate, while its margins – corresponding to ventral ectoderm in insects – now meet at the dorsal midline.

In the meantime, some ideas just refuse to lie down and die. Fifty years after the comprehensive demolition of Gaskell's theory, arthropod ancestry formed the basis of a truly extraordinary jackdaw of a paper by Raw (1960) in which ideas from Berrill (1955), Hubrecht, Patten and others are welded onto a Gaskellian chassis.

Raw holds that vertebrates originated from a 'prot-arthropod' [*sic*] that had but lately deviated from annelid stock and was yet to evolve compound eyes (a convenient act of goalpost-moving that subverts one of the most egregious difficulties of Gaskell's scheme).

Driven by an overtly Lamarckian[56] desire to exploit the rich detritus streaming down to the sea from inland, these creatures moved to freshwater where they evolved into vertebrates (cf. Berrill). The dorsal tubular nervous system was formed by the coalescence of perhaps three pairs of cords (echoes of Hubrecht's nemerteans); the gut was formed by the ventral overgrowth of the arthropod food grooves by the limbs, which fused to become gill arches (a nod to Patten); the notochord formed because it had to.

Raw's ideas are as strange as those of Gaskell, and suffer from precisely the same theoretical difficulties, so do not merit further discussion here. Intrepid readers can seek out the paper for themselves, and make their own judgement.

Should they do so, they will be rewarded by a paper by Sillman (1960) which appears in the same issue of the *Journal of Paleontology* immediately following that of Raw. Sillman goes right back to the days of Lamarck (1809) with an assertion that vertebrates are more closely allied with cephalopod molluscs than with echinoderms.

Sillman questions Garstang's ideas on the echinoderm alliance on the grounds that it is the adult animal, not the larva, that should give the clue to ancestry, and that adult echinoderms are fundamentally unlike vertebrates. The obvious question – that cephalopods are equally alien – is treated with almost cabbalistic tentacle-waving:

> In this connection the ten upper extremity digits of the tetrapod ... seem to resemble in more that [*sic*] a fortuitous manner the design of the squid with its ten tentacles doing the same thing in a different way. (1960, pp.543–544)

Paedomorphosis is questioned as a force in evolution, and as with all ideas based on protostome ancestry, Sillman is forced to assert the degeneracy of tunicates and cephalochordates.

Sillman's claims for a close embryological similarity between vertebrates and cephalopods are dubious: they include the assertion that cleavage in echinoderms is radial and the yolk homolecithal, but cleavage is incomplete (and the yolk telolecithal) in both cephalopods and vertebrates. One can see how these questions arise: they stem from a questioning of an over-dogmatic insistence of the fundamental distinctions between protostomes and deuterostomes. Nevertheless, none of Sillman's proposed character similarities stand much detailed scrutiny.

Other contributions are similarly erroneous: Dillon (1965) suggests that the hydrocoel of enteropneusts and echinoderms survives in chordates in another guise, that of the dorsal tubular nerve cord, by homology of the hydropore with the neuropore and subsequent rearward extension as far back as the neuroenteric canal. However, he seems to be confusing the hydrocoel with the axocoel or protocoel (as appropriate). This may explain his failure to appreciate that the hydrocoel, in contrast with the protocoel, is a paired structure, with the left half assuming the greater importance in echinoderms.

Space does not permit an exhaustive discussion of every single idea, sensible or otherwise, about vertebrate origins. One of the more interesting recent reviews and probably also, sadly, one of the most difficult to obtain, is that of Ivanova-Kazas (1990). This deals succinctly with ideas proposing derivation from nemerteans, annelids, arthropods, carpoid echinoderms (see below), hemichordates, dipleuruloid larvae and neotenous chordates, and concludes that all chordates could probably be derived from a hemichordate similar to modern enteropneusts, though in all likelihood more generalized and primitive, and with a dipleurula-like larva.

Three other ideas are worthy of mention before moving on, if only because of their methodological interest. Gutmann (1981) stresses the primacy of functional morphology as a key to phylogenetic reconstruction. His conception of vertebrate origins stems from the idea that the metameric coelom that forms the hydrostatic skeleton of annelids is more primitive than the trimeric coelom of deuterostomes. The vertebrate ancestor, therefore, if not an annelid by name, is

> ... a metameric, coelomate worm-like animal with a complex set of circular, transverse and longitudinal body muscles. The coelom plus the complex body musculature formed the hydrostatic skeleton (1981, p.63)

All else follows from this: the closest chordate to this model, and therefore the most primitive, is the amphioxus[57]. Tunicates and enteropneusts are seen as highly derived and specialized offshoots. Pterobranchs are derivatives of enteropneusts, echinoderms of pterobranchs.

Løvtrup (1977) links vertebrates not with echinoderms but with annelids and molluscs. His views are interesting chiefly in that they are cladistic to an austere degree. He goes further than the 'transformed' school of cladistics that eschews all evidence but the purely neontological (see in particular Rosen *et al.*, 1981), discounting also characters which he sees as 'non-morphological'. His thesis is thus built largely from biochemical and physiological features, and is clothed in what Jefferies (1986) calls 'quasi-Euclidean' garb, complete with axioms and theorems derived therefrom.

Most of the problems with Løvtrup's scheme lie with a number of errors of fact and interpretation. But the central error concerns the method itself. One of its central axioms boils down to the simple assertion that 'non-morphological' characters carry more phylogenetic weight than morphological ones. As Jefferies (1986) points out, the fact that such an axiom is not self-evident renders worthless any theorem on which it is based.

As if in reaction to phylogenetic reconstruction by the cladistics of which Løvtrup is perhaps the most extreme exponent, Gans (1989) seeks to address the issue through the formulation of nested series not of clades, but of 'scenarios'. These consist of tests of proposed phylogeny in the light of their supposed environmental and physiological consequences. Despite Gans' assertions to the contrary, scenarios seem to be very like the kind of just-so stories so deprecated by cladists (Forey, 1982). The phylogenetic conclusions vary little from those of Garstang, Berrill and their successors, who also used a scenario-based form of reasoning even if they did not distinguish it with a special name.

At this point, mention should be made of two other papers co-authored by Gans and Northcutt (Gans and Northcutt, 1983; Northcutt and Gans, 1983). Both stress the importance of the neural crest in the origin of vertebrates, to the extent that the vertebrate 'head' is a neural-crest-generated neomorph. In Northcutt and Gans (1983), they note that all the organ systems of vertebrates possess derived features that arise, embryonically, from neural crest and epidermal placodes; developmental modifications from epidermal nerve-nets associated with a shift from a filter-feeding to a predatory mode of existence. Distinctively vertebrate hard tissues arose in association with the organs of special sense, and only later became associated with mechanical support. The idea of a 'new head', discussed further in the next chapter, has largely been refuted by work on the amphioxus (Holland *et al.*, 1992a), and although the elaboration of neural-crest tissue in response to a more active lifestyle as described by Gans and Northcutt has a certain appeal, it is still reminiscent of Garstangian story-telling.

Gans and Northcutt's ideas allow for a smooth transition back to the mainstream, of which Denison (1971) provides a succinct summary, setting down scenario-like guidelines for inferring the condition of the vertebrate ancestor. These guidelines might easily be summarized as a cladogram.

The earliest known vertebrates, known with certainty from the Ordovician, lacked jaws and paired fins. Earlier still, vertebrates presumably lacked a mineralized skeleton, which accounts for their lack of preservation, although they would have had a well-developed head with organs of special sense, derived from neural crest.

Later research suggests that the conodont animal may represent this grade of organization (Sansom *et al.*, 1992) although doubt has been expressed about the presence of neural-crest-derived tissues in these creatures (Forey and Janvier, 1993). To judge from the amphioxus – which has segmented muscles but no neural crest – the common ancestor of vertebrates and the amphioxus had muscular segmentation but no neural crest.

Earlier still, the common ancestor of urochordates, cephalochordates and vertebrates lacked either segmented muscles or organs of special sense, although it would have been a chordate by virtue of a notochord, dorsal tubular nerve cord, paired gill slits and endostyle. This creature may have lain close to the ancestry of hemichordates, although the latter presumably lacked a notochord (whether by secondary loss or primitive deficiency thereof).

The common ancestor with echinoderms was yet not much further off, although this must have lacked gill slits and a central nervous system. It would, however, have exhibited radial cleavage, deuterostomy and have had a tripartite coelom derived from invagination of the archenteron. In short, a deuterostome.

What did this common ancestor of deuterostomes look like? Denison (1971) suggests that it would have been bilaterally symmetrical, and yet motile: despite the tendency towards sessility in its descendants, no trace of a sessile past can be found in vertebrate structure or ontogeny. It was provided with cilia for locomotion and feeding, and may have lacked tentacles (this last feature implying that the pterobranch lophophore is convergent on the lophophore of brachiopods, bryozoa and phoronids, and does not reflect shared common ancestry). The creature would have been microphagous rather than a filter-feeder. Although Denison (1971) does not say so, I think that this ancestor would have looked rather worm-like, although whether its preferred habitat was the plankton or the sea-bed is impossible to say.

From his discussion on the dual 'somatico-visceral' nature of vertebrate anatomy (of which more, see next chapter), Romer (1972) concludes that the vertebrate ancestor was most emphatically sessile and probably also attached to the substrate. He speculates that it would have been a simple filter-feeding creature consisting of a digestive system and little else, and suggests the pterobranchs as perhaps the closest animals now living to this ancestry. Enteropneusts (fairly sessile animals for all that they are unattached) are still very 'visceral' in nature, but possess gill slits for filtration instead of a lophophore[58], and (arguably) the rudiments of 'somatic' organization in the form of what seems to be the beginnings of a dorsal tubular nerve cord.

The adult tunicate is essentially a large sac, perforated with gills and perfectly suited to its role as a sessile filter-feeder – it is the acme of its line of evolution 'beyond which little further evolutionary progress would seem probable' (1972, p.146). The tadpole larva is thus an escape route from an evolutionary dead end, its adoption of 'somatic' structures such as a notochord and a tail its passport to salvation.

Romer follows Garstang's line that the tadpole larva is, as it has always been, the larval form of a sessile adult which, in the vertebrate lineage, became transformed through paedomorphosis into a new adult form of a radically different kind.

He is less certain, though, about Garstang's scheme for the origin of the tadpole larva, from tornariae or auricularia-type larvae. Instead he pushes the problem far back in time, noting the views of Whitear (1957) and others that the tadpole larva might have been an ancient chordate invention antedating the origin of tunicates *per se*.

The amphioxus, though, presents more problems. Although superficially an intermediate between the tadpole larva and a vertebrate – and has a notochord and tubular nerve cord – it lacks even the rudimentary brain and sense organs possessed by a tadpole larva, has a nephridial type of excretory system and an 'excessive multiplication' of gill slits. Romer sees it as an aberrant and possibly degenerate offshoot: like Berrill, Romer sees it as 'laggard' which 'fell by the wayside' and has somehow 'failed'. The asymmetries of its development are also held as evidence of its peculiarity.

The earliest true vertebrates, though, had taken matters in hand. Although primitive jawless fishes still subsisted by pharyngeal filter feeding, they had elaborate brains, sense organs and somatic musculature for 'transporting this feeding device to favorable [*sic*] spots for its activity' (1972, p.149).

2.3 GISLÉN AND CARPOID ECHINODERMS

As shown above, Garstang's paper of 1928 has attracted much comment, and is still widely cited today. A paper published two years later that has received less prominence is that of Swedish echinoderm specialist Torsten Gislén (1930) who, in contrast to Garstang, highlighted the asymmetries of chordate and echinoderm structure and development, particularly with reference to the extinct carpoids, to forge a somewhat different view of the possible course of vertebrate ancestry[59].

What interested Gislén in particular was the apparent bilaterality of chordates imposed, it seemed to him, on an underlying asymmetry evident only in the development of certain forms, especially the amphioxus. 'We shall see that in the *Chordonia*', he writes:

> there are transitory remnants of an asymmetrical stage in the development. There is, however, *a strong tendency towards symmetrisation* and bilaterality which sooner or later sets in and in the *Vertebrata* results in a practically complete symmetry. (p.200, original emphasis)[60]

2.3.1 Asymmetric carpoids

This process – the struggle to re-impose bilateral symmetry on a secondarily non-bilateral pattern – is not unusual in echinoderms. It is seen, for example, in animals as varied as spatangid echinoids and holothurians, so it is not unreasonable to expect the same thing to have happened in Palaeozoic echinoderms such as the carpoids, in the amphioxus and in other deuterostomes.

In this spirit, Gislén emphasizes that no assumption of ancestor–descendant relationship should be read into his comparisons of modern deuterostomes with carpoids: rather that the resumption of bilaterality was a tendency that found varying expression in several distinct but related lineages[61].

One tendency in many modern echinoderms is the reduction of the food-gathering system of ambulacra that converge on a central mouth. In various living and fossil ophiurans, holothurians, crinoids and echinoids, ambulacra may be overgrown by calcite plates to produce a canal that may be secondarily transformed into a duct for nerve tissue. At the same time, the original mouth – at the vertex of the ambulacra – may become redundant, its function doubled by the anus. The result is a sac-like gut (which may also serve for gas exchange), and an ambulacral nervous system that can be used for innervation of many parts of the body.

Gislén follows an earlier suggestion by Jaekel (1918) that the reduction of the ambulacra in several carpoids (to the point of absence in mitrates) might have followed the same course.

This is fine as far as it goes, but this idea raises problems of interpretation when applied to cornutes such as *Cothurnocystis* and *Ceratocystis*, which each bear a single row of large openings on one face of the theca, of a kind not seen in either cinctans or mitrates. Gislén scorns attempts to interpret these either as gonopores (Jaekel, 1918) or new 'mouths' (Bather, 1925). The former idea would imply serial repetition of organs within the theca, which Gislén regards as unlikely.

The latter is more complex, and Gislén takes a great deal of care in his arguments against it. Ultimately, it goes against the grain of Gislén's own ideas of the elimination of ambulacra by overgrowth: were the pores of *Cothurnocystis* new 'mouths', they should appear more open early in ontogeny, and close up as the animals got older. On the contrary, young and small animals have a smaller number of pores, which they augment as they get older and larger. Thus, were these pores accessible to a buried ambulacrum, they indicate increasing size and extent in that ambulacrum with age – which is quite the opposite of what one would expect.

Gislén's solution is to propose that these openings represent excurrent openings for a system of respiratory diverticula (similar to those found in some holothurians) that would filter the incurrent water coming in through the (old) anus, concentrating the pulp of solid particles for digestion or egestion, as appropriate. These slits would presumably arise by coalescence of the intestine with the body wall, and subsequent perforation – something that is not without precedent, as Gislén demonstrates with an example from nudibranch molluscs.

Of course, these slits appear on only one side of the remarkably asymmetrical cornute theca, and the asymmetries of echinoderms in general demand explanation in terms of underlying coelomic organization[62]. As I showed in Chapter 1, the fundamental deuterostome pattern of three paired coeloms arranged front-to-back is substantially modified in echinoderms, in which the right protocoel (axocoel) and right mesocoel (hydrocoel) are reduced or absent.

At the turn of the century, researchers suggested that this arrangement recalled a situation in which primitive echinoderms became fastened to the substrate by the right anterior corner. If so, this primitive stalk cannot be homologous with the crinoid stalk, derived from the right somatocoel. 'We can be fairly certain that the stalk which we find in the oldest Cystideans is homologous with the stalk in the younger Crinoids[63]' writes Gislén:

> This has therefore only contained processes from the somatocoele [*sic*]. It seems to me it is simpler to explain the asymmetries of Pelmatozoans by assuming that their ancestral form either lay on its right side or became fixed to the substratum with this side, than by supposing that they became fixed by the right anterior end. (p.216)

In this passage, Gislén sought only to clarify the origination of the Pelmatozoan stalk to accord with known coelomic arrangements. But it is tempting to read into these lines a harbinger of Jefferies' idea of 'dexiothetism', in which asymmetries in echinoderms and vertebrates can be ascribed to the habit of an ancestral form that (in Gislén's words) 'lay on its right side'. Of course, the 'ancestral form' in Jefferies' conception was not necessarily the ancestor of Pelmatozoans only, but something altogether more primitive than either echinoderms or vertebrates – more like the pterobranch *Cephalodiscus*.

After this discussion of the peculiar asymmetries of fossil and modern echinoderms, the oddities of the amphioxus (Chapter 1) assume phylogenetic significance.

Why, for example, do the gills of the amphioxus originate in such a lopsided fashion? Why is it that the first gill slit develops in such a skewed, distended way? Why does the larva swim in such a curious manner, cork-screwing itself through the water? With the long tradition of echinoderm – indeed deuterostome – asymmetry in mind, adaptive necessity seems a hollow, catch-all answer.

Gislén emphasizes that the small asymmetries of the adult are trivial compared with those familiar from larval development. Indeed, much of the later development of the animal is about Symmetry Regained. For although the adult amphioxus[64] tends to rest lying either on one side or the other, it does not do so on one side consistently, as would a carpoid – constrained to do so by its very shape.

Gislén then sets about studying the spiral motion of the swimming amphioxus, thought to be the best way to direct food into the larval mouth, situated as it is on the left side of the head. Very young larvae rotate in an anticlockwise (right-to-left) sense when seen from the posterior end, this rotation being driven by cilia. In older larvae and adults, though, cilia are less important than the somewhat asymmetrical pre-oral hood, which tends to drive rotation in the *opposite direction*, that is clockwise (left-to-right). Between is a stage during which the larva hangs motionless. It is in this stage that the most pronounced asymmetries of the larva appear, develop and fade.

If spiral motion is there to help an asymmetrical larva feed, why is the animal motionless when this asymmetry is most pronounced? Why, if the mouth is a fundamentally left-sided structure, does the direction of rotation change during ontogeny? Why, if the left-sided larval mouth would benefit from a current generated by a left-to-right rotation, does this mode appear only after the animal loses this feature? And why do the larvae of ascidians swim in the same spiral fashion, even though they are incapable of feeding and do not display the same degree of asymmetry as amphioxus larvae?[65]

Asymmetry is nothing more than a feature of deuterostomes: a quirk that runs in the family, a relic from an earlier time when adults were still in a state of Symmetry Lost. There is no need to drag out selection as an excuse. Indeed, to do so denies the value of asymmetry as an important seam of evidence concerning vertebrate ancestry.

Once one accepts that the asymmetries of the amphioxus are 'historical reminiscences' (p.229), one should be able to find trace of them in other deuterostomes, too. Here Gislén goes a little overboard, finding asymmetry wherever he looks, from vertebrates down to enteropneusts and pterobranchs. Never mind that the case for asymmetry is somewhat weak in the latter two cases – given the chance, Gislén would have found asymmetry in every lamp-post and manhole cover.

Only then does Gislén tie in modern asymmetries with carpoid morphology. He compares (Figure 2.3a) the obverse, gill-bearing face of *Cothurnocystis* with a larval amphioxus pictured lying on its dorsal surface, the left-sidedness of its gill-slits evident. In other words, carpoid obverse means amphioxus left. The large left-skewed mouth of the larval amphioxus corresponds to the old carpoid anus – now, of course, doubling as a mouth.

The pharynx in both cases is perforated on the left side of the amphioxus by the primary gill slits (the secondary ones, of course, appear later in ontogeny, and come to rest on the right side). These primary gill slits correspond to the *only* row of gill slits that cornute carpoids ever have – right-side slits do not appear. The 'new' anus of the amphioxus is a posterior gill slit, moved back along the gullet with the extension of the body.

The fate of the 'old' mouth is somewhat different. In the amphioxus, it becomes the junction between the larval gullet and the posterior end of the spinal cord, an embryonic feature called the neurenteric canal. In Gislén's terms, the transition between gut and nerve cord is quite logical, as it has a precedent in the neural function of crinoid ambulacra once they are roofed over[66].

What of the mitrates, superficially more symmetrical than cornutes? Gislén compares their simple, paired gill-slits with the simple, paired slits of appendicularians[67]. In particular, he compares the swollen, flexible proximal portion of the mitrate stem with the same region of the appendicularian tail, which contains a swollen nerve ganglion. In Jeffries conception, the junction between body and tail was to assume

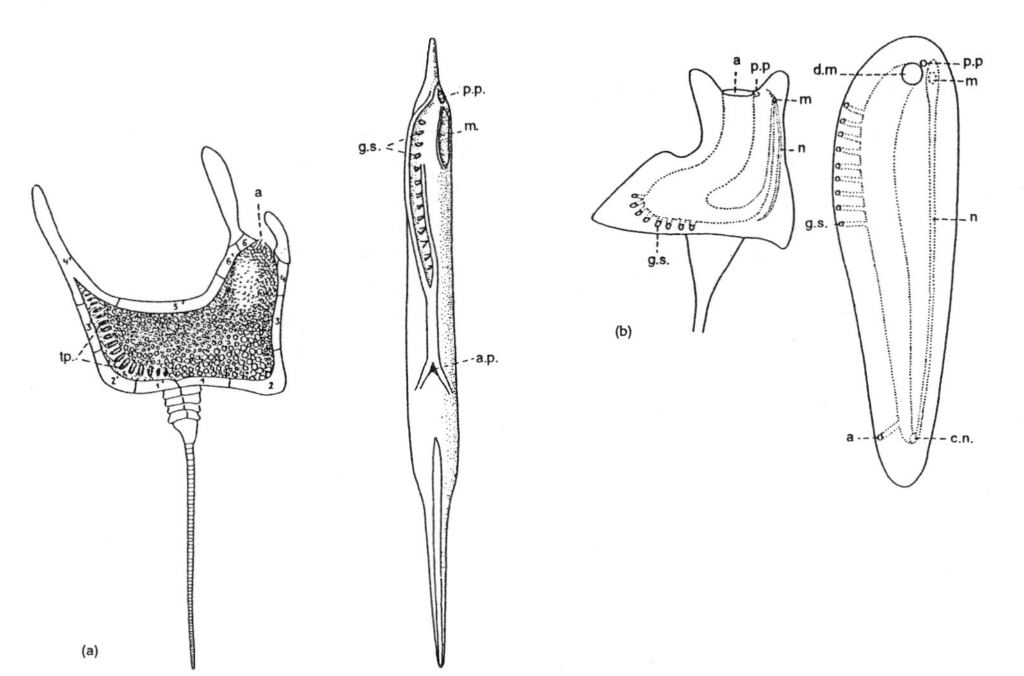

Figure 2.3 Gislén's (1930) comparisons between carpoids and the amphioxus. (a) The obverse face of the cornute *Cothurnocystis elizae*, compared with an amphioxus larva seen in ventral view. Note the left-sided mouth (m) in the amphioxus: we are invited to compare the gill slits (g.s.) of the amphioxus with the pores in the carpoid (t.p.). (b) From the same paper, a reconstruction of the inside of a carpoid showing the pharynx perforated with gill slits (g.s.), the original anus (a), now an intake, the original (now subtegminal) mouth (m) and the original ambulacrum (now a nerve tube) (n). This is compared with a schematic of an acranian larva. Note the reconstruction of the 'new' anus (a) from a posterior gill slit, the old anus, now the definive mouth (d.m.), and the original mouth (blastopore) (m).

particular significance, and contained the definitive chordate brain.

Much of Gislén's paper (between p.250 and approximately p.276) concerns the origin and homologies of the notochord, and is the weakest part of the work. Bearing in mind that the notochord develops as an endodermal outpocketing beneath a nerve tube that invaginates from the exterior, Gislén can hardly help comparing this with the hydrocoel in the arm of an echinoderm, situated beneath an ambulacrum that closes up and assumes the function of a nerve cord. This may seem fanciful when read today (and perhaps as much then). Much follows from the homology between notochord and hydrocoel: internal connections between the front of the notochord and the hypophysis, thence to the exterior via the structures known as Seessel's Pocket or Rathke's Pouch in embryonic vertebrates, are held to correspond with the internal links between the hydrocoel and protocoel and thence to the exterior (via the protocoel pores) in echinoderms and hemichordates.

To go further, Gislén argues that the notochord is homologous with the *left* hydrocoel, and that the subnotochordal rod of lower vertebrates is in fact the reduced right antimere of the notochord, derived from the right hydrocoel. The fact that the subnotochordal rod is absent altogether in the amphioxus speaks for the unusually asymmetric nature of this animal when compared with vertebrates.

Lastly, Gislén makes some interesting comments about the status of segmentation in chordates, which lead us nicely on to Chapter 3. Gislén is quite explicit in his view that first, chordate metamery originated in the tail; second, this metamery is superimposed on a fundamentally trimeric structure; and third, metamery of muscles, that is somites, has no fundamental connection with the serial repetition of gill slits. This is because metamery is a characteristic of the tail, but gill slits are features found in the head (Figure 2.4).

> In case [primitive vertebrates] are ... to be derived from ancestors similar to Carpoids, the muscular segmentation ought to have emanated from the proximal part of the carpoid tail, i.e. from the post-branchial part of the body. (p.281)

Then, as evidence:

> In a young fish-embryo the muscular segmentation appears only in the posterior part of the body of the embryo. It is not until later that the somites become as if pushed forward towards and over the head which thus at earlier stages is muscularly unsegmented. (p.281)

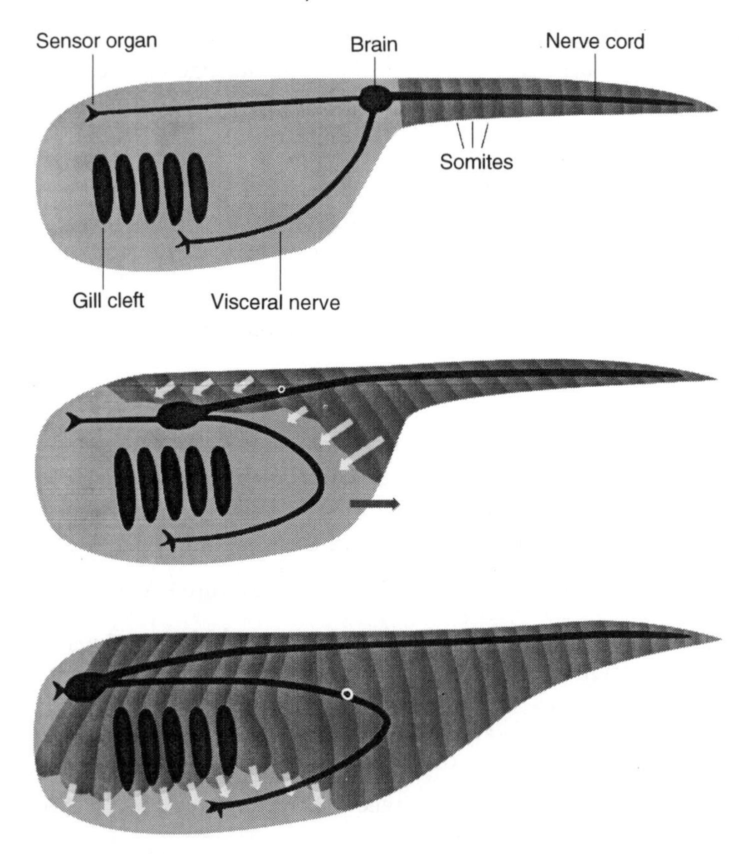

Figure 2.4 The transforming tadpole. This diagram illustrates how a tunicate-tadpole-like animal might be transformed into a vertebrate by the overgrowth of posterior somitic structures. Top: the animal is divided into an anterior 'head' containing the pharynx and viscera, unsegmented but for the seriation of gill slits; and a posterior locomotory 'tail', containing nerve cord and notochord. A ganglion ('brain') at the front of the tail section receives innervation from the head, from organs of special sense at the anterior, and from the viscera. Middle: the 'tail' domain extends forwards, over the top of the anterior 'head' domain. This has the effect of pushing the brain to the front of the animal, shortening the sensory nerve trunks but leaving long, looping tracts in the viscera (the vagus nerve being a relic). On reaching the pharynx, the tail somites intercalate with the gill slits, but the ventral region of the 'head' remains largely unsegmented. Bottom: this animal is essentially a modern vertebrate. The animal at the top, though, is hypothetical, although it bears strong resemblances to a tunicate tadpole larva, Romer's 'somatico-visceral animal' (compare with Figure 3.2) and (especially in respect of the neuro-anatomy) Jefferies' reconstruction of a mitrate carpoid. Indeed, the transformation shown here is essentially the same as that process which Jefferies invokes to turn a mitrate into a modern vertebrate. The idea seems fanciful, though it explains several features of vertebrate anatomy, such as the distribution of neural-crest-derived ('tail') tissue, the layout of nerves, especially in the viscera, and the distribution of *Hox* genes. (Picture by the author, redrawn by Sue Fox.)

Examples include the amphioxus and the hagfish *Bdellostoma*, in which the relationship between the positions of gill slits with respect to particular somites changes during development. This fundamental dislocation is everywhere stressed:

> All this points towards the conclusion: *The muscular segmentation has proceeded from the posterior part of the body and has gradually been pushed forward over the animal whose anterior part before this was muscularly unsegmented.* (p.283, original emphasis)

Furthermore, this segmentation is restricted to the dorsal side:

> We find a myomery only on the dorsal side even in very small embryos of *Amphioxus* ... Thus the dorsal part of the animal is myomerized, the ventral, on the contrary, is muscularly unsegmented. As only the dorsal part of the somatocoele has been myomerized we must imagine that the muscle-bundles have advanced over the obverse face of the Carpoid. (pp.283–284)

As for serial repetition of the gill-slits (branchiomery):

> We can state that branchiomery in the Vertebrates does not correspond originally to the neuro- and myomery (cf., for instance, the development of *Bdellostoma*). In the cases where it does so (at early stages), as in Acranian larvae, I consider that it is a secondary phenomenon, a result of the tendency towards correlation and symmetrization which can be traced so universally in the ontogeny of the Vertebrates. (p.284)

2.3.2 Fearful asymmetry

Gislén's paper is evidence of an inventive mind at work. At the same time, these ideas are (for the most part) rooted in real evidence: everywhere Gislén cites current work to support his fusion of palaeontology and experimental embryology. This puts Gislén's scheme in a different league from, say, that of Gaskell, whose ideas were easily bruised by even the gentlest application of external evidence. And yet Gaskell's ideas were a matter of discussion in the serious scientific arena for twenty years, whereas Gislén's have assumed a far lesser prominence. Why?

Several reasons spring to mind. First, Gaskell's ideas came from a time when it was still fashionable to think of creatures in terms of the Naming of Parts, as static structures that served as canvases for the drawing up of homologies. Such fluidity as existed came in the form of recapitulation, evolutionary progress by the appendage of new stages on the end of ontogeny. Little thought was given to how these new stages might have arisen, or how creatures such as morphological intermediates lived and breathed. For all their protestations to the contrary, the embrace of darwinism by Gaskell's followers was simple-minded and misplaced – they failed to appreciate the meaning of adaptation and how it might interact with morphology.

Gislén's scheme is very much in this old tradition. It is strongly recapitulative, in the sense that Symmetry Broken and Symmetry Regained are traceable in the larvae of modern amphioxus and vertebrates, and in the adult stages of the venerable carpoids. Apart from a short discussion on the motion of carpoids, morphology is discussed in terms of phylogenetic history rather than current adaptive need.

By 1930, though, such modes of thought were falling from favour. Garstang's work, published earlier, adopted adaptation – not recapitulation – as the explanation for most, if not all, morphological features, especially in larval forms. After all, adaptation was underpinned by a real mechanism (natural selection) whereas recapitulation remained a description of a pattern divorced from the process whereby that pattern might have originated. Garstang's work clearly represented the spirit of progress, and was modified and discussed by researchers such as Berrill who tended to stress the current adaptive character of morphological features, and their functional rather than phylogenetic importance. At the same time, palaeontological evidence was seen as an archaism, palaeontologists as disciples of now-discredited recapitulation[68]. By Berrill's time, discussion of topics such as vertebrate ancestry was unfashionable enough for Berrill to apologise for his temerity, as a self-confessed 'voice speaking from the wilderness' (1955, p.1) in writing a book about it.

Second, and more pragmatically, Gislén's work was published in a fairly obscure journal, probably off the usual reading lists of the dwindling numbers of people who would have cared enough to seek it out. And once found, Gislén's treatment of echinoderm palaeontology and vertebrate embryology with equal facility puts some demands on the reader. Without some appreciation of both echinoderm morphology and vertebrate embryology, Gislén's paper is stony ground indeed.

Third, as we have seen, the entire topic tends to attract its share of thinkers whose originality remains unfettered by such considerations as experimental verification, and whose sometimes odd ideas thus do not lend themselves to testing. This explains why such as Bateson turned his back on the subject with evident disgust. At first sight, Gislén's paper seems to fall very much within this sad and eccentric *oeuvre*, and its message is only apparent after a second or third reading. It is easy to see why few would have bothered attempting a first.

I should like to think (if only half-seriously) that there is another reason for Gislén's obscurity. For just as Garstang promoted adaptation, Gislén saw asymmetry as the key to vertebrate evolution. One could argue that asymmetry is a phenomenon that people find psychologically uncomfortable. Everywhere regularity and symmetry are seen as signs of beauty, desirability and health (Enquist and Arak, 1994). Striking asymmetries are regarded with fascination and horror. How disturbing, then, to look at the egregiously irregular *Cothurnocystis* and see within its fearful asymmetry the seeds of time sprouting to produce one's own self.

One who did cite Gislén was the palaeontologist William King Gregory (1936) in a paper about 'organic designs'. Gregory held that most, if not all morphological diversity was reducible to two principles: the repetition of similar parts (polyisomerism) and the subsequent emphasis of particular copies of these parts (anisomerism). Intriguingly, these principles may be as applicable to the classification of homeobox genes as to structural morphology, if not more so (Chapter 3).

In reviewing vertebrate origins, Gregory figures various carpoids in the company of the ostracoderm fish *Drepanaspis* (Figure 2.5), and asserts several times that morphological transformation could have turned an echinoderm, via the carpoid state, into a vertebrate. For example:

> [Certain groups of Carpoid were] performing some remarkable experiments in the modification of a quinqueradiate symmetry into a new dorso-ventral asymmetry and a partial bilateral symmetry, and that some of them approach in general patterns to the 'dorsal' and 'ventral' shields of *Drepanaspis* ... (p.321)

Gregory dismisses all other ideas about vertebrate origins as unlikely: but he does not cite Garstang. He had caught up ten years later, in a review paper for the *Quarterly Review of Biology* (1946).

For its brevity, accessibility and range, this review should be required reading, even if Gregory's sidelong style suggests that reading it today should be accompanied with a generous pinch of salt. It is doubly interesting, though, in that it places carpoids in context with Garstang's idea of paedomorphosis.

In the supposedly paedomorphic origin of vertebrates from a tadpole larva, it is usually assumed that the adult form of this ancestor was a sessile creature similar to a tunicate or pterobranch. But this assumption does not necessarily follow – in Garstang's conception, all that matters is that the vertebrate form appears as a paedomorphic offshoot from a larva, rather than a recapitulative addition to the adult form[69]. Once that is accepted, the precise form or habit of the former adult hardly matters (Whitear, 1957). If so, the 'tunicate' tadpole larva could just as easily have been an interpolation in the life cycle of a carpoid echinoderm as a tunicate.

This idea allows for a much more vigorous, long-lived tadpole larva, once common to the ontogenies of chordates and carpoids, but retained today only in tunicates, abbreviated and modified as a simple habit-selection device. Thus the teasing possibility of a homology between carpoid stalk and tunicate-tadpole tail, and a note on the asymmetries of amphioxus and how they are paralleled in *Cothurnocystis* – with the reminder that phylogenetic transformation between the two, though seemingly great, 'is perhaps no greater than the ontogenetic metamorphosis of a caterpillar into a butterfly' (1946, p.361).

And, lastly, a brief paragraph suggesting that the similarities between carpoids and some ostracoderms was, perhaps, more than skin deep. His arguments, though, are unconvincing: apart from the fact that carpoid skeletons are calcitic rather than phosphatic, there seems no reason why Gregory's readers should prefer carpoids over (say) eurypterids, were one to judge ancestry of the vertebrates on the criterion of physical similarity with ostracoderm fishes.

And apart from that, the search for discussion of the possible chordate affinities of carpoids yields scant pickings. Perhaps more people thought about them than were prepared to commit themselves on paper. Crowson (1982), for example, remembers (in another context) that he

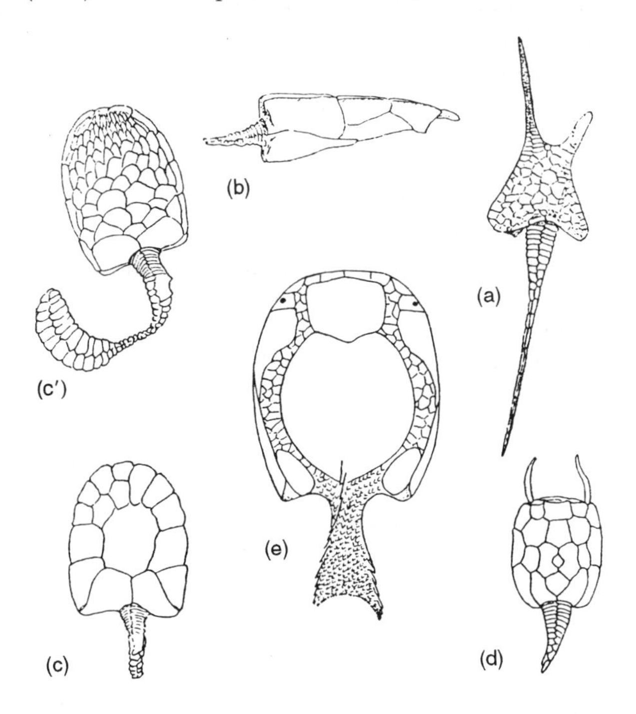

Figure 2.5 The superficial resemblance between carpoid echinoderms and extinct armoured agnathans, as pointed out by Gregory (1936). (a) *Dendrocystoides scoticus*, (b) *Lagynocystis pyramidalis*, (c, c') *Mitrocystella barrandei*, 'lower' and 'upper' surfaces, (d) *Placocystites forbesi* and (e) *Drepanaspis*. All the animals shown are carpoids, except (e), which is an agnathan (reproduced by permission of Cambridge University Press).

> ... first came to suspect these fossils [carpoids] ... as possible ancestors of Vertebrata around 1950, and I even suggested this idea to students at that time, though not venturing to put my ideas into print. (1982, p.252)

Reasons for his support include the acceptance of the bipartite tadpole-like division of the vertebrate body (similar to Romer's 'somaticovisceral animal' described in the next chapter) with concomitant rejection of the idea of the Balfourian scheme of segmentation and the recognition of an unexplained asymmetry in carpoids as well as the heads of modern chordates.

The review by Denison (1971), although far less colourful than that of Gregory (1946), is arguably more sensible. By that time Gislén's ideas had been adopted and developed by Jefferies (1967, 1968a,b), and Denison focuses his attention on Jefferies' ideas rather than those of Gislén. Jefferies' contribution is discussed in detail in Chapter 4, but for now, though, it is enough to note that whereas Gislén had noted the similarities between carpoids and vertebrates, and drew back from any assertion that the former might have been ancestral to the latter, Jefferies (1967) interpreted carpoids (or, at least, cornutes and mitrates) quite explicitly as chordates. Denison, therefore, is bound to criticize Jefferies' ideas in this light, and raises several important technical points concerning Jefferies' interpretation. His critique is presented in full in Chapter 4. His conclusion is that it is simplest to regard carpoids as early echinoderm offshoots (Andrews, 1984) that resemble chordates through convergence or by virtue of their common deuterostome heritage. It follows from this that echinoderms adopted a calcite skeleton before settling down to pentamerous symmetry.

For the very latest in ideas concerning vertebrate origins, one should turn to Peterson (1994), who, like Denison, finds much to criticize in what has now become known as the calcichordate theory. Peterson suggests that the solution to the 'impasse' of vertebrate origins will not be in the acquisition of morphological novelties, but in the molecular changes that underlie them. These, in part, are the subject of the next chapter.

NOTES

29 For example, to say of Willmer (1975) that he proposed the nemertine worms as vertebrate ancestors 'solely because no one else had' (Thomson, 1987b) is unfair, given that the nemertean idea has roots going back more than a century.

30 Gregory also notes that the 'popular emphasis on the irrevocability of degeneration may have delayed a wider acquaintance with von Baer's principle of "paedogenesis" and Kollman's "neoteny"', taken up by Garstang.

31 See Hall (1992) for an illuminating discussion on the conflict between Cuvier and St Hilaire and its impact on subsequent biological thought

32 Many authorities, irrespective of the views on vertebrate origins to which they subscribed, often held the organization of the nervous system as central to the entire argument. Particularly telling examples are seen in the paper by Gaskell *et al.* (1910), discussed below.

33 Such comparisons, though, were quite common. Gaskell, like Patten, was a noted student of early vertebrate anatomy and, like him, was drawn into making comparisons between arthropods, ostracoderms, and fossils such as *Bothriolepis*, now known to be a placoderm, a member of an extinct class of jawed vertebrate. Even Edward Drinker Cope (more famous, these days, as a tireless seeker after dinosaurs) was drawn into the debate. Gregory (1946) reports his assignment of *Bothriolepis* as a member of a new order of tunicates, distinguished by the straightness of its gut relative to the (looped) urochordate gut. Hence he classified *Bothriolepis* as an Antiarch (*anti* = opposite, *archus* = anus). The name survives, as the order of placoderm fishes to which *Bothriolepis* belongs.

34 As in the annelid theory, the transformation from arachnid to vertebrate demanded the creation of a new gut from an infolding of the ventral surface. The ventral mid-line of many arthropods (particularly crustaceans) is used as a food-groove, so one might imagine how this could be enclosed by paired series of appendages which met at their tips to form a kind of pharyngeal tunnel. And, as luck would have it, the gaps between these appendages make convenient gill slits.

35 Again, the alimentary function of the pharynx in tunicates and the amphioxus would be dismissed as degenerate, but its filter-feeding function in young ammocoetes could, presumably, be explained away.

36 In the Linnean report, it is noticeable that Gaskell and his supporters are given to somewhat more intemperate language than their critics. Could it be that the arthropod idea was (to coin a phrase) on its last legs, such that its exponents were forced to take refuge in strong language?

37 The evolution of the hermit crab was one of MacBride's favourite examples of recapitulation at work, and one much criticized by Garstang (1922).

38 Gaskell's theory ended up pilloried by A. S. Romer in *The Vertebrate Body* (1970) as an 'amusing variant of the arachnid theory' (p.26).

39 Everywhere Goodrich stresses the clean lines and perfect simplicity of the amphioxus. He does not discuss how these features are marred by the asymmetries that dominate its ontogeny and, to an extent, alter its structure as an adult. Again, his identification of the 'simple' nephridia of the amphioxus with the flame-cells of flatworms is now known to be mistaken, and they can be regarded instead as vertebrate-like glomeruli writ small (Ruppert, 1994). However, we can forgive Goodrich all these foibles for his recognition that protochordates might be primitive rather than degenerate, and could therefore yield useful information about vertebrate ancestry.

40 ... and one of the more notable victims of the hoax discovery of Piltdown, which was at the time of the Linnean meeting still two years in the future.

41 cited as the Address of the Chairman of the Zoology Section to the American Association for the Advancement of Science meeting of 1901, printed in *Science*, 1910, p.7.

42 Consult Hall (1992) and Gould (1977) for discussions on recapitulation and its place in biological thought.

43 Paedomorphosis is often confused with neoteny, which has deeper roots – Gregory (1946) cites Kollmann (1882) as a source, but goes on to use the terms interchangeably (Gregory, 1946, p.354). The distinction may be purely a historical one. Here I refer to neoteny as the ability for larval forms to reproduce sexually, as in the axolotl. Paedomorphosis, though, as advanced in opposition to recapitulation, casts neoteny in terms of a circumstance that may exert an evolutionary influence, far beyond the practical circumstances of particular species in which neoteny is seen as an instance.

44 Those salamanders that retain larval external gills, and remain in the water where they become sexually mature.

45 Newts and salamanders.

46 It is probably wrong, too. Evidence from molecular phylogeny (Wada and Satoh, 1994) and sperm morphology (Holland *et al.*, 1988) suggests that larvaceans branched from close to the base of the tunicates. The implication is that the ancestor of the tunicates as a group was free-living, which contrasts with Garstang's view that larvaceans are neotenous doliolids or ascidians.

47 Bone (1960) argues that the processes are rather different, and one cannot hold up the paedomorphic origin of the larvacea as evidence for the origin of the vertebrates in a similar fashion. Larvaceans originated by paedomorphosis from another free-swimming group, the doliolid thaliaceans, which Garstang presumed to have originated from ascidians by paedomorphosis. Yet it is evident that metamorphosis does take place in doliolids, and that the adult form of (say) *Pyrosoma* is quite different from that of a tadpole larva. Rather, doliolids are better thought of not as paedomorphic ascidians but as *adult* ascidians that have undergone metamorphosis but which have failed to settle. Instead of elaborating on their larval phases, doliolids suppress their tadpole larvae: adult motility requires no separate, motile larval phase to ensure its dispersal. Thus, larvacea did not arise by paedomorphosis from a sessile ancestor, but from one that was already motile. As noted elsewhere, the motile ancestry of larvaceans, and indeed all tunicates, is supported by work by Wada and Satoh (1994) and Holland *et al.* (1988).

48 Incidentally, the second and third reasons were that Garstang often sketched his ideas in light verse that some might have thought flippant; and that his ideas conflicted with Haeckelian ideas which were then much more popular than they are nowadays, a problem that Garstang himself identified, as I noted above.

49 This circumstance is discussed by Hardy in his foreword to Garstang's *Larval Forms* (1951). Berrill may be largely responsible for the rehabilitation of Garstang's ideas.

50 For example, most tunicates have a notochord that contains just 42 cells. The number is rarely different, and then only by one or two. The exception is the

larvacean tunicates which have only 20 notochord cells, or thereabouts. There is only so much that can be done with so few cells.

51 Lacalli *et al.* (1994) make the same point in support of Garstang's scheme.

52 Berrill's discomfiture about the apparently non-nervous state of the nerve cord would be relieved by subsequent ultrastructural studies. Katz (1983), Crowther and Whittaker (1992) and others have now shown that within the nerve cord, the bulk of which consists of non-neuronal glial cells, lie two parallel, ventrolateral nerve tracts containing axons.

53 But there is a curious statement on page 129 of *Origin*:

> And among the most abundant of all marine organisms are their [i.e. the ascidians'] immediate descendants, the thaliacean and larvacean tunicates. This statement I realize contains the conclusion that the pelagic tunicates have evolved from the sessile tunicates, and not the other way round as Garstang (1928) and others have at time supposed ... (1955, p.129)

This statement is at odds with Garstang's clear contention that pelagic tunicates evolved from sessile ones (see in particular Garstang, 1928, pp.54–55). I can only suppose that Berrill was either referring to Garstang's pre-paedomorphic ideas of the 1890s, or that this statement is a simple editorial error of transposition.

54 The amphioxus is incapable of regulating the influx of water even when the medium is only mildly brackish, around 90% sea water. This fact, according to Binyon (1979) raises questions about the functions of the animal's numerous excretory 'flame' cells. Following Berrill (1955), Binyon supposes that the amphioxus is therefore a 'freshwater reject' rather than a creature primitively marine and whose solenocytes might perform some other function.

55 Even if Garstang's ideas about paedomorphosis may not have stood the test of time, at least in respect of larvacean ancestry, his simple, original idea that ciliated nerve tracts might be transformed into a dorsal, tubular nerve cord has received steady experimental support (see Katz, 1983; Crowther and Whittaker, 1992, and references therein). Note, however, that the collar cord of enteropneusts is also ciliated (Nielsen, 1987), a finding that adds strength to the enteropneust case.

56 This is made explicit when on p.499 Raw states that 'Lamarckism seems an important factor in evolution, as important perhaps as Darwinian natural selection. It is difficult to see how evolution could have occurred without some degree of inheritance of acquired characters in geological time'.

57 Interestingly, the amphioxus was first described not as a chordate, but a slug: its first appearance in the systematic literature was as *Limax lanceolatus* Pallas 1778. Its true position was recognized by Costa, who called it *Branchiostoma* in 1834 and by Yarrell, who called it *Amphioxus* in 1836 (details from Sedgwick, 1905).

58 Romer comments that the enteropneust gill series 'makes for a much more efficient method of filter feeding than the antique lophophore system' (1972, pp.144–145). Were that strictly true, it would be hard to account for the success of the lophophore in brachiopods, phoronids and bryozoa, let alone pterobranchs and, most recently, lophenteropneusts: deep-sea lophophore-bearing enteropneusts.

59 This paper would probably have been consigned to the pot-pourri of miscellaneous ideas of varying soundness as discussed above, but for the fact that it provided the central inspiration for Jefferies' own extensive and much-discussed canon, itself the subject of Chapter 4: such is the justification for according Gislén's work the prominence it receives here.

60 Jefferies kindly allowed me to see his own copy of Gislén's paper, complete with his (Jefferies') own annotations. Jefferies cross-references this statement to a passage on page 234 that reads 'By assuming that the *Acrania* started from forms similar to [the cornutes] *Ceratocystis* or *Cothurnocystis* it seems to me that we gain an explanation of a large number of the asymmetries touched upon above as well as of organizational peculiarities'. That an explanation for asymmetry should be a requirement of any theory of vertebrate ancestry is an important motivation of Jefferies' work.

61 This point is made again on p.241: 'As should be evident from the above', writes Gislén:

> I thus regard the *Acrania* as derived from a Carpoid type, similar to *Cothurnocystis*. I purposely say similar, so as not to be misunderstood, for I will by no means assert that *Cothurnocystis* is any direct ancestor; it is presumably of too late a paleontological [*sic*] date for this.

I draw attention to this with reference to a frequent criticism of Jefferies' ideas, that his scheme of carpoid–vertebrate relationship is at variance with the stratigraphic record. This criticism stems from a misreading of cladograms as 'family trees'. It also demonstrates the present obscurity of Gislén's paper, in which a statement concerning this problem is quite clearly and unambiguously made.

62 However, Gislén suggests that although only one set of slits is apparent, carpoids could still have been more symmetrical internally. Indeed, the respiratory diverticles are stated to have broken through in pairs in the more symmetrical mitrates. That these are described as 'earlier more symmetrical' (p.220), presumably earlier than cornutes, denies Gislén the opportunity to postulate a cornute–mitrate transition as an example of bilateral symmetry regained. This is ironic given the importance Gislén attaches to the process – as well as its later prominence in Jefferies' later ideas about 'mitrate organ-pairing'.

63 Gislén adds a footnote here that reads 'I am not perfectly convinced that the stalk in the *Carpoidea* is homologuous [*sic*] with that in other *Pelmatozoa*. It may be possible that it developed somewhat further back than in other Echinoderms' (1930, p.216). The term 'further back' presumably means morphological placement rather than phyletic occurrence. Nevertheless, the ambiguity of this statement – and Gislén's reticence on the reasons for his own lack of conviction – may have contributed to Jefferies' occasional changes of mind concerning the homologies of the carpoid stalk. Jefferies has annotated this particular footnote with a marginal highlight, as if to draw attention to it, but makes no further comment on the copy in my possession.

64 And, as Gislén showed by experiment, the ammocoete larva of *Lampetra fluviatilis planeri*. As recently as 1994, Hirakow and Kajita noted that early amphioxus larvae lie upon one side, and that on one occasion 'it was found that 71% (27 of 38) of larvae lay upon the left side'. The significance of this is uncertain.

65 The relationship between swimming and feeding in the larval amphioxus has been a source of debate for decades. It now seems set for resolution thanks to the simple expedient of filming larval amphioxus at various stages of development (Gilmour, personal communication). It now seems clear that the distended first gill slit is used as a mouth, although it is rather inefficient at trapping the microscopic algae on which the larvae subsist. This may explain why amphioxus in the wild thrive in highly eutrophic waters where algae grow in sufficient density to compensate for the poor grazing ability of the larva: it may also explain the difficulty of culturing amphioxus in the clean, clear waters of laboratory aquaria (N. D. Holland, personal communication).

66 Jefferies' reconstructions of the carpoid visceral plan differ considerably: his scheme does not set much store by the link between nerves and gut, and does not reconstruct the hindgut in the same way as does Gislén.

67 ... although he makes no statements about their phylogenetic relationship with cornutes. In contrast (see Chapter 4), Jefferies holds that mitrates evolved from cornutes, evolving many new structures on the right side as Symmetry Regained, and that chordates evolved from mitrate ancestors. Furthermore, he suggests that mitrate gill slits are not so much single slits as atriopores, and that the animals would have had several pairs of internal gills, similar to the amphioxus.

68 cf. Garstang's comment: 'It is, however, the palaeontologists who are the real defenders of the Biogenetic stronghold. With them the Law is a faith that inspires to deeds, while to the embryologist it is merely a text for disputation' (1922, p.92).

69 As Garstang himself says (1922, p.84), 'The Coelenterate, Coelomate, Protochordate, Gnathostome, and Tetrapod are successive grades of differentiation both in the ontogeny and phylogeny of a Frog: but at none of these grades does the ontogeny recall the form and structure of a possible adult ancestor.'

3
Head to head

Fillet of a fenny snake
In the cauldron boil and bake;
Eye of newt, and toe of frog,
Wool of bat, and tongue of dog,
Adder's fork, and blind-worm's sting,
Lizard's leg and howlet's wing,
For a charm of powerful trouble,
Like a hell-broth, boil and bubble.

Shakespeare, *Macbeth*

3.1 HEADS AND TAILS

No theory of vertebrate origins can claim to be truly synthetic unless it addresses the subject of the segmentation of the vertebrate body, and the origins of the vertebrate head.

The two topics are closely connected. By segmentation, I refer to the presence of serially repetitive structures along the body axis. For example, hemichordates and chordates have a series of gill openings arranged along the body, each with its own supporting structures, innervation and musculature that seem to have derived from a common model. The body muscles of cephalochordates and chordates are likewise divided into a series of 'blocks' or somites, again, each one with its own supporting structures, innervation and so on. Other parts of the body, such as the excretory system, the brain, the cranial and spinal nerves, the limbs and so on are constructed at least in part along the lines of serial repetition.

This kind of organization is called metameric segmentation, in which the body is divided into a number of segments or 'metameres', which although designed according to a common scheme, may differ from one another in detail, in structure and function: a bit like the different storeys of the same office block. At the same time, metameres must interact to form a functional whole.

Metameric segmentation occurs to a greater or lesser extent in several major groups of animals, most notably the arthropods and annelids (thus giving rise to hypotheses of vertebrate ancestry from arthropods or annelids, as seen in the last chapter).

Not all segmentation is metameric: the segments of cestodes (tapeworms of the phylum Platyhelminthes) and the attached polyps of cnidarian medusae, for example, are not segments of a single individual but the unsegmented embryonic stages or primordia of several whole, distinct individuals, each attached to a common stalk.

Metamerically segmented animals tend to be active and free-living, and move about by means of directed deformation of the trunk and tail regions. Earthworm movement is efficient because the fluid-filled coelom is divided into segmental compartments, allowing for regional variations in hydrostatic pressure which, in essence, propel the animal. Many-legged arthropods direct the limbs in each segment with reference to those in other segments, so that the animal as a whole moves along in a coordinated way (witness the wave-like undulation of millipede legs). This coordination is arguably facilitated by the division of the body into regional centres, or segments.

Active chordates move by throwing the body and tail into 'S' shaped curves, achieved by waves of contraction and relaxation spreading through segmentally arranged muscle blocks. One could argue, therefore, that metameric segmentation evolved in several metazoan lineages in parallel.

3.2 THE METAMERIC VERTEBRATE

For more than a century, vertebrates have been seen as, essentially, segmented from stem to stern: a kind of amphioxus made complicated. Modern ideas of segmentation in vertebrates, particularly as regards the head, are associated with the work of Francis Maitland Balfour on the anatomy of the head region of sharks[70].

One of Balfour's contemporaries, though, cautioned against laying too much stress on segmentation as a phylogenetic character, as serial repetition occurs in most animals in varying degrees (Bateson, 1886). This assertion is made to counter, explicitly, ideas that vertebrates were linked with annelids. But vertebrates were more likely to have derived from an unsegmented ancestor, and as discussed above, Bateson finds a small measure of sympathy with the idea, developed by Hubrecht (1883), of a link between vertebrates and nemerteans. Perhaps surprisingly in hindsight, this link also reportedly (Bateson, 1886, p.555) found some favour with none other than Francis Maitland Balfour.

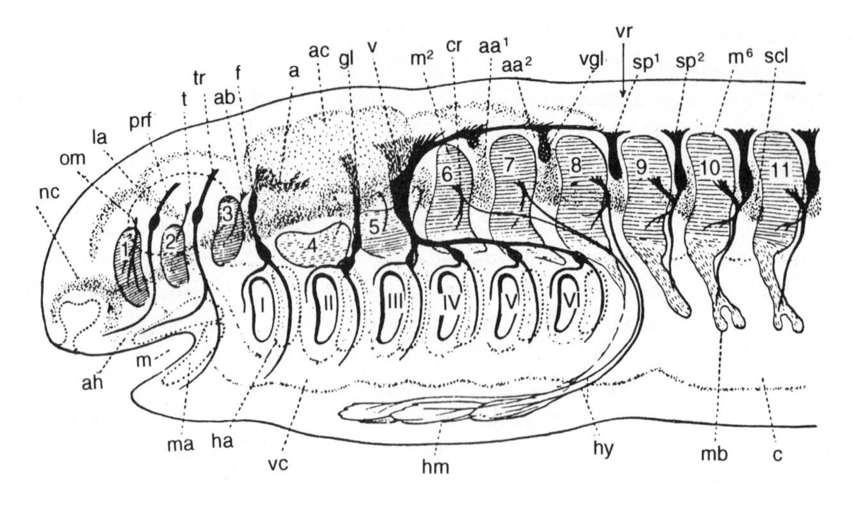

Figure 3.1 The canonical picture of head segmentation in the young dogfish (from Goodrich, 1930).

The anatomical studies of Balfour's school, notably those of E. S. Goodrich and Gavin De Beer, support this contention, that vertebrates are primitively segmented from one end to the other (Goodrich, 1918, 1930; De Beer, 1922, 1937). In a drawing that has now become famous (Figure 3.1), Goodrich (1930) illustrates the regular alternation of cranial nerve roots and gill arches that characterizes the young dogfish, indicative of an inherently segmented head. Of course, the alternation is far from regular, but this can be accommodated in a scheme in which the embryonic layout of a modern vertebrate differs substantially from its ancestor. The earliest vertebrates would have lacked jaws: these would have developed from an anterior set of branchial arches, with much modification of more posterior branchial arches, innervation and musculature.

3.3 ROMER'S SOMATICOVISCERAL ANIMAL

The Balfourian view of the chordate body plays down the fact that some parts of the body are more segmented than others, and different regions are segmented in different ways.

In many ways, the simplest kind of chordate metamerism is that of the amphioxus. The body is dominated by the somites, each clearly demarcated from the next by a sheet of tissue, and which together extend from the anterior end to the tip of the tail. The notochord acts as a purchase against which the muscles can act, and extends all the way from one end of the animal to the other. Although the division into somites is more marked dorsally than ventrally, they effectively form 'rings' around the animal. Needless to say, many other structures in the body are serially repeated, such as nerves, nephridia and gill slits.

Tunicates are, at least at first sight, a complete contrast. They do not display metameric segmentation to anything like this degree, with the possible exception of the seriation of the gill slits[71]. Instead, it is more useful to consider the tunicate tadpole larva, at least, as bipartite: the front end, containing the viscera; and the hind end, devoted to motility with its notochord, muscle cells and nerve cord.

Given that its function is locomotory, it is surprising that the tail of the tunicate tadpole (and that of adult appendicularian tunicates) does not appear segmented. In the case of the tadpole, at least, this could be a simple consequence of a body plan cast on very few cells (Berrill, 1955: see Chapter 2). In which case the unsegmented condition could be a primitive feature: tunicates evolved from a sessile ancestor, and the motile larva was an evolutionary innovation. Segmentation became a necessity in those paedomorphic, obligate swimmers among tunicate larvae that evolved into cephalochordates and vertebrates.

On the other hand, the unsegmented condition of the tadpole tail could be a consequence of degeneracy contingent on smallness. Perhaps the ancestors of tunicates were fully motile, even as adults, and had segmented tails. But given that the animals today are motile only for a very short period in their lives, efficient and sustained locomotion (in search of prey or mates, for example) is less of a priority than finding a suitable settlement site as quickly as possible. Natural selection might therefore have favoured a simple, unsegmented tail without the elaborate repetition of structure that segmentation involves – especially as the animal contains too few cells to allow corralling large sheets of them into somites. This idea is consistent with the view that the sessile habit of adult tunicates is a specialization of the group that evolved relatively late in its history, and also with the view that tunicates need not be more distant relatives of vertebrates than is the amphioxus.

The segmentation or otherwise of the tail of the tunicate tadpole larva has long been a matter for debate. Lankester (1882) was for it; Seeliger (1894) was against. Recent ultrastructural (Crowther and Whittaker, 1994) and genetic work (R. Di Lauro, personal communication) suggests that Lankester might have been right after all, in that the tail of the tunicate tadpole larva was originally built upon segmental lines, with the implication that overt segmentation was secondarily lost. This carries the further implication that tunicates descend from motile ancestors, and that the larvacea are in fact primitive – a complete contrast to Garstang's view.

As I discussed in Chapter 2, Garstang and Berrill did not discuss the origins or evolution of segmentation in chordates in anything more than the sketchiest terms, perhaps because of the assumptions that, first, tunicates were phylogenetically more primitive than cephalochordates; and, second, that segmentation could be explained, simply, as an adaptation to locomotion developed in the ancestry of cephalochordates and chordates which, because of its complexity, would most parsimoniously be considered as having evolved in common.

However, the presumption that the amphioxus is a metameric animal *sine qua non* tends to hide its other features, notably its larval asymmetries. It thus falls into the role of vertebrate sister group rather easily. But the fact that vertebrates and arthropods have a metamerism sufficiently similar – superficially – to have anyone regard the two groups as close relatives suggests that this assumption is weak[72].

Nevertheless, were one to suppose that the tunicate tail *were* once segmented, the amphioxus would seem to represent the posterior end of the whole tunicate tadpole, a segmented tadpole tail without a head. The adult tunicate represents the unsegmented front end of the tadpole larva, and is therefore completely unlike the amphioxus: the two represent different parts of the same animal, the first, the front end, the second, the hind end. Vertebrate segmentation is something of a compromise between the two. Considered in this way, the vertebrate is an elaborate version of what Romer (1972) called the 'somaticovisceral animal', divided into two regions, rather like the hypothetical tadpole larva with a segmented tail (Figure 3.2)[73].

Romer (1972) viewed segmentation as a pattern imposed on a more fundamental, bipartite division of the chordate body, the 'somatic' and

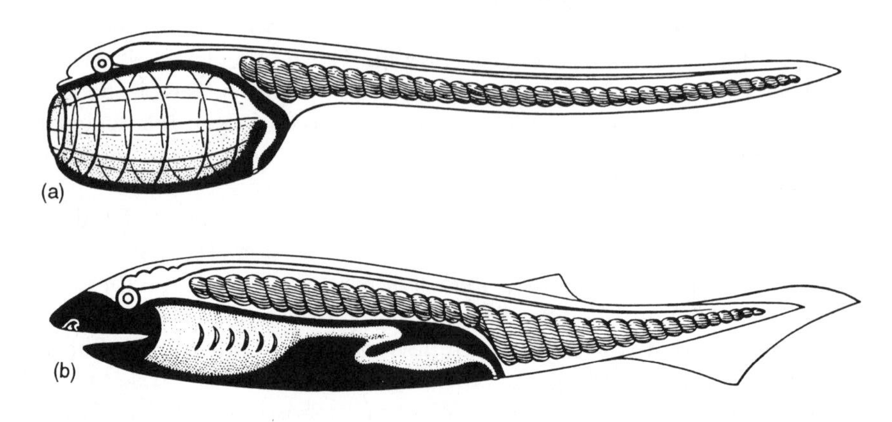

Figure 3.2 Romer's somaticovisceral animal. This picture illustrates the contrast between 'somatic' (clear) and 'visceral' (black) components of the vertebrate body. In (a), the somatic animal lies posterior to the visceral animal except that sense organs and the nerve cord extend forward and dorsally (contrast with the scheme outlined in Figure 2.4). In (b), the ancestral chordate, the two parts have overlapped to a considerable degree (from Romer, 1970).

the 'visceral'. 'In many regards', he writes:

> the vertebrate organism, whether fish or mammal, is a well-knit unit structure. But in other respects there seems to be a somewhat imperfect welding, functionally and structurally, of two somewhat distinct beings: (1) an external, 'somatic,' animal, including most of the flesh and bone of our body, with a well organized nervous system and sense organs, in charge, so to speak, of 'external affairs,' and (2) an internal, 'visceral,' animal, basically consisting of the digestive tract and its appendages, which, to a considerable degree, conducts its own affairs, and over which the somatic animal exerts but incomplete control.

He continues;

> A consideration of what evidence can be gathered from our knowledge of living vertebrate relatives among chordates or hemichordates suggests that this dichotomy has a historical background; that the remote chordate ancestor, as exemplified by pterobranchs, acorn worms and tunicates, was essentially a small, sessile, simply built animal that included in its structure little but a food-gathering and digestive apparatus; that in the developmental stage of more advanced chordates there was added to this, for better distribution of the young, a locomotor 'unit' with muscles, supporting structures, and brain and sense organs; that this newly acquired 'somatic animal' was at first restricted to the larval period, and resorbed for adult existence; but that, eventually, in progressive chordates, the new somatic unit was retained throughout life. (1972, pp.121–122)

Romer marshalls his evidence that wherever organ-systems or tissues cross the somaticovisceral 'divide', they assume differences of character consonant with separate origins. For example, the muscles of the gut and the vascular system (the heart and blood vessels) are smooth, derived from the splanchnic mesoderm. Those of the outer parts of the body are striated, derived from the somitic mesoderm as a fundamental feature of segmentation:

> These structures [i.e. the somites] give the first and major indication of such segmentation as a vertebrate possesses (and, in fact, the segmentation of vertebrae, ribs and spinal nerves occurs as a fundamental adaptation to this primary segmentation). (1972, pp.122–123)

Unlike the trunk, the head appears to contain relatively little in the way of somites. Some anterior trunk myotomes send extensions round the back of the gill region to form part of the throat and tongue musculature, and there are three pairs of anterior outliers that give rise to the three pairs of (striated) eye muscle, innervated by cranial nerves III (oculomotor), IV (trochlear) and VI (abducens). So, one can winnow the somatic from the visceral on the basis of the striation of their constituent muscles. But this could say more about present circumstance than the history of vertebrate body form.

However, there is an exception that proves the rule: the striated muscles that move the jaws and gill-bars of fish (which persist in the neck, ear and face in land vertebrates) derive not from myotomes, but from splanchnic mesoderm, just like the smooth muscle that lines the gut and blood vessels. In functional terms, there is no reason why somite-derived muscle from the trunk could not have done the same job (after all, somitic mesoderm makes it up as far as the tongue, and behind the orbits). These pharyngeal muscles seem to be a set of visceral muscles, specialized to operate structures associated with the front end of the gut. They are striated simply because striated muscles would do the job better than smooth ones, in spite of their visceral origin.

Bones and cartilages, like muscles, are mostly mesodermal in origin, either from the sclerotome (the axial skeleton and the back of the braincase) or the lateral plate (ribs and limbs). But the skeleton that supports the throat (gill bars) the face, the sensory capsules, the skull roof and parts of the front and undersurface of the braincase are ectodermal in origin. They originate, of course, in neural crest, that quintessentially vertebrate tissue.

Again, there is no reason why these structures could not have arisen from ordinary mesenchyme, as does the axial skeleton – the fact they do not, says Romer, suggests a fundamentally different origin, and more evidence for a fundamental division between somatic and visceral:

> All in all, it is clear that the development of the visceral skeleton from mesectoderm [i.e. neural crest] is part of an ancient and basic pattern in vertebrate development and history. And this fact, combined with the likewise distinctive nature of the visceral muscles associated with this visceral skeleton [see above], bring forcibly to mind the prominence of the pharynx in the lower chordate relatives of the vertebrates. Of special interest is the fact that in amphioxus, tunicates and even acorn worms, in which there are no other skeletal structures of any sort, cartilage, or procartilage, is developed to support the gill region. The visceral skeleton is ancient in its origins – the oldest skeletal structure, in chordate history, far antedating the mesodermal skeletal elements evolving in the 'somatic' animal. (1972, p.131)

This sentiment goes back to Bateson (1886). The problem with this interpretation is that although many lower chordates (and enteropneusts) have distinctive pharyngeal skeletons, only vertebrates have neural crest. It could be that the migration of neural crest into the branchial region during development runs along very old rails, replacing branchial mesenchyme which, although not ectodermal in origin, is different from the mesenchyme characteristic of skeletal tissues in the 'somatic' part of the animal.

Much of Romer's argument concerns nerves, the ways in which they join the spinal column, and how they innervate the rest of the body. Romer follows the classification of nerves into four types[74], which are segregated in space as well as differentiated in function.

Motor cell bodies are concentrated in the ventral edge of the spinal cord, whereas sensory axons aggregate more dorsally, their bodies being housed in ganglia external to the cord itself. The dorsoventral sequence runs from (1) to (4) (see note 74), so that visceral origins are grouped, along with sensory or motor function. The same sequence holds for the brainstem, which 'strongly suggests that the somatic-visceral contrast is an ancient one in the history of the vertebrate nervous system' (1972, p.135).

The peripheral nervous system is even more interesting. Whereas somatic motor nerves innervate their target muscles directly, visceral motor nerves must do so via an intermediary. Before it reaches its target, a preganglionic visceral motor nerve penetrates a ganglion of visceral-nerve cell bodies. Here the preganglionic fibres make contact with postganglionic nerves that carry the signal to its destination. These postganglionic nerves may patch into the diffuse nerve nets that patrol various parts of the intestine, and which may use different neurotransmitters from those found in the central nervous system. In many ways, visceral ganglia are the border checkpoints between two very different countries.

Romer speculates that the deep-lying nerve networks, associated with the gut, represent the primary nervous system of the 'visceral' animal. The autonomic nervous system, with its elaborate relay system, is the attempt by the 'somatic' animal to make contact and exert power over it.

Textbooks show that sensory nerves enter the spinal cord through the dorsal root, whereas motor nerves (somatic and autonomic) leave through the ventral root. This may be true in mammals, but the routing of autonomic nerves has had a chequered past: even in mammals, a few autonomic fibres exit via the dorsal root. The proportion is greater in fishes, and apparently indiscriminately dorsal or ventral in lampreys. In fact, four of the cranial nerves in lampreys – V (trigeminal), VII (facial), IX (glossopharyngeal) and X (vagus) – lack separate ventral motor roots altogether, and spinal nerves in the amphioxus all appear to be dorsally rooted – there are no separate ventral roots for the motor nerves at all. This may stem from a day when dorsal and ventral routes were not lined up, but staggered, so that a ventrally placed, somatic motor root could penetrate a myotome, whereas a root containing the somatic sensory, visceral sensory and autonomic nerves could weave its way between the segments. This root would be dorsal, to keep well clear of the myotome.

Interestingly in this context, although all the roots in the amphioxus appear to correspond to a mixture of sensory and motor fibres as found in lampreys, there appear to be no correspondents with the purely sensory dorsal roots as seen in vertebrates generally. Based on this, Fritzsch and Northcutt (1993) propose that cranial and spinal nerves in vertebrates are independently derived serial homologues of elements in an amphioxus-like ancestral pattern, separation of which was facilitated by the evolution of the neural crest.

Romer speculates that the peripheral nervous system was originally bipartite – deep, diffuse nerve nets in the viscera, as yet 'unaware' of the central nervous system above sending somatic motor nerves to the myotomes and receiving somatic sensory input from nerves coming in between the segments. Only later on would the hand of colonialism extend downwards, first as scouts on missions to monitor conditions (visceral sensory), then as administrators to direct operations (autonomic system). With the system of visceral ganglia, the form of government is more like a protectorate than a colony.

The distinct but subtle demarcation of the nervous system is also seen in the head, in the disposition of the cranial nerves. Despite the complexities imposed by the presence of the brain and organs of special sense, the cranial nerves that are clearly ventral, such as III (oculomotor), IV (trochlear), VI (abducens) and XII (hypoglossal), are all somatic motor nerves that innervate myotome derivatives; or clearly dorsal, such as V (trigeminal), VII (facial), IX (glossopharyngeal), X (vagus) and XI (spinal accessory), all of which contain components from the other three classes of nerve; somatic sensory, visceral sensory and visceral motor (and, as discussed above, these nerves lack ventral roots altogether in lampreys). The separation between dorsal and ventral nerves, rather than their agglomeration into shared roots, may tell of a more primitive, 'visceral' organization than that seen in the trunk.

There is a further difference. Whereas the orderly arrangement of trunk nerve roots is, to an extent, dictated by the myotomes, in the head it is the visceral skeleton of the gills that seems to call the tune.

This is an important distinction, as it contrasts with the Balfourian scheme in which the arrangement of nerves in the head is interpreted as having been dictated by somites, just as it is in the trunk. In his critique of Goodrich's views (Goodrich, 1918), Romer writes:

> It may happen by chance that during development some one gill bar and its musculature may lie below some one specific myotome and its derived musculature. But there is no *a priori* reason to think

> that the two segmental systems – one basically mesodermal and related to the 'somatic' animal, the other basically endodermal, 'visceral' in origin – have any necessary relationship to one another. Gill slits and somites arise quite independently of one another in the embryo. And phylogenetically, one may note that a gill slit segmentation is highly developed in acorn worms, in which the somite system is not developed at all. (1972, p.141)

Bateson would have approved. He saw a clear separation between the segmented trunk and a head that is essentially unsegmented but for the branchial skeleton, which is seen, in any case, in the unsegmented enteropneusts. His sentiment presages that of Romer almost word for word:

> No doubt the cranial nerves may, by arbitrary divisions and combinations, be shaped into an arrangement which more or less simulates that which is supposed by some to have been present in the rest of the body, but little is gained by this exercise beyond the production of a false symmetry (Bateson, 1886, p.562)[75]

In phylogenetic terms, then, the Romerian chordate precursor was essentially a visceral animal, with an epidermal and a visceral nerve net, but no central nervous system as such. The story of chordate evolution is, in Romer's view, the story of the evolution of the somatic component, and its gradual domination over the visceral.

Recent work on the development of vertebrates tends to support the existence of an old division between (visceral) 'head' and (somatic) 'trunk'.

Work on mice and chick embryos shows that, at a particular point in embryogeny, the hind portion of the brain, contiguous with the spinal cord, is divided, by visible constrictions, into bulges or segments called rhombomeres. These bulges are more than just superficial: the nerve cells in each rhombomere develop from a population that is specific to that rhombomere (Lumsden and Keynes, 1989). The compartmentalization of rhombomeres is morphologically analogous to the segmental compartments of insect bodies.

These bulges continue down the spinal cord into the trunk, where they are called myelomeres, rather than rhombomeres, which strictly refers to the anterior part of the hind brain called the rhombencephalon. Differences between myelomeres and rhombomeres are revealed in the way they relate to surrounding tissues, at least in work on the domestic chick (Storey *et al.*, 1992).

Chick rhombomeres seem to have an inherently segmental character, in that they influence the segmental specification of other tissues nearby, such as those destined to form parts of the branchial arches. Chick myelomeres, on the other hand, are moulded by the arrangement of somites on either side. Until the somites rein them in, the young nerves seem to emerge from the spinal cord more or less at random (Lim *et al.*, 1991).

In chicks then, the nerve cord is patterned according to different sets of rules depending on the region through which it passes, supporting the idea that the head and trunk are discrete structures, each with its own developmental tradition[76].

Graphic confirmation of a sharp divide between head and trunk comes with the discovery that mice lacking functional copies of a gene called *lim1* are born perfectly normal, with beating hearts and complete notochords – but cleanly lacking all parts of the head anterior to the otic vesicle (Shawlot and Behringer, 1995). This finding is interpreted as confirmation of Spemann's classic embryological work on the presence of head-specific factors in special parts of the early embryo, 'organizers', to induce body axes when transplanted (De Robertis, 1995). But could it be that *lim1* is a genetic marker for an ancient division between heads and tails?

3.4 A NEW HEAD

The vertebrate body plan, then, does not seem to be governed by any intrinsic segmentational scheme that pervades the entire animal. The head, in particular, can be viewed as a distinct and separate structure, not derived by the simple elaboration of the front end of an amphioxus-like animal[77].

The clearest (and most extreme) view along these lines is that of Gans and Northcutt (1983), who propose that vertebrates have 'a new head', derived largely from neural crest and epidermal placodes and their interactions with other tissues. Gans and Northcutt show that many of the features we consider as distinctive of vertebrates relate to neural crest, and, in their view, a distinct head region (Chapter 1). Tunicates, the amphioxus and other animals lack neural crest, and so do not have the distinctive vertebrate head.

3.5 HEAD SEGMENTATION

As noted above, though, the work of Balfour and his school suggests that the vertebrate body plan is rather different. Rather than being divided into, essentially, a 'head' and a 'tail', or 'soma' and 'viscera', Balfourian 'segmentationists' see the body, including the head, as inherently segmented from one end to the other, and assert that it can be derived by the elaboration of an amphioxus-like ancestor.

In other words, the anterior end of the amphioxus corresponds with the anterior end of the vertebrate head, and not with some point in the middle such as the boundary between the hindbrain and midbrain.

The evolution of the neural crest has allowed the vertebrate head to become more complex than the front end of an amphioxus, although the two are positionally equivalent. This conflicts with the Gans and Northcutt notion of a 'new head' which, in the vertebrate, has no homologue in the amphioxus.

Much of this 'segmentationist' work has been confirmed and even extended by electron-micrographic, embryological and histological studies (reviewed by Schaeffer, 1987). It is clear that although the head lacks significant amounts of somitic mesoderm, the mesoderm in the head is divided into somite-like units or 'somitomeres' (Meier and Packard, 1984; Jacobson, 1988).

In this scheme, amphibians have about 6 to 9 head segments, but the number of somitomeres after the first (which underlies the prosencephalon, the anterior portion of the brain) is doubled in amniotes: each somitomere may represent a subdivision of an ancestral amphibian segment[78]. Moreover, the presence of segmentation could be homologous with the presence of segmentation in the anterior parts of amphioxus, even if there is no clear correspondence between particular segments. The fact that the vertebrate head is segmented supports the view that it can be derived by simple elaboration of the front-end of an amphioxus-like animal.

Support for this model comes from work on the zebra fish *Brachydanio rerio*; segmentation of the neural tube as well as the brain is defined from an early stage and is not influenced by the somites (Eisen *et al.*, 1986). Gene expression studies in the amphioxus (Holland *et al.*, 1992a, 1994a; Garcia-Fernàndez and Holland, 1994) suggest that the anterior region of the animal is indeed cognate with the vertebrate head.

These results clearly differ from the work on chicks, which demonstrates a marked difference between the way segmentation is regulated in the head and trunk, respectively (see above), and may reflect differences between development in non-amniote vertebrates (such as fish and frogs) on the one hand, and amniotes (of which the chick is one) on the other.

So which is right? Is the vertebrate a fundamentally bipartite creature, divisible into 'somatic' and 'visceral' parts? Or is vertebrate somitic segmentation a primitive feature, shared with the amphioxus? Recent genetic work has the potential to shed light on these questions.

3.6. GENETIC PATTERNS[79]

This fly's portrait (Figure 3.3) is to geneticists what the Mona Lisa is to art historians – an icon that evokes mystery as well as being something of a *cliché*.

Figure 3.3 A fruit-fly *Drosophila*, with an extra pair of wings, a result of a mutation in the *bithorax* complex of homeotic genes.

It is a specimen of the fruit fly *Drosophila melanogaster*, with two pairs of wings instead of the usual one. As the result of a genetic mutation, the third segment of the thorax, which usually bears a set of small balancing organs, has been transformed into a copy of the second thoracic segment, which usually bears a pair of wings.

The result is a fly with two copies of the second thoracic segment, and therefore two pairs of wings. The process whereby one structure assumes the 'identity' of another is called homeosis, from a Greek word meaning 'becoming like'. This mutation, which is called *bithorax*, is said to be 'homoeotic', in that through its agency, one structure (in this case, the third thoracic segment) becomes like another (the second segment). Another well-known homeotic mutation is *antennapedia*, in which small leg-like structures develop on the head in the place of antennae. These mutations are manifested as anomalies in the positioning of structures (legs, wings) with respect to particular segments. However, they do not affect the formation of the segments themselves, and so tell us nothing about the process of segmentation *per se*.

The genes which, when defective, lead to homeotic mutations, are now known to encode transcription factors – proteins that regulate the activities of other genes (either by binding to specific regions of DNA or by assisting that binding process), which may be structural or regulatory.

But homeotic genes also control the regulation of other homeotic genes, as well as being regulated themselves by other sets of transcription factors, encoded in their turn by other sets of genes. Thus is introduced a complicated series of feedback loops that allow the tight control of the timing and order of transcriptional regulation – in other words, to maintain the integrity of the developmental programme as a whole.

3.6.1 Recipe for a fly

The 'big picture', therefore, of fruit-fly development seems to be of a network or 'cascade' of genetic interactions that result in the formation of a mature fly from an egg, segments and all – and a fly in which there is a place for everything, and everything in its place, time after time.

The process starts even before fertilization. The expression of around 30 maternal genes impresses upon the egg the anterior–posterior and dorsoventral body axes of the future fly. Later on, other genes known as 'gap' and 'pair-rule' genes determine the position of segments in the developing grub, and the polarity of the segments (that is, which end of the segment is anterior, and which posterior, relative to the whole grub)[80]. These genes restrict and determine the boundaries of segments. The actual 'identity' of the segments, manifested by the nature of the structures that it will produce (legs, wings, mouthparts and so on) is determined by the homeotic genes.

Homeotic genes, therefore, do not govern the formation of particular structures, but perform a rather more abstract task: that is, they ensure that structures develop in their proper places consonant with segmental identity (or position with respect to the whole body axis), irrespective of the nature or function of the particular structures concerned[81].

3.6.2 The homeobox

Many homeotic genes share a distinctive 180-base-pair section of DNA that has come to be known as the homeobox (McGinnis *et al.*, 1984a,b; Struhl, 1984; reviewed by Akam, 1984, and Slack, 1984).

This sequence encodes a stretch of 60 amino acids that is folded into three tightly wound helices of a particular kind (called 'alpha' helices), I, II and III, arranged in line and connected by loops (see Riddihough, 1992 and references therein). The loop between II and III allows them to form a hairpin-like ('helix-turn-helix') structure such that helix III can anchor itself to specific sequences of nucleotides in the major groove of the DNA double-helix. This anchorage is the key to the function of proteins encoded by homeotic genes – they bind to the promoter regions of other genes, thus exercising control over their expression.

Homeobox genes, therefore, form a distinct group, membership of which means, by definition, the encoding of DNA-binding proteins with a particular kind of DNA-binding structure (the homeobox). Indeed, a wide variety of genes is now known to contain homeoboxes or homeo-box-like sequences (for a review, see Manak and Scott, 1994). Because these sequences denote involvement in DNA binding and thus the regulation of gene expression in the animal, it is no surprise that many of the genes concerned are involved in development. Their mutations are implicated in developmental defects ranging from embryonic abnormalities to cancer, particularly when some error of position or timing is implied.

Not all genes involved in development are homeobox-containing genes, of course. Many of them encode cell-surface receptors, or vital parts of the inter- or intracellular signalling pathway, whereby cells communicate and instruct one another concerning gene expression. Given that much of developmental elaboration involves the communication of signals by this route (the end result of which might, of course, involve gene transcription through the agency of a homeobox-containing gene), mutation in these genes can lead to developmental defects or cancer.

3.6.3 The homeobox cluster

The *antennapedia* and *bithorax* mutations of *Drosophila* are caused by defects in two of a closely related set of homeobox-containing genes. Several other genes belong to this group, each restricted in its domain of expression, and denoted in many cases by characteristic mutations in adulthood.

In addition to sharing many details of their structure apart from the homeobox (suggesting that they may have been generated by the duplication and subsequent divergence of a small number of ancestral gene sequences), they are all grouped together in two large clusters BX-C and ANT-C, collectively known as the HOM cluster.

Furthermore, the order in which the genes are arranged on the chromosome is the same as the order along the anterior–posterior axis in which they appear to be expressed in the segments of the animal: in general, the more 'downstream' the gene in the cluster with respect to the direction of transcription, the more anterior its anteriormost expression boundary[82].

Now, this may be nothing more than a coincidence, but it is conceivable that this correspondence really does have something to do with the positioning of structures along the anterior–posterior axis of the animal. It could be that natural selection has ensured that a set of genes of such fundamental importance to development is always kept together, and in a particular order, so that genes in the HOM cluster regulate one another as well as more distant genes, thus ensuring the proximity of each to each. Experimental evidence for this idea, though, is equivocal (González-Reyes *et al.*, 1990). Nevertheless, homeobox-containing genes may have several 'enhancers', each designed to interact with factors specific for the tissues in which the homeobox-containing gene is to be expressed (see, for example, Gould *et al.*, 1990; Whiting *et al.*, 1991). These other factors could, of course, be the products of other homeobox genes in the same cluster.

That the relationship between structure and gene order is more than coincidental is supported by the remarkable correspondence of sequence between certain genes in the HOM cluster and homeobox-containing genes in mice (known as *Hox* genes), as well as in their characteristically restricted domains of expression (Awgulewitsch *et al.*, 1986) and the one-for-one order in which these genes are arranged along the chromosomes in each case (Figure 3.4). Moreover, both arrangements correspond to the position at which each gene is expressed along the anterior–posterior body axis.

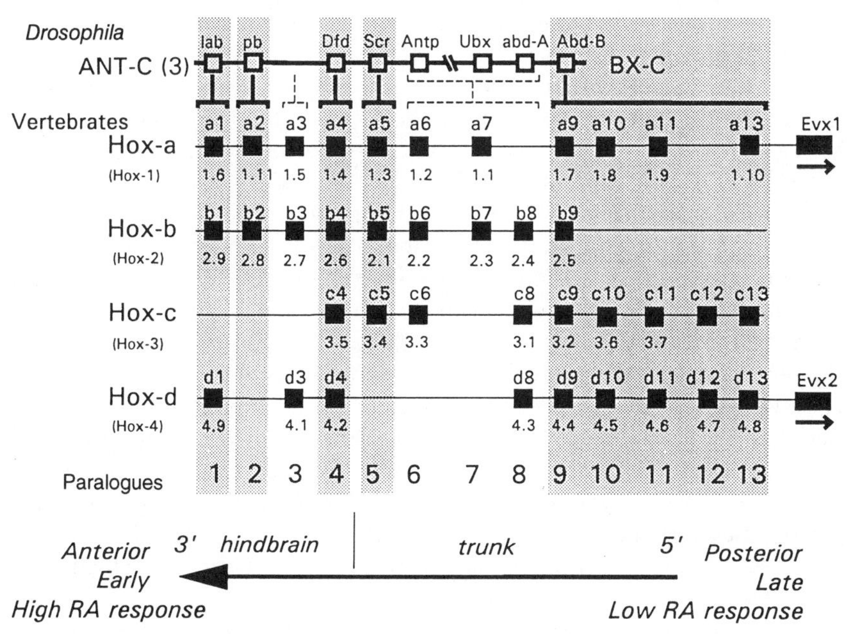

Figure 3.4 The relationship between the ANT-C and BX-C clusters (collectively HOM) and the four *Hox* clusters in mammals (from Krumlauf, 1994). Paralogues are lined up, one after the other. This diagram is a useful guide to the recently adopted nomenclature for the *Hox* genes (above each gene e.g. Hox-b, Hox-bl) and how it relates to the old (below each gene e.g. Hox-2, Hox 2.9).

Of course, there are substantial differences in organization between fly and mouse genomes. First, *Hox* genes in mice are more closely packed than their cognates in *Drosophila*. For example, the whole of the mouse *Hox-2* cluster – a group of about 11 genes – is about the same size as the *antennapedia* gene.

More importantly, mice have not one cluster but four, each on a different chromosome. Many of the homeobox-containing genes in each *Hox* cluster are closely related to genes in corresponding positions (and regions of phenotypic expression) in the other three clusters: so that a *Hox* gene in one cluster has up to three 'paralogues', one in each of the other clusters (Figure 3.4). There are 13 sets of paralogues (called 'paralogous groups'), with up to four members in each set, one from each chromosomal cluster. However, no cluster has the full complement of 13 genes. The paralogous groups are numbered from 1 to 13, where 1 refers to the anterior end of the animal (which in structural terms, confusingly, is the 3-prime rather than the 5-prime end of the cluster) The clusters evolved, presumably, through gene duplication and subsequent divergence from a small set of 'ancestral' genes in a single cluster (Holland *et al.*, 1994b; Ruddle *et al.*, 1994)

The relationship between particular mouse and fly genes is less clear. Although *Hox* genes are similar to HOM genes in a general sense, few one-for-one correspondences exist between individual fly and mouse genes. The fly gene *Abdominal-B* (*AbdB*), expressed in the posterior part of the animal, corresponds to not one but five sets of *Hox* genes (paralogous groups 9–13), also expressed in the hind parts of the mouse. The genes *abdominal-A* (*abdA*), *ultrabithorax* (*Ubx*) and *antennapedia* (*Antp*) are all similar to one another, and similar in a general sense with three cognate paralogous groups 6–8; *sex combs reduced* (*scr*) may be homologous with cognate group 5, and *proboscipedia* (*pb*) with cognate groups 2 and 3.

Overall, though, it seems as though the homeobox-containing gene clusters in mice and flies evolved from a much simpler arrangement in a common ancestor with a small number of homeobox–containing genes, perhaps five or six: reading from the 5 to the 3 direction, one might call these genes *AbdB*-like, *abdA/Ubx/Antp*-like, *Scr*-like, *Dfd*-like, *pb*-like and *lab*-like. This arrangement may have been derived from a cluster of about three genes early in metazoan evolution. These genes would have corresponded, broadly, with *AbdB*, *abdA/Ubx/Antp/Scr/Dfd* and *pb/lab* (Schubert *et al.*, 1993).

The one-to-one correspondences that do exist, however, are striking. *Deformed* (*Dfd*) and *labial* (*lab*) correspond, respectively, with genes in cognate groups 4 and 1 in such detail that their promoter regions[83] are interchangeable. Awgulewitsch and Jacobs (1992) showed that an enhancer sequence from *Dfd* generated expression of a reporter gene within the anterior expression domains of its cognate *Hox* genes in mice. The converse has also been demonstrated – mouse *Hox* (Malicki *et al.*, 1990) and human *HOX* genes (Malicki *et al.*, 1992) can specify segmental identity when transplanted into flies. Other remarkable correspondences of expression between mammals and flies are summarized by Manak and Scott (1994) in a review on the evolutionary conservation of homeobox gene function.

In mice as well as flies, then, HOM/*Hox* expression is confined to certain regions of the animal, notably excluding the anterior part of the head. In vertebrates, successive genes in the *Hox* cluster, reading from cognate groups 13 down to 1 (or from 5-prime to 3-prime) have expression patterns that extend, respectively, from the posterior end to well-defined, successively more anterior boundaries. As in *Drosophila*, these boundaries are segmentally defined, and tend to coincide with the constrictions that demarcate the boundaries between the rhombomeres of the hindbrain, a remarkable parallel with the strict (and more overt) 'compartmentalization' seen in *Drosophila* (Wilkinson *et al.*, 1989b; Lewis, 1989; Murphy *et al.*, 1989; Fraser *et al.*, 1990; Lawrence, 1990)[84]. The further anterior one travels along the body axis, the fewer *Hox* or HOM genes are expressed in a given segment. In *Drosophila*, HOM expression domains tend to be confined to particular regions or segments, although there are some areas of overlap.

In addition, the expression of a *Hox* or HOM gene tends to conform to a rule known as posterior prevalence, that wherever the expression domains of two genes overlap, the phenotype is often determined by the one with the more posterior domain of expression[85].

In *Drosophila*, for example, *Ubx* is expressed in the back part of the third thoracic and the front half of the first abdominal segment, whereas the domain of expression of *Antp* is more anterior, and extends throughout the third thoracic segment. When *Ubx* function is lost through mutation, *Antp* fills the gap, and thoracic structures appear in the first abdominal segment. Normally, the more posterior *Ubx* masks the expression of *Antp* in the first abdominal segment.

Posterior prevalence operates in mice, too: a copy of *Hox-d4*[86], normally expressed at the base of the skull, when driven by regulatory regions of the more anteriorly-expressed *Hox-a1* transforms what would have been occipital and basicranial parts of the skull into vertebrae (Lufkin *et al.*, 1992). *Hox-a1* itself is normally expressed far further forward, although its expression pattern is complicated by the effects of *Hox-a3*, expressed slightly more posteriorly and affecting different groups of cells (Lufkin *et al.*, 1991).

As for the head, so for the trunk; Kessel and Gruss (1991) show that varying patterns of *Hox* expression dictate the precise patterning of the vertebrae in mice, according to the rule of posterior prevalence. Mice with mutations in *Hox-b4* show homeotic transformations of the axis vertebra to the more anterior atlas, also in accordance with posterior prevalence (Ramírez-Solis *et al.*, 1993). Similarly, mice in which the thoracic *Hox-c8* has been deleted show defects consistent with posterior prevalence, notably the attachment of the 8th pair of ribs to the sternum, and the appearance of a 14th pair of ribs from the (normally ribless) 1st lumbar vertebra (Le Mouellic *et al.*, 1992). Ectopic over-expression of a human homologue of the normally more anterior *Hox-c6* has similar phenotypic effects in mice (Jegalian and De Robertis, 1992)[87]. And the list goes on: Krumlauf (1992a), McGinnis and Krumlauf (1992) and Manak and Scott (1994) review posterior prevalence in more detail, discussing its rules and exceptions.

Intriguingly, anterior boundaries of expression are not spaced evenly along the body axis in vertebrates: in mice, they cluster in the hind-brain area, and correspond initially to boundaries between successive rhombomeres. In mouse rhombomeres, the members of a paralogous group share the same expression pattern in a general way[88]: perhaps the placement of all these boundaries, in the hind-brain area, marks the boundary between the 'head' and 'trunk' of a bipartite animal (Lonai and Orr-Urtreger, 1990; cited by Holland, 1992).

It has been suggested that the compartmentalization of rhombomeres, and the controlled specification of structure in each case, is connected with a unique 'combination' of *Hox* expression patterns imposed by these staggered anterior boundaries. This combination is sometimes referred to as a 'code' whereby cells receive their cues to develop in a certain way in accordance with their position along the body axis (see Krumlauf, 1992a for a discussion of the concept of *Hox* codes).

When neural crest cells migrate from the dorsal edge to form the face and branchial region (see Meier and Packard, 1984, for the classic embryological study of neural crest migration), they take this *Hox* expression pattern with them, so it should be possible, in principle, to 'match' branchial somites with particular rhombomeres by virtue of their *Hox* 'codes' (Figure 3.5)[89]. Neural crest cells in rhombomeres 2, 4 and 6 populate the first, second and third branchial arches, respectively; crest cells in rhombomere 3 and 5 die soon after emergence, responding to signals from adjacent rhombomeres (Graham *et al.*, 1993). In this way, *Hox* patterns that originate in the neural crest are imposed on a wide range of neural-crest-derived and other tissues.

Interestingly, neural crest tissue in the hind-brain is organized into streams by mechanisms intrinsic to the neural epithelium, before it enters the branchial region. This contrasts with the trunk, in which segmental accumulation of neural crest cells is governed by the local mesodermal environment (Graham *et al.*, 1993).

Similar intrinsic mechanisms also determine the expression of *Hox-a2*: although the most rostrally expressed *Hox* gene, its expression boundary in the neural tube is staggered with respect to that in crest-derived cells, in a way that contradicts the idea of *Hox* coding. Although the gene is expressed in rhombomere 2, it is not expressed by neural-crest cells derived from this region once they get to the first branchial arch. However, the gene is expressed in both neural tube and crest-derived cells at the level of rhombomere 4 (Prince and Lumsden, 1994). When discussing the mechanisms underlying this finding, Prince and Lumsden (1994) note that disjunct expression is more common in the trunk, except that in the case of *Hox-a2*, the decision to switch off expression of the gene in the first branchial arch is intrinsic to the pre-migratory cell population, not imposed by pre-existing branchial tissue.

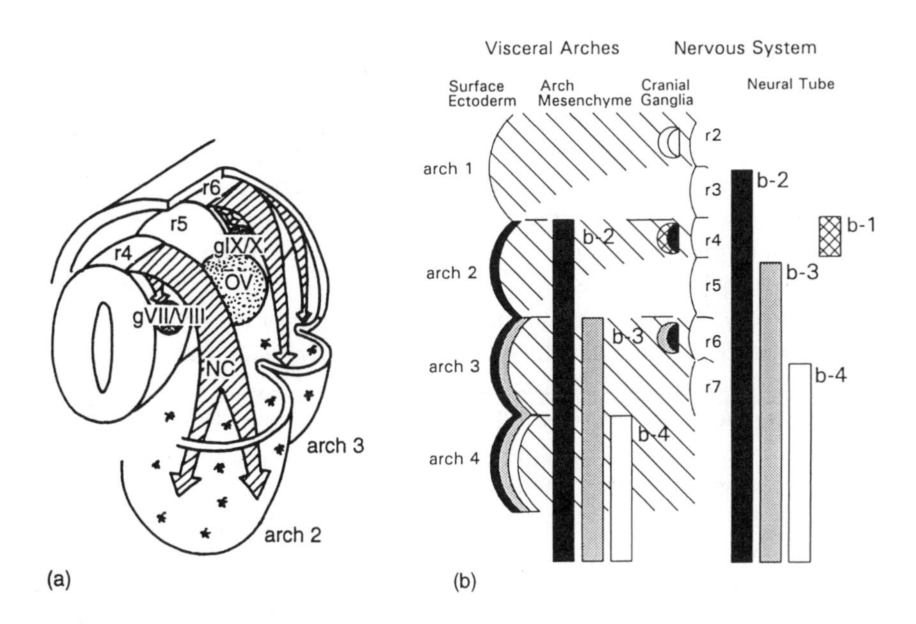

Figure 3.5 Diagram showing migration of cranial neural crest and patterns of *Hox-2* (Hox b) expression. (a) illustrates the process in a mouse embryo between 8 and 9 days after fertilization. Mesenchymal neural crest (large hatched arrows labelled NC) migrates into the branchial arch, followed by neurogenic crest (small hatched arrows, culminating in the dark circles, which represent ganglia (gVII/VIII for facial–acoustic sensory complex; gIX/X for glossopharyngeal/vagus complex), arch 2, arch 3 = branchial arch, OV = otic vesicle, r = rhombomere. (b) The *Hox-2* (Hox b) code in the hind-brain and branchial arches. The areas of the neural plate where the neural crest is generated and the branchial arch into which it migrates are shown by the cross-hatching. Note that expression boundaries in the rhombomeres vary with a two-segment periodicity, with the exception of *Hox-2.9* (Hox-b1), which is a *labial* subfamily member, like *Hox-1.6* (Hox-a). Because of the presence of regions with little or no crest (r3, r5), the expression pattern in the branchial arches is offset from that in the neural tube. According to the revised nomenclature for Hox genes, Hox-2.8 = Hox-b2, Hox-2.7 = Hox-b3, Hox-2.6 = Hox-b4 (from Hunt and Krumlauf, 1991b).

The expression and functions of *Hox* genes can now be demonstrated experimentally using strains of so-called 'knockout' mice in which specific genes are damaged or deleted by a technique called 'gene targeting', in which selected genes are displaced by homologous recombination with introduced segments of DNA containing null mutants of the genes (Wright and Hogan, 1991). Chisaka and Capecchi (1991) used this method to create a line of transgenic mice carrying a null mutation for *Hox-a3*, which has an anterior limit of expression between rhombomeres 4 and 5, just anterior to the otic vesicle. Mice homozygous for the null mutation die at birth, with a spectrum of abnormalities: they lack the thyroid and parathyroid glands, have defects in a variety of tissues in the throat, and may also have defects in the heart and arteries, as well as craniofacial abnormalities.

This research demonstrates that a *Hox* gene, initially expressed in the central nervous system, exerts a wide-ranging influence through adjacent neural-crest-derived tissues. However, the somewhat disjunct range of disorders suggests that the missing gene is, in part, being 'covered' by its paralogues, so 'true' homeosis may not have been demonstrated. Technical improvements have allowed the obvious solution – the simultaneous knock-out of two or more paralogues. Indeed, Condie and Capecchi (1994) show that simultaneous knockout of *Hox-a3* and *Hox-d3* has a synergistic effect, producing deficits more marked than the sum of either produced in separate strains.

On the face of it, these results tend to support the 'segmentationist' view of a fundamental metamerism in vertebrate structure, in that *Hox* gene expression is the agency responsible for the correspondence between gill arches and cranial nerves, rather than their juxtaposition by happenstance.

On the other hand, the migration of *Hox*-coded neural crest tissue from neural tube to craniofacial structures might reflect nothing more than the secondary imposition of segmentation on a fundamentally bipartite plan, reflected by the staggered boundaries of *Hox* gene expression domains in the hindbrain. After all, *Hox* expression and neural crest are both 'somatic' inventions, imposed on a pre-existing 'visceral' pharyngeal chassis as found in enteropneusts, which have pharyngeal slits but not neural crest.

Second, the versatility of *Hox* coding decreases as one moves anteriorly, because progressively fewer *Hox* domains extend that far forward. There comes a point, within the hind-brain, where no more *Hox* genes are expressed. If there are serially repetitive structures anterior to this (other than those which can be derived from the migration of neural crest cells), they must be specified by other genetic systems (see below).

Given that this anterior boundary of *Hox* expression corresponds (roughly) to the boundary between 'head' and 'trunk', and the fact that *Hox* expression becomes richer and more varied towards the posterior end of the axis, one might suggest that regional specification by *Hox* genes is a feature of the 'somatic' part of the animal, and is alien to the 'head' or 'visceral' part. The 'somatic' part has moved forward and over the top of the 'head' part during evolution, thinning in potency with increasing distance from the tail end, as shown both by the progressive anterior poverty of *Hox* gene domains, as well as by the phenomenon of posterior prevalence.

This idea receives added support from studies of the trunk. As shown above, *Hox* expression is a determinant of vertebral as well as cranial structure, in that it ensures that the correct type of vertebra develops at the appropriate point along the backbone. There is a difference, though. As we saw, vertebrates seem to have four clusters of Hox genes, such that a gene in a given cluster has paralogues in the other three clusters. In the head, a given gene has a domain of expression that is identical with that of each of its three paralogues (Hunt *et al.*, 1991a). This is not the case for the trunk, where paralogues may have different domains of expression[90] (Kessel and Gruss, 1991).

What does this mean in terms of evolution? Holland (1992) suggests that the degree of similarity between the large-scale structures, functions and expression domains of related genes is a measure of phylogenetic relatedness, in the same way that similarity of gene sequence can be used to infer phylogenetic pattern. For example, it is reasonable to suppose that the four sets of *Hox* genes in mice are *prima facie* evidence for multiple gene duplication events somewhere in animal phylogeny, after the ancestral mouse diverged from that of *Drosophila*. Once genes are duplicated, one would suppose that each copy would, on its own account, acquire its own mutations, and its sequence would gradually diverge from that of its fellows and from the common ancestor. The same may be true of function and expression domains, as each copy comes under the various influences of extraneous genetic factors (such as transcriptional control). The longer that each paralogue is independent, the more chance it has to acquire its own peculiarities.

The fact that the paralogues of *Hox* genes in the head have congruent expression domains, whereas those in the trunk do not, may suggest that *Hox* expression has been a feature of the trunk longer than the head. This, in turn, may be further evidence that the control of segmental specification by *Hox* genes is a characteristic 'somatic' feature that has subsequently been adopted by the 'head'.

In flies, as in mice, HOM expression is absent in the more anterior parts of the head. But the concept of a bipartite vertebrate ancestor is meaningless when applied to *Drosophila*, so how do the two patterns of expression come to be so similar? It seems likely that homeobox genes once fulfilled functions quite different from segmental specification, and have been co-opted subsequently for this purpose, independently in the ancestries of *Drosophila* and the mouse.

Common ancestry is part of the reason, as there seems to be evidence that orderly clustering of homeobox-containing genes is a feature of metazoan organization in general[91], and certainly antedates the divergence of the insect and vertebrate lineages (and, by implication, those of protostomes and deuterostomes).

One such cluster has been isolated from the nematode worm *Caenorhabditis elegans*. This unassuming creature has four HOM-like genes, *ceh-13*, *ceh-15*, *mab-5* and *ceh-11*, arranged in that 3 - 5 order on a single chromosome, and (as far as is known) expressed, correspondingly, from anterior to posterior, just like HOM genes in *Drosophila* or *Hox* genes in mice. The four genes appear to resemble *Drosophila* genes *labial*, *Deformed*, *antennapedia* and *Abdominal-B*, respectively. There are estimated to be at least 60 homeobox-containing genes in *Caenorhabiditis* (Bürglin *et al.*, 1989), although not all of them will fall into the HOM class.

The discovery of this cluster (Bürglin *et al.*, 1991) caused a stir (Kenyon and Wang, 1991), especially as its function was not clear: *Caenorhabditis* has neither a discrete head nor any segmentation apart from the most superficial kind of annulation, so even if the cluster is involved in the definition of spatial domains, it need not be involved in segmental specification. Furthermore, *Caenorhabditis* development follows a strictly predetermined pattern such that the ancestry of every adult cell can be traced back to the egg. Mouse and fly development are very different. In mammals, certainly, the determination of structure in embryogeny depends on the interaction between cells[92], rather than the identities of the cells *per se*. However, the *Caenorhabditis* genes appear to be involved with regional specification (mutations in *mab-5* result in the loss of structures in the posterior part of the body) and conform to posterior prevalence.

Homeobox-class genes have been isolated from a number of animals considered to be primitive, such as flatworms and cnidarians. A gene related to the *Drosophila* homeobox-containing pair-rule gene *even-skipped* has been isolated from a coral, *Acropora formosa* (Miles and Miller, 1992). Although a role in segmentation can be ruled out in this simple diploblastic animal, there is as yet no clue as to what function it performs.

However, there is some evidence that homeobox genes in the hydra *Chlorohydra viridissima* act in concert to determine axial structure, and play a part in the animal's legendary ability to regenerate its structure after amputation. Schummer *et al.* (1992) report the cloning of four *Chlorohydra* homeobox-containing genes which they call *cnox1* to *-4* (short for 'cnidarian homeobox'). Of these, the homeoboxes in *cnox1* and *cnox2* resemble those of *Drosophila* HOM genes, particularly *labial* and *Deformed*. The homeobox of a third, *cnox3*, is similar to that of *distal-less*, a gene which, in *Drosophila*, acts in concert with *wingless* to regulate limb development (Cohen, 1990)[93]. The genes are expressed in the regenerating crown of tentacles after amputation, *cnox1* peaking before *cnox2*, *cnox2* before *cnox3*. The *cnox4* gene is somewhat different, and resembles the *Drosophila muscle-specific homeobox* (*msh*), a motif found in a different class of homeobox-containing gene (see below). In contrast, two homeobox genes in another non-segmented animal known for its regenerative powers, the turbellarian flatworm *Dugesia tigrina*, do not seem to be involved in regeneration. Instead, they seem to be implicated in specific cell or tissue determination (Garcia-Fernàndez *et al.*, 1991, 1993).

Given that the results from *Caenorhabditis* and other simple, non-segmented animals suggest that the role of homeobox genes in overt segmental specification is likely to have been secondary, the original function of homeobox-containing genes is obscure. McGinnis and Krumlauf (1992) pick one among many: that they may have played a part in the patterning of metazoan nervous systems, but became co-opted for various other uses, and the link with metameric segmentation in arthropods and vertebrates was achieved independently in each case. Primordial links with the nervous system have also been invoked to explain expression patterns of the gene *engrailed* in a number of phyla (Patel *et al.*, 1989a, b; see below), and a variety of genes involved in neurogenesis in *Drosophila* have now, in addition, been implicated in mesoderm differentiation (Corbin *et al.*, 1991).

3.6.4 Beyond *Hox*

There are several different classes of homeobox-containing gene, many of which are involved in development, and which are present in a wide range of organisms (even unicells and plants). The unifying feature, the homeobox (or features like it) encodes for the DNA-binding motif of a protein involved in transcriptional regulation.

In addition, there are a number of other genes, found in many organisms, which although they do not necessarily contain homeoboxes, have equally profound influences on development, and which may interact with homeobox genes.

What follows is a very brief tour of a selection of some of those genetic systems that may be of importance in tracing the ancestry of the vertebrates (much of which is reviewed in detail by Holland, 1992, and Manak and Scott, 1994) – more of a taster than a comprehensive survey, as this kind of information dates quickly, and the reader will be able to fill in quite a few gaps.

The *engrailed* (*en*) gene in *Drosophila* contains a homeobox. It is expressed in the posterior portion of each developing segment, and helps define and maintain the position of segment boundaries. Later, it is expressed in parts of the central nervous system. A second gene called *invected* lies adjacent to *engrailed*. The two are very similar and have similar expression patterns: they also share an intron not seen in vertebrate, brachiopod and echinoderm versions of *engrailed*, suggesting that *invected* is the result of an arthropod-specific duplication event. Vertebrates from hagfish upwards seem to have at least two *engrailed*-like genes, although the lamprey and invertebrates such as sea-urchins, brachiopods, grasshopper *Schistocerca americana*, crayfish *Procambarus clarki*, lobster *Homarus americanus*, leech *Helobdella triserialis* and roundworm *Caenorhabditis* may have just one each (details and citations in Patel *et al.*, 1989a,b and Holland, 1992). Expression of *engrailed* in the grasshopper, crayfish and lobster, both with respect to neurogenesis and the stripe-like patterns in the body, are indicative of its segment–polarity function, similar to that in *Drosophila*. Later work revealed segmental *engrailed* expression in the polychaete annelid *Eisenia foetida* (Patel *et al.*, 1989b), and also that the leech has a tolerably segmental pattern of *engrailed* expression (Lans *et al.*, 1993; *contra* Patel *et al.*, 1989b), as does the chiton, a primitive mollusc (Jacobs and De Salle, 1994). But in vertebrates, at least, *engrailed* expression has nothing to do with segmentation of the body.

The genes are expressed in a wide variety of tissues in vertebrates, but all vertebrates examined express the gene or genes in the junction between mid-brain and hind-brain, anterior to the expression domain of the *Hox* cluster. Indeed, this is the only common region of expression for *En-1* and *En-2*, the mouse versions of the genes. As with other animals, *engrailed* expression is linked with neurogenesis and the developing neural system. Holland (1992) suggests that this conservation reflects an ancestral expression site in the mid-brain–hind-brain boundary, with the duplication and diversification in the vertebrate lineage offering opportunities for co-option of function in other tissues. The origin of vertebrates seems to have been marked by the duplication and diversification of *engrailed*-like genes primarily involved in neuronal development, with their subsequent co-option for a variety of purposes.

As its name suggests, *even-skipped* (*eve*) is a 'pair-rule' gene, one of several that lay down the segmental organization in the *Drosophila* embryo. A homeobox-containing gene, it is expressed at later stages in parts of the nervous system and also in the extreme posterior end of the embryo. At least two homologues are known from mammals, *Evx-1* and *Evx-2*, and one, *Xhox-3*, from the frog *Xenopus laevis* (Ruiz i Altaba and Melton, 1989a, b). In humans, the two *Evx* genes seem to be associated with the front end of two of the four *Hox* clusters. Based on this prediction, they should be expressed more posteriorly than any of the *Hox* genes, and this seems to be borne out in embryo mice.

The *wingless* (*wg*) gene does not contain a homeobox, although its function seems to be, in part, to control the expression of genes that do. The product of the gene in *Drosophila* is a diffusible signalling molecule which interacts in a complex way with the product of the homeobox-containing *engrailed* and other genes to specify the position of boundaries between embryonic compartments (see Ingham and Martinez-Arias, 1992, and Jessell and Melton, 1992, for reviews). Components of the *wingless* signal pathway in *Drosophila* turn up in vertebrates in other contexts: a vertebrate homologue of the gene *hedgehog* (*hh*), known as *sonic hedgehog*[94] is involved in the patterning of the limb (Riddle *et al.*, 1993; Smith, 1994; Fietz *et al.*, 1994; Blair, 1994). In flies, embryonic lethal mutants of *wg* result in an embryo without segmental boundaries, as well as changes in the nervous system. Less lethal mutations leave the flies to survive to adulthood, occasionally with defects such as the absence of wings, hence the gene's name.

In 1982 a gene was described in mice which, when mutated by the insertion of the mouse mammary tumour virus (MMTV), led to cancer. By virtue of its role in insertional mutation, it was named *int* (Nusse and Varmus, 1982). Later it was realized that *int* and *wg* were homologues, and this is reflected in the name of the ever-growing family of which they are the archetypes, the *Wnt* (*wingless*+*int*) family. Apart from mice and *Drosophila*, *Wnt* genes have since been found in all classes of vertebrate, including agnathans, as well as in annelids, echinoderms and the nematode *Caenorhabditis*.

Since their discovery, *Wnt* genes have attracted attention by virtue of the striking nature of the mutations with which they are associated. Defects in the mouse gene *Wnt-1* (as *int* was renamed) range from the lack of the anterior part of the cerebellum (with consequent motor dysfunction) to the absence of the entire cerebellum and parts of the mid-brain, the rest of the central nervous system remaining eerily undisturbed (McMahon and Bradley, 1990). Mice with these latter defects die at birth.

Wnt-1 is expressed in the most posterior part of the mid-brain in both mouse and zebrafish embryos, and also in the spinal cord. The spinal cord is unaffected by *Wnt-1* deficiency, which is presumably covered by other genes expressed in the region (some of which, such as *Wnt-3* and -*3A*, belong to the *Wnt* family). However, the gene is not expressed in the anterior part of the hind-brain from which the cerebellum is derived. The absence of the cerebellum in *Wnt-1* mutation may be a consequence not so much of the lack of the gene itself but of disruption of the expression of a homologue of *engrailed* (McMahon *et al.*, 1992). It could be that both genes are required to reinforce the expression of the other, just as in *Drosophila*.

Seven out of ten *Wnt* genes studied by Parr *et al.* (1993) show restricted, sharply defined expression patterns in the brain, perhaps explaining the mutant phenotypes observed in earlier studies. *Wnt-3*, *-3a* and *-7b* observe sharp boundaries in the fore-brain, whereas in the spinal cord, *Wnt-1*, *-3* and *-3a* are expressed dorsally, *Wnt-5a*, *-7a* and *-7b* more ventrally, with *Wnt-4* expressed both dorsally and in the floorplate.

Expression of *Wnt-1* and other members of the family in places where they may not normally occur ('ectopic' expression) has the equally remarkable property of inducing duplicate body axes in embryos of the frog *Xenopus* (Sokol *et al.*, 1991). However, remarkable as these effects are, ectopic expression cannot reveal whether axial specification is the bailiwick of *Wnt* genes *au naturel*, or if the ectopically expressed genes are simply substituting for different, unknown factors. Several different *Wnt* genes are expressed naturally in specific regions during *Xenopus* development, as they are in the development of every animal investigated, but do not seem directly involved in axial formation. It could be that as yet undiscovered *Wnt* genes are responsible, and that ectopically expressed *Wnt* genes are sufficiently similar to substitute for their activities.

The diversification of the *Wnt* gene family has been shown to track vertebrate phylogeny. In an analysis of 55 partial and 17 previously published *Wnt* sequences from a variety of vertebrates and invertebrates, Sidow (1992) shows that *Wnt-1* through *-7* were present in the common ancestry of arthropods and deuterostomes. However, duplication events involving *Wnt-3*, *-5*, *-7* and *-10* seem to have occurred on the lineage to jawed vertebrates, after the divergence of echinoderms – after which the rate of molecular evolution in vertebrate *Wnt* genes seems to have slowed dramatically. This may reflect constraints imposed by increasing complexity of an interactive regulatory network.

Vertebrate *Wnt* genes are generally expressed in dorsal, neural contexts, whereas in *Drosophila*, *wg* is expressed in more ventral, ectodermal settings. As discussed in Chapter 2, differences like this have been used in support of an arthropod/annelid inversion (Arendt and Nübler-Jung, 1994; Nübler-Jung and Arendt, 1994) as well as the auricularia theory (Lacalli, 1995).

Genes in the *Wnt* family (and others) initiate and regulate mesoderm induction or segmental specification through downstream agents such as homeobox genes. Another interesting determinant of anterior–posterior specification inasmuch as it concerns the notochord is called *brachyury*, required in the mouse for proper notochord formation (summarized by Ruiz i Altaba, 1991). Mice with progressively subnormal amounts of *brachyury* product show increasingly severe truncations of posterior regions of the body. The gene has been cloned in *Xenopus* and the zebrafish *Brachydanio*, where it seems to be required for notochord and early-mesoderm formation in response to induction. Intriguingly, *brachyury* is also expressed in posterior parts of the *Drosophila* larva. In summary, *brachyury* seems to be essential for the specification of the posterior parts of the body, and also for posterior extension of mesodermal structures such as the notochord (Yamada, 1994). Indeed, *brachyury* expression is a common theme in vertebrate gastrulation, and may serve as a marker for the definition of stages within this process common to all vertebrates (De Robertis *et al.*, 1994). Two copies have recently been cloned in the amphioxus (Holland *et al.*, 1995), where it appears to serve a similar function as in vertebrates. It has also been cloned from tunicates (Yasuo and Satoh, 1993, 1994) where it has the explicit function of specifying the fates of cells destined to become notochord. It has also been found in echinoids, where it is expressed in mesenchyme (Harada *et al.*, 1995).

Pet-lovers have long suspected that there is a tendency for animals with white fluffy coats to be hard of hearing. This has some basis in fact: symptoms of deficit in mice with a mutation called *splotch* include patchiness of pigmentation (hence the name) as well as defects in the formation of the ear[95]. Precisely why colour and hearing are connected is unclear, but both are related to the migration of neural crest cells during development, and with the activity of a gene called *Pax-3* (Epstein *et al.*, 1991).

This gene appears to be defective in a human condition called Waardenburg's syndrome that accounts for up to 3 in 100 cases of hereditary deafness; it, too, is associated with imperfect pigmentation, notably a blaze of white scalp or hair on the forelock. Other symptoms of *splotch* mice go far beyond ear and skin defects – they include gross neural tube defects such as spina bifida, and Epstein *et al.* (1991) suggest that defects in the human homologues of *Pax* may be responsible for at least some of the one in every thousand babies born with either spina bifida or anencephaly.

Pax-3 is one of a number of genes in the large and varied 'paired-box' group, found in a wide variety of organisms (Burri *et al.*, 1989). These genes have a distinctive DNA sequence, the 'paired' box in addition to a homeobox, although some just have the paired-box (Bopp *et al.*, 1986, 1989; reviewed by Manak and Scott, 1994): *splotch* mice have a genetic lesion actually within the paired box of *Pax-3* itself, a deletion of 32 nucleotides (Epstein *et al.*, 1991).

The archetypal *Pax* gene is the *Drosophila* developmental gene *paired*, one of the 'pair rule' genes along with others such as *fushi-tarazu* and *even-skipped*. However, possession of a paired box is no clue to function: neither *fushi-tarazu* nor *even-skipped* have it, although it is found in the segment-polarity gene *gooseberry* (Bopp *et al.*, 1986).

In vertebrates, *Pax* genes are expressed in the dorsal part of the neural tube and the hind-brain, at the stage in embryogenesis when the otic vesicle is forming. The expression domain of at least one member of the *Pax* family, the zebrafish *pax[zf-a]*, extends further forward, into the forebrain (Krauss *et al.*, 1991a), where *Pax* genes seem to be involved with brain regionalization in the early embryo – like *Hox* genes, their expression boundaries tend to coincide with morphological 'landmarks' such as nerve bundles (Krauss *et al.*, 1991b). Overall, however, they seem to be involved in the axial specification of structures connected with the neural tube: whereas defects in *Pax-3* result in failure of neural tube closure, defects in *Pax-1* (which is also expressed in the mesoderm) result in large-scale malformation of the entire vertebral column.

Two genes of special interest are expressed in what corresponds to the head of the very early embryo of *Drosophila*. One, *orthodenticle* (*otd*), is expressed in the second and third segments of the head: the other, *empty spiracles* (*ems*), in the second, third and fourth, in the last of which it overlaps the domain of *labial*, the most anteriorly expressed of the HOM genes. Both *ems* and *otd* contain homeoboxes, but they are somewhat different both from each other and that of the HOM genes[96].

One difference concerns the part they play in *Drosophila* development, for they are both 'gap' genes (Cohen and Jürgens, 1990; Finkelstein and Perrimon, 1990; both reviewed by Ingham, 1990), which in the trunk, at least, act to define the regions of expression of the HOM genes. Unlike HOM genes, they are direct targets for maternal genes that determine anterior–posterior polarity in the egg. The fact that no HOM genes are known to be expressed in the anterior portion of the *Drosophila* head has led some to suggest that *ems* and *otd* double as homeotic genes, although this has not been demonstrated unequivocally.

Surprisingly given their divergence of sequence, homologues of these genes are also found in the vertebrate head (reviewed by Holland *et al.*, 1992b). In the mouse, there are two homologues each of *otd* and *ems*, respectively *Otx1* and *Otx2*, and *Emx1* and *Emx2*, expressed in the forebrain in a nested set reminiscent to the pattern of *Hox* genes in the hindbrain, except that expression boundaries tend to extend rearwards from the front end, rather than forwards from the rear end (Simeone *et al.*, 1992; Figure 3.6). *Emx1* is expressed in a small patch in the posterior part of the roof of the telencephalon, the most anterior part of the brain. *Emx2*, in contrast, is expressed in most of the roof of the telencephalon (overlapping the *Emx1* domain at the front and at the back) as well as the front part of the floor of the diencephalon, the next most anterior compartment of the brain.

The expression domains of the *otd*-class genes are more extensive. *Otx1* is expressed in the roof of the brain from a point just in front of the most anterior edge of the *Emx2* domain, right to the posterior boundary of the mesencephalon roof, as well as a patch on the floor of the diencephalon coincident with the *Emx2* domain there, but extending further backwards. The domain of *Otx2* is similar except that it extends slightly further forwards in both the telencephalon roof and the diencephalon floor. Taken together, they form a 'nest' of expression domains that can be arranged, in order of extent, *Emx1*< *Emx2*< *Otx1*< *Otx2*, with a clear focus in the roof of the telencephalon.

The evidence suggests that the duplication and nested domains of expression of these genes provide a combinatorial 'code' for the specification of structure in the fore- and midbrain, as *Hox* genes do in the hindbrain and trunk. However, even the code provided by the *Emx* and *Otx* genes is too coarse for the variation evident in the vertebrate brain: as Holland *et al.* (1992b) suggest, expression patterns in other genes may refine the pattern[97].

A gene in *Drosophila* called *muscle-specific homeobox* (*msh*) has homologues in a variety of vertebrates, and is expressed in a wide variety of tissues and organs such as the teeth, eyes, limbs and heart. The only connection is that all the tissues concerned are uniquely vertebrate elaborations. Given that the mouse has at least three *msh*-like genes, and the zebrafish *Brachydanio rerio* at least four (Akimenko *et al.*, 1995), and *Drosophila* and the ascidian *Ciona intestinalis* have only one each, it could be that duplication and diversification of *msh* and other genes parallels that of *Hox* genes in the early evolution of vertebrates (Holland, 1990, 1992; Holland *et al.*, 1994b).

Some of the genes involved in *Drosophila* dorsoventral patterning are, intriguingly, related to those which, in mammals, determine the forma-

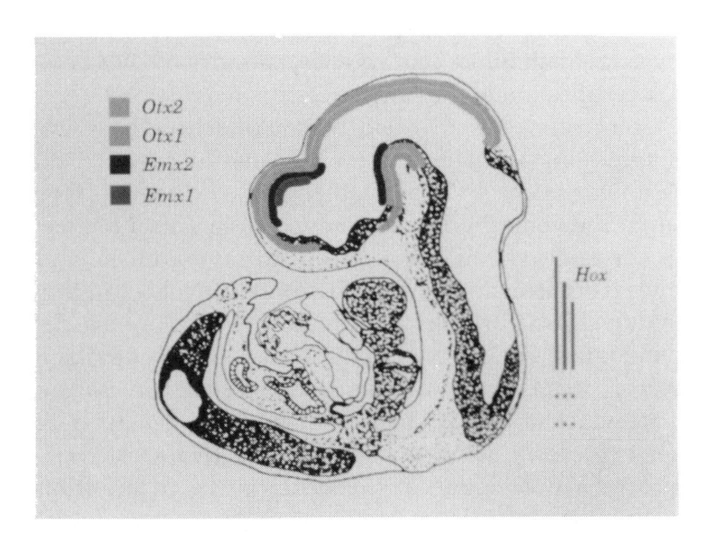

Figure 3.6 Mammalian expression domains of *Emx-1* and *Emx-2*, *Otx-1* and *Otx-2* in the developing central nervous system of mice embryos 10 days after fertilization. Note their position far anterior to *Hox* expression domains. Te = telencephalon, Di = diencephalon, Mes = mesencephalon, Met = metencephalon, My = myelencephalon (from Simeone *et al.*, 1992).

tion of bone: the fly gene *decapentaplegic* (*dpp*) resembles the genes for human bone morphogenetic proteins 2 (*BMP2*) and 4 (*BMP4*), and *tolloid* is related to *BMP1* (Shimell *et al.*, 1991). They are all part of a family that contains at least 15 members, the *DVR* genes (see review by Lyons *et al.*, 1991). The name (*decapentaplegic-Vg-related*) is a contraction of *decapentaplegic* and the *Xenopus* gene *Vg-1*, involved in gastrulation, the first two members of this functionally diverse family to be identified. The *DVR* genes are in turn members of a large family of genes that encode hormone-like cell-signalling molecules, the transforming-growth-factor-β (TGF-β) family.

As it turns out, BMPs are involved in more things than bone formation (hair and feather formation, and possibly limb formation, for example); *dpp* is part of a complicated regulatory system involving homeobox-containing genes as well as segment-polarity genes such as *wingless* – and not only in dorsoventral patterning; and *Vg-1* seems to be part of the general genetic underpinning of mesoderm induction in *Xenopus*. From this, Lyons *et al.* (1991) conclude that *DVR* genes are genetic middlemen that tend to get involved wherever there are tissues undergoing inductive interactions.

As with other types of gene surveyed here, there is probably a greater diversity of *DVR* genes in vertebrates than in *Drosophila*. Although *Drosophila* tends to substitute alternative gene transcripts for alternative genes, the examination of primitive chordates and other deuterostomes might reveal, as with other family genes, episodes of *DVR* gene duplication and diversification in the early ancestry of vertebrates.

More recent work has revealed correspondences between *dpp* and BMPs as interesting as that between HOM and *Hox* genes. In short, vertebrate homologues of *DVR* and other genes which in *Drosophila* are expressed in the ventral ectoderm are found in dorsal neuronal tissue, and vice versa. *Dpp* itself is expressed dorsally and, in concert with other genes (Ferguson and Anderson, 1992) is involved in the differentiation of dorsal structures. In the frog *Xenopus*, in contrast, *BMP-4* (one of its homologues) is expressed ventrally, and has ventralizing activity (Fainsod *et al.*, 1994). To give another example, the *Drosophila* gene *short gastrulation*, which appears to be an antagonist of *dpp* and necessary for its proper expression, is expressed ventrally (François *et al.*, 1994). It appears to be a homologue of the gene *chordin* in the frog *Xenopus*, which expresses a potent dorsalizing factor (Sasai *et al.*, 1994; Holley *et al.*, 1995) in response to the activity of the homeobox gene *goosecoid* (Cho *et al.*, 1991). In turn, *goosecoid* (which we have encountered in relation to the paired-box family) is a potent contributor to the formation of the body axis, and is found in the organizer region of the dorsal blastopore lip in *Xenopus* ('Spemann's Organizer' of classical embryology) and in homologous regions in other vertebrates, such as 'Hensen's Node' in the chick (De Robertis *et al.*, 1994). For a more recent review, see De Robertis and Sasai (1996).

For all their appeal, these links should be tinged with caution. Peterson (1995b) points out that *BMP-4* is just one of two or more vertebrate homologues of *dpp*, which seem to have resulted from the same kind of genetic duplication event in vertebrate history that produced the *Hox* clusters. It could be that the various members of the *BMP* family have diverged in function since the duplication event. In which case, it would be hard to decide whether the ventral expression of *BMP-4* is primitive for the family in general, and representative of the function of the *BMP* ancestor immediately after the inversion event. *BMP-2*, for example, seems to be expressed dorsally, just like *dpp*. If this condition is primitive, it gives the lie to the inversion story. However, it could be that the example of the *BMP* genes may turn out to be inapposite, as the products of both *BMP-2* and *BMP-4* seem to compete for the same receptors (Graff *et al.*, 1994), which may be evidence for a similar range of activity. Nevertheless, if the pattern of duplication, divergence and co-option (Holland, 1990; Holland *et al.*, 1994b) is true in general, then Peterson's criticism must stand.

Arendt and Nübler-Jung (1994) use genetic evidence like this in support of the idea that sometime early in their history, chordates experienced an inversion of the dorsoventral axis with respect to that of arthropods. This inversion would have happened very early in embryogenesis. Perhaps significantly, Arendt and Nübler-Jung (1994) seem to regard vertebrates as representative of chordates and even deuterostomes more generally – perhaps this remarkable pattern of genetic inversion is simply a symptom of the origins of deuterostomy, rather than of vertebrates.

3.7 HEAD TO HEAD

The *DVR* story aside, what does our knowledge of these genetic systems really tell us about vertebrate evolution in particular, rather than that of deuterostomes in general? As mentioned above, Holland (1992) suggests that one might be able to discern antiquity of function from the disparities of current usage. For example, the fact that the expression boundaries of paralogous *Hox* genes in mice correspond in the hind-brain, but not in the spinal cord, suggests that *Hox* genes are relative newcomers to the hind-brain, and have had less chance to diversify there than in the trunk, where they have been long-time residents.

More broadly, it is clear that although the functions and expression domains of *Hox*/HOM genes may vary from animal to animal, the fact that they seem generally associated with the specification of the anterior–posterior axis argues that this is an ancient, perhaps primordial function of the group.

Holland (1992) and Holland *et al.* (1994b) note, as have others (see, for example, Pendleton *et al.*, 1993), that the origin and evolution of vertebrates is connected with an increase in the number of homeobox and other families of genes[98], and therefore with the potential for the co-option of some of the newly copied genes for a variety of functions not necessarily connected with axial specification (Holland, 1990), but which are nonetheless characteristic features of vertebrates, including eyes, ears, limbs and craniofacial structures: indeed, the whole panoply of neural-crest-derived structural complexity. Animals without these features only have one set of *Hox*/HOM genes. Such echinoderms as have been studied only seem to have a single set, and the same seems to be true of tunicates and amphioxus (Ruddle *et al.*, 1994; Garcia-Fernàndez and Holland, 1994).

And as we have seen, the origin of vertebrates also seems to have been associated with the duplication and diversification of homologues of genes and gene families outside the *Hox*/HOM group, such as *engrailed, wingless, orthodenticle, empty spiracles, even-skipped, caudal, distal-less, msh, dpp* and many others without particular relationship to development (Holland *et al.*, 1994b).

The expression of some of these, such as *engrailed* and *wingless*, does different things in *Drosophila* from vertebrates. Others, how-ever, show a remarkable conservation of function. Yet others, such as the *DVR* group, seem to do both – involved in the specification of the dorsoventral axis in flies and vertebrates, but having acquired an additional purpose in the development of bone, a unique feature of vertebrates.

Along with the *Hox*/HOM group, homeobox-containing genes in the *otd* and *ems* groups are also concerned with axial specification, only in the fore-brain rather than the hind-brain and spinal cord. Holland (1992) suggests that rather than the acquisition of a 'new head', vertebrates exploited the potential offered by a more varied palette of genes to elaborate what was already there, and develop new axial domains corresponding to the fore-brain.

The amphioxus provides important evidence to support this argument. Peter Holland and his colleagues have just started to investigate the developmental genetics of what, after all, is the closest thing we know to a headless vertebrate. Their exciting results, though built on what we know about *Drosophila* genetics, are therefore far more relevant to our present concerns.

Early work produced a homeobox-containing gene from the amphioxus which they call *AmphiHox3*. The sequence of this gene is similar to *Hox-b3* (Holland *et al.*, 1992a). *AmphiHox3* is expressed in posterior mesoderm (but not in the somites) and in the nerve cord as far anterior as a point just in front of a particular pigment spot, coinciding with the boundary between somites four and five. There is a substantial portion of the animal, anterior to this point, in which the gene is not expressed. Genes in the cognate *Hox* group are expressed in the mouse hind-brain as far forward as the boundary between rhombomeres four and five.If the similarity between *AmphiHox3* and *Hox-b3* and its paralogues stretches as far as domains of expression, this suggests that a large portion of the nerve cord in the amphioxus, anterior to the *AmphiHox* domain of expression, appears to correspond to the vertebrate mid- and fore-brain, at least in broad terms. The results negate Gans and Northcutt's idea of a 'new head' (Gans and Northcutt, 1983) or that the entire brain of vertebrates is an elaboration with the small cerebral vesicle at the extreme anterior end of the amphioxus. The neural crest may be a vertebrate innovation, but the head itself is not.

What this result cannot tell us is whether there is a one-for-one correspondence between particular regions of the neural tube of amphioxus and particular parts of the vertebrate mid- or fore-brain. It could be that whole stretches of the vertebrate brain want for homologous structures in the amphioxus. Later work that produced more homeobox genes (Holland *et al.*, 1994a) was able to answer this question. Analysis of what turned out to be the amphioxus homeobox cluster (Garcia-Fernàndez and Holland, 1994) in parallel with comparative anatomy (Lacalli *et al.*, 1994) shows that the neural tube of the amphioxus has the equivalent of a hind-brain (though greatly elongated) and the diencephalon, the part of the fore-brain from which originate the paired eyes of vertebrates. There does not seem to be anything cognate with the telencephalon (the foremost part of the fore-brain) or the mid-brain.

All this may suggest that the head was built from within a structure that was fundamentally segmented, and that bipartite division is illusory. However, certain differences can be explained by both views. For example, it is still possible to view the amphioxus as a fundamentally bipartite animal in which segmentation has been imposed from behind on a weakly cephalized chassis. (Jefferies, 1973, interprets the mitrate *Lagynocystis* as a stem cephalochordate in which the 'head' and 'tail' regions are sharply defined.) Why, for example, do the expression domains of *Hox* genes not extend all the way to the front of either vertebrates or the amphioxus, but fade out forwards in the hind-brain? Why are the expression domains of the genes of the *ems* and *otd* class confined to the front end, and fade out backwards in the mid-brain? Are these expression domains markers of a fundamental division into Romer's 'somatico-visceral animal'?

Later work (Garcia-Fernàndez and Holland, 1994) indicated that the single amphioxus *Hox* cluster bears little resemblance to the *Drosophila* cluster, but a strong general similarity to just one of the four mammalian clusters (*Hox-c*). The implication is that the amphioxus is primitive with respect to vertebrates, the nearest thing there is to a vertebrate ancestor, just before the duplication event – were it degenerate, we would expect to see the decadent pseudogenetic remnants of three other *Hox* clusters, but we do not. This simple fact weakens theories that demand the degeneracy of the amphioxus, of necessity. It is this fact, perhaps more than any other, that lays the shades of Gaskell and Patten to rest.

Work on the *Hox* genes of other deuterostomes is less clear-cut. Among the echinoderms, it seems that echinoids, at least, have a single cluster, but many of the genes homologous to the anteriorly expressed *Hox* genes in vertebrates have yet to be isolated (Ruddle *et al.*, 1994).

Hox genes have been isolated from a variety of ascidians and the larvacean *Oikopleura* (Ruddle *et al*,. 1994; Holland *et al.*, 1994b; Di Gregorio *et al.*, 1995; Saiga, personal communication). It is very likely that urochordates in general have a single *Hox* cluster in which most individual constituent genes have diverged in sequence from their vertebrate or *Drosophila* homologues to make them hard to isolate.

3.8 TAILS

This molecular work allows some resolution of a few long-standing problems. First, all metazoa seem to share a cluster of homeobox genes related to the *Drosophila* HOM cluster and the vertebrate *Hox* clusters. These genes always seem to have had some association with the specification of anterior–posterior polarity, even in non-segmented animals, as well as with the formation of the central nervous system. Nevertheless, the upstream regulators of these genes seems to differ from group to group. In *Drosophila*, for example, the expression of HOM genes is regulated by the 'gap' and 'pair rule' genes. This does not seem to be the case in vertebrates, in which 'gap' and 'pair-rule' homologues do different things. This suggests that the cluster is associated, particularly, with the 'phylotypic stage' of development (Slack *et al.*, 1993; Duboule, 1994), which occurs after gastrulation but before the elaboration of distinctive adult structures.

It also suggests that the co-option of HOM/*Hox* genes for the metameric segment specification in particular (as opposed to axial specification in general) has happened at least twice, independently: once by the annelids and arthropods, once by the chordates. This conclusion weighs against a direct relationship between annelids/arthropods and chordates. Deuterostomes other than vertebrates and the amphioxus have HOM/*Hox* clusters, but are not segmented.

Second, the overall pattern of *Hox* gene expression in vertebrates suggests that segmentation is very much a property of the posterior region, which has since been imposed, from behind, on a primitively unsegmented 'head'. This forward growth resulted in the intercalation of branchial slits, somites and cranial nerves viewed as characteristic of vertebrate segmentation on the Balfourian model.

Third, the origin of the vertebrates was accompanied by large-scale duplication and subsequent diversification of genes in general, not only *Hox* genes. This idea was explored in an influential book by Ohno (1970), and its various ramifications for structural novelty, genetic regulation and so on have been much discussed (Lundin, 1993; Holland *et al.*, 1994b; Ruddle *et al.*, 1994; Bird, 1995). This may have been a two-stage process (Holland *et al.*, 1994b). The origin of vertebrates was accompanied by the initial duplication of the single cluster into two, and further duplication to produce four clusters may have occurred at about the time of the evolution of jaws. The number of clusters in lampreys and hagfish is not known, but it may be two or three (Holland *et al.*, 1994b; Ruddle *et al.*, 1994). This allowed for the elaboration of characteristically vertebrate features such as the neural crest, and the structures derived therefrom.

Fourth, renewed consideration of the molecular genetics and anatomy of the amphioxus shows that it is a close but primitive relative of the vertebrates, rather than a degenerate and distant one. This conclusion rules out all theories in which the amphioxus is cast as a degenerate vertebrate. This also applies, albeit more tentatively, to urochordates: although the urochordate *Hox* cluster seems highly divergent and degenerate, the cluster does nevertheless seem to be single. Were urochordates degenerate vertebrates, they would have the remains of more than one such cluster.

Fifth, the vertebrate head is not a neomorph, but a neural-crest-derived elaboration on a more or less well-defined anterior bucco-pharyngeal region evident in non-vertebrate chordates, particularly the amphioxus.

In the past twenty years, molecular genetics has allowed the solution to problems (such as the origins of segmentation and the head) that had puzzled zoologists for the past two hundred, whence the disgust and defection of such talents as Bateson. How ironic, then, that the trend towards experimental biology started by Bateson and other refugees from *Balanoglossus* should culminate in the very answers they sought for so long, so fruitlessly? For in Bateson's time, the reason for the appearance of a repetitive structure was 'as yet unknown, and the laws that control and modify them are utterly obscure' (Bateson, 1886, p.546). He continues:

> But in view of what has been adduced it is surely not too much to say that enough of their mode of working can be seen to enable us to realise that they are at least powerful enough to have produced anatomical features of high importance, and further that the metameric segmentation of the Vertebrata is distinctly of the kind

which they elsewhere present must be admitted; but there is no evidence to show that this result differs in kind from that which occurs on a smaller and more restricted scale in almost all animals. Whether the repetitions which occur in the Annelids and Arthropoda are also the products of this force in a still higher degree cannot yet be certainly stated.

NOTES

70 Balfour, a keen mountaineer as well as anatomist, met his death on 19 July 1882, while attempting the ascent of the Aiguille de la Belle Etoile on the south face of Mont Blanc in the Alps. On his return he was to have assumed a professorship awarded him by the University of Cambridge in recognition of his prodigious talent. He was 31 years old.

71 Bateson (1886) followed the view that tunicate gill slits were, essentially, repetitions of a single gill slit as seen in appendicularians (then seen as primitive), and possibly homologues of pharyngeal structures in nemertines. But if appendicularians are regarded as advanced (Garstang, 1928) then the ancestral form with a single gill slit would be a pterobranch.

72 Thus, the derivation of vertebrates from segmented arthropod or annelid ancestors, as proposed by Patten, Gaskell and others (Chapter 2), presupposes that segmentation is a shared feature, something that Garstang did not consider. Even though these ideas are unworkable for other reasons, they do try to address the origin of vertebrate segmentation. They come to the natural (and, in their own way, most parsimonious) conclusion that vertebrates and arthropods share segmentation through common ancestry rather than functional convergence. Again, Bateson (1886) exposed the fallacy of these ideas as they concerned annelids, although they could be applied with equal force to the relationship between chordates and arthropods, and with rather less force (because of independent comparisons) to the relationship between vertebrates and cephalochordates.

73 There is nothing new under the sun, though. R. P. S. Jefferies reminds me that the belief of a clear distinction between head and tail in the ancestry of vertebrates goes back to the work of Froriep in the 1880s.

74 They are (1) somatic sensory (afferent) nerves, which convey information to the brain gathered from sense organs, both the external sense organs and the proprioceptors in the muscles and tendons. These nerves monitor conditions in the outside world and the body's attitude towards them; (2) visceral sensory (afferent) nerves, which convey information to the brain gathered from sensory receptors in the gut and other viscera; (3) visceral motor (efferent) or autonomic nerves, which convey impulses from the central nervous system to the visceral muscles and glands, and (4) somatic motor (efferent) nerves, which convey impulses from the central nervous system to the somatic musculature.

75 Presumably the 'some' in this quotation alludes to Balfour (who died four years earlier) and his school.

76 In a useful short review of some of the examples I discuss here, Thorogood and Hanken (1992) suggest that a simple, bipartite model may be too simplistic. They suggest that the vertebrate body is divided into at least three regions, the trunk, the hindbrain and the anterior head. The hindbrain region may be seen as a region in which the old-fashioned head and trunk regions have met, and in which they interact, posing special problems. (For further discussion and citations on the genetic basis of the head–trunk divide, see Hunt and Krumlauf, 1991a,b).

77 For a comprehensive treatment of the development and evolution of the vertebrate head, consult Hanken and Hall's three-volume treatise *The Skull* (Hanken and Hall, 1993), the successor to Gavin de Beer's *The Development of the Vertebrate Skull* (1937), rooted in turn in the Balfourian tradition. The two works are compared by Holland (1993).

78 Closer inspection suggests that the situation in amphibians may be a special feature of that group, and may not represent the primitive craniate situation. For, despite many differences in early embryogeny, the pattern of somitomere formation in fishes perhaps bears closer comparison with birds, reptiles and mammals than amphibians (Martindale *et al.*, 1987; Gilland, 1989; Gilland and Baker, 1993).

79 Much of the background to the genetic patterning of development (and many additional references) can be found in the following reviews, which are just a few among many: Slack (1984), Struhl (1984), Ruddle *et al.* (1985), Gehring (1985, 1987), Akam (1987), Akam *et al.* (1988), Ingham (1988, 1990), Gaunt *et al.* (1988), Keynes and Stern (1988), Doe and Scott (1988), Holland and Hogan (1988), Holland (1988, 1990, 1992), Lewis (1989), Lumsden and Keynes (1989), Lonai and Orr-Urtreger (1990), Lumsden (1990), Keynes and Lumsden (1990), De Robertis *et al.* (1990), Kessel and Gruss (1990), Ruiz i Altaba (1991), Cohen and Jürgens (1991), Finkelstein and Perrimon (1991), Gaunt (1991), Lyons *et al.* (1991), Melton (1991), Rosenfeld (1991), Hunt and Krumlauf (1991a,b), Wright (1991), Marx (1992), McMahon (1992), Gurdon (1992), Krumlauf (1992b), McGinnis and Krumlauf (1992), St Johnston and Nüsslein-Volhard (1992), Ingham and Martinez-Arias (1992), Holland *et al.* (1992b), Noden and Van De Water (1992), Nusse and Varmus Lawrence and Sampedro (1993), Kenyon (1994), Lawrence and Morata (1994), Krumlauf (1994), Manak and Scott (1994), De Robertis *et al.* (1994), Holland *et al.* (1994b), Ruddle *et al.* (1994) and Carroll (1995).

These reviews vary greatly in focus, tone, content and intended audience, so the reader should feel free to pick and choose. Obviously, the more recent ones are the most reliable, but they also tend to be complex and focus on increasingly particular issues. An easy way into this formidable body of work is to look up the 1994 supplement of the journal *Development*, which contains several of the reviews listed here, in a convenient and accessible format. Some of the older reviews might be read as simple introductions, or simply for their historical interest. Ingham (1988), for example, gives a good account of the basics of early development in *Drosophila*, even though progress has rendered much of the detail obsolete.

Research into the genetic basis of development is currently moving at fantastic speed, so by the time you read this, many of these reviews will have aged, not always gracefully. This book cannot hope to survey even a small fraction of the published work, much of it in any event being of little relevance to the matter in hand. The only remedy is to monitor the learned journals for reviews of the latest work.

80 Although the segments of the adult do not correspond exactly to those of the larva, many of the same genes are involved in segmental specification and identity in both larval and adult tissues.

81 For a classic early ('pre-homeobox') review of the homeotic genes of *Drosophila* from the viewpoint of a pioneer of the field, see Lewis (1978).

82 The term 'downstream' when applied to genetic interactions is ambiguous in that it also implies position in a series of events, each one of which depends on the one before. An equivalent and less troublesome term when referring to gene order is '3-prime' (3′), which refers to the conventional number of the hydroxyl group in the deoxyribose moiety of a nucleotide that is at the trailing edge, with respect to the direction of transcription. The leading edge, 'upstream' is epitomized by the '5-prime' (5′) hydroxyl group.

83 A promoter region is a part of a gene to which transcriptional factors bind, switching the gene on or off. The sequence of a promoter region is tailored for such binding events, and has no meaning when 'translated' into a protein sequence.

84 The same is true for other kinds of regulatory gene. Some genes known to encode the 'zinc-finger' type of DNA binding protein are also expressed in step with rhombomere compartmentation, and thus *Hox* expression. An example is *Krox-20* (Wilkinson *et al.*, 1989a).

85 These correspondences in expression as well as structure are surprising, in that the mammalian means for controlling the expression of homeotic genes do not correspond to the hierarchy of maternal, 'gap' and 'pair-rule' genes seen in *Drosophila*.

86 The nomenclature of vertebrate homeobox genes reflects the cluster ('a' to 'd') to which the gene belongs, as well as its position (equivalent to paralogy group) within that cluster (Scott, 1992). Thus *Hox-d4* is the member of cognate group 4 in cluster 'd', and is expressed more anteriorly than members of cluster 5, more posteriorly than those of cluster 3, and so on. The previous name for this gene, *Hox-4.2*, reflected the cluster membership (cluster 4) but the subordinate or trivial number ('2') bore no relationship either to the cognate group or the positioning of the gene with respect to other genes in the same cluster or in other clusters. Here I use the revised nomenclature throughout, even though the original papers to which I refer may have used the pre-1992 nomenclature. I mention this should the reader refer to older papers and be confused by the nomenclature found therein.

87 Oddly, however, mice in which the expression of *Hox-c8* is driven by regulatory sequences of the more anteriorly expressed *Hox-a4* do not exhibit the anterior expression of posterior structures. Rather, anterior structures come to the fore, and the phenotype is similar to that of *Hox-c8*-null mutants

(Pollock *et al.*, 1992). Clearly, the correct expression of homeobox genes depends on a carefully struck balance between surfeit and deficit.

88 Nevertheless, there are differences in both spatial and temporal detail, and between the neural tube and surrounding tissues (see Gaunt, 1991, for a general discussion, and Prince and Lumsden, 1994, for a good example).

89 See Hunt and Krumlauf (1991a,b), for reviews, and Hanken and Thorogood, 1993, for a contrasting view based on current models of cranial morphogenesis.

90 So much is true for the neural tube. In practice, the axial positions of expression boundaries of the same gene in different tissues (such as neural tube, lateral-plate mesoderm or surface ectoderm) tend to be offset from one another.

91 Slack *et al.* (1993) suggest that the possession of a homeobox-gene cluster is a diagnostic feature of multicellular animals which they call the 'zootype', in that it was responsible (at least in part) for the sophisticated genetic control necessary to regulate an integrated yet multicellular organism. Duboule (1994) further suggests that such genetic clustering is intimately involved with the timing of development such that the embryogeny of all vertebrates goes through a characteristic 'phylotypic' stage in which they all look quite similar, irrespective of differences imposed by gastrulation or subsequent adult form.

92 Such as the creation of craniofacial structures from the interaction of neural-crest-derived tissues with underlying head tissues.

93 The murine homologue of *Distal-less*, *Dlx*, is expressed in a restricted region of the diencephalon and telencephalon in the forebrain (Price *et al.*, 1991). Given that hydroids have neither legs nor brains, the murine expression pattern (in the broadest sense of anterior–posterior axial specification) is closer to the hydroid than the fly. The involvement of *Dll* in *Drosophila* limb specification is arguably a secondary, arthropod co-option.

94 'Sonic the hedgehog' is a cartoon character from a video game popular around the time of the isolation of this gene. The humour of some developmental geneticists is clearly on a par with their cultural sophistication which, in other respects, hardly progresses beyond childrens' literature: one vertebrate member of the *hedgehog* family has been named *tiggywinkle*.

95 See Noden and Van De Water (1992) for a review on the genetics of mammalian ear development.

96 Another gene, *caudal* (*cad*), resembles *ems*, and its expression domain overlaps the most posterior HOM gene, *Abd-B*. Both *cad* and *ems* have about 50% sequence similarity with the HOM gene *antennapedia*. Homologues of *cad* are known in vertebrates. As in *Drosophila*, they seem to play a part in regional specification of posterior parts of the animal in a way analogous to HOM and *Hox* genes in the head and trunk (see McGinnis and Krumlauf, 1992; Manak and Scott, 1994, for reviews and further references).

97 Examples may be mouse homologues of the *Drosophila* gene *Distal-less* (*Dll*). Like *ems*, it contains a homeobox but is not in the HOM cluster. *Dll* is expressed in the head of *Drosophila*, and its murine homologues seem to be expressed in a segmentally restricted pattern in the diencephalon (see

Holland, 1992, and McGinnis and Krumlauf, 1992, for a review and further references).

98 The fact that *Drosophila*, an advanced metazoan with a distinct head, has only one set of HOM genes compared with four in the mouse is not a mark of primitiveness in flies with respect to vertebrates. HOM expression in flies is controlled in a different way from the régime that pertains in mice, as evidenced by the different functions of (for example) *wingless* and *engrailed* in each group – *Drosophila* tends to use differentially spliced transcripts fromthe same gene rather than transcripts from different genes. It is more likely that the mouse shares more of a transcriptional 'background' with the amphioxus than it does with *Drosophila*, such that more (and more diverse) copies of *Hox* genes are required for the elaboration of the brain and associated structures. Therefore, it is perfectly possible for flies to make do with one set of genes whereas mice need four. And of course, the fact that the elaboration of structure is correlated with duplication and diversification of genes does not necessarily mean that diversification is the proximate cause of structural complexity in general – but only in the vertebrate group.

4

Jefferies' Calcichordate Theory

Left. The *left* side of anything is frequently considered to be unlucky, of bad omen (*cf.* AUGURY; SINISTER), the *right* the reverse.

Brewer's Dictionary of Phrase and Fable

4.1 CARPOIDS REVISITED

Jefferies' Calcichordate Theory is the subject of this chapter. In this section I shall describe the general features of carpoids in a little more detail than I did in Chapter 1, with some discussion of how their features have been interpreted by Jefferies and other authors. I shall also review the main features of Jefferies' ideas, before going on to discuss his research in depth later in the chapter.

As introduced in Chapter 1, there are essentially four kinds of carpoid; cinctans, solutes, cornutes and mitrates. When Jefferies first studied their relationships, they were conventionally grouped together as the echinoderm subphylum Homalozoa, usually on the basis of bilaterality or asymmetrical shape, and the primitive absence of pentameral symmetry (Nichols, 1972). Such definitions, based on absences and primitive features, suggest that the Homalozoa is an unnatural, paraphyletic assemblage: as the only completely extinct echinoderm subphylum, it also numbers living representatives among its unique absences (Figure 4.1).

Within the Homalozoa, the cinctans comprised the sole order within the class Homostelea; the solutes comprised the sole order of a different class, Homoiostelea, whereas cornutes and mitrates constituted two separate orders within the class Stylophora. In *The Fossil Record*, Jefferies *et al.* (1967) mention a fifth order, the digitates, but suggested that these really belong with another group, the eocrinoids.

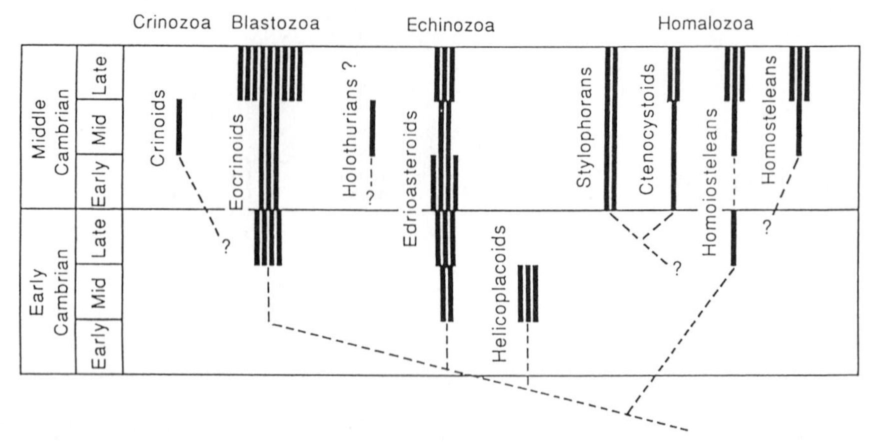

Figure 4.1 The usual picture of early echinoderm classification and origination. Carpoids are grouped as Homalozoa, subdivided into Stylophora (cornutes and mitrates) with Ctenocystoids as possible relatives, Homoiostelea (solutes) and Homostelea (cinctans). (From Sprinkle, 1992.)

The classification today is somewhat different. The traditional categorical ranks such as Class and Order have broken down, and the advent of cladistics has exposed much inherent paraphyly. *The Fossil Record 2* (Benton, 1993) is a very different creature from its predecessor, and is explicit in its debt to Jefferies' work.

4.1.1 Cinctans

The cinctans are restricted to the Middle Cambrian. They are relatively large for carpoids, being up to 40 or 50 mm long. *Trochocystites bohemicus* was the first to be described (Figure 4.2a). It looks, if anything, like a miniature tennis racquet. The 'body' or theca of the animal is disc-shaped, with a many-plated integument on the dorsal and ventral faces, bounded and edged by a 'frame' of large, marginal plates. These plates are drawn out posteriorly into a panhandle called the stele, which represents the 'handle' of the racquet. If flexible at all, this appendage would have warped up and down, rather than side-to-side.

The ventral integument is featureless by comparison. The dorsal surface, though, bears what looks like the typical echinoderm anal 'pyramid' near the front, at the left. The front edge of the animal bears two ambulacra, which converge on a mouth situated just to the right of centre. Another large opening, situated right at the front, is covered by what looks like a moveable plate, the operculum. It is conventionally regarded as the anus, but Jefferies (1990) argues that this function is fulfilled by the pyramid-like structure, and interprets the large anterior opening as a gill

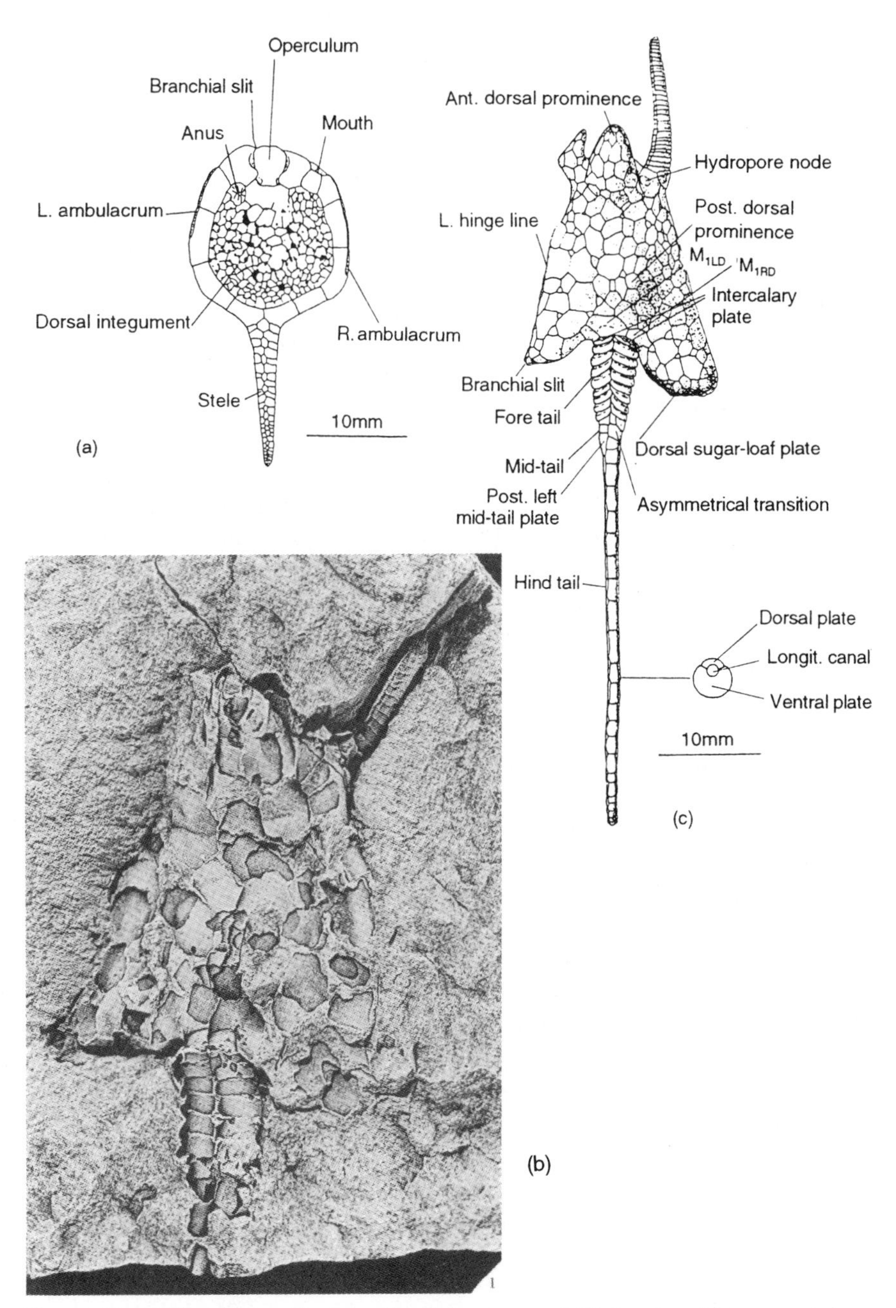

Figure 4.2 (a) The cinctan *Trochocystites bohemicus* and (b) fossil and (c) reconstruction of the solute *Dendrocystoides scoticus* (from Jefferies, 1990, with Jefferies' interpretations).

slit. Friedrich (1993) accepts this interpretation and also records a hydropore–gonopore behind and to the right of the mouth.

4.1.2 Solutes

Solutes are known from the Middle Cambrian to the Lower Devonian. They are extremely irregularly shaped, as shown by the drawings of *Dendrocystoides scoticus* (Figure 4.2b, c). Solutes are large for carpoids – they may be more than 10 cm long[99]. The theca is an irregular potato-like blob, flattened or more usually globular, and often uniformly covered with a continuum of variously sized polygonal plates. Various openings have been interpreted as mouth, hydropore, gonopore, anus, gill-slit and so on, but with the exception of the mouth and the anal pyramid, these openings are often hard to discern.

The theca bears two appendages, one at each end. The shorter one is covered by rings of plates which, on one surface, are separated by a long mid-line groove (and presumably could be opened along this groove). This structure may be interpreted as a 'feeding arm' bearing an ambulacrum, protected by moveable cover plates. The longer appendage of solutes, which without preconception can be called the 'stele', is roughly divisible into three sections. Proximally (nearer the theca) it is relatively thick and covered by thin imbricating rings of plates, leaving a large internal cavity that could have been filled with muscle; distally, the stele is long and thin, with plates arranged biserially – sometimes the series are at right and left, but sometimes at dorsal and ventral. There is space for a small canal down the inside of the stele. The proximal and distal sections grade into each other in the rather indistinct middle section.

The disposition of the plates in the stele, especially in the mid-section, is generally irregular with almost no left–right asymmetry. However, bilateral symmetry is evident in the proximal and distal parts of the appendage in dorsoventrally biserial forms. The function of the long appendage is somewhat mysterious although Jefferies (1990) suggests that the proximal section may have been flexible and muscular, driving the movements of the stiff, rod-like distal section, and that the stele as a whole was locomotory, pulling the animal across the sea floor. However, all authorities agree that the plates making up the stele were too strongly interlocked for them to have unfolded to reveal an ambulacrum (see Philip, 1979 for discussion and references).

4.1.3 Cornutes

Cornutes are known from the Middle Cambrian *Ceratocystis perneri*, to the end of the Ordovician. Even a cursory glance marks them out as odd indeed. They are smaller than solutes, the theca being a couple of centimetres across and generally flattened as in cinctans rather than globular. As in cinctans, the upper and lower surfaces are rather different in character from each other, and the various parts seem better demarcated from one another than is the case in solutes. Generally, though, the theca is irregular in shape, sports varying numbers of spines and other excrescences and varies greatly between species.

In most if not all cornutes, though, an area on one face (the 'obverse' of Bather, the 'dorsal' of Jefferies) of the theca is perforated with a set of openings which may or may not be arranged in a series. These have been interpreted as mouths (Bather, 1925), gonopores (Jaekel, 1918) or gill slits (Gislén, 1930; Jefferies, 1968a,b). In some forms such as *Cothurnocystis elizae* these slits are well-defined by a framework of plates; in others they are seen, simply, as holes between plates.

Like the solutes, cornutes have a stele that is divisible into three sections, although these sections are better defined than in solutes, and the appendage as a whole is clearly bilaterally symmetrical. The proximal section consists of a large interior space ringed with thin imbricating plates, sometimes interspersed with thinner, plated integument that may have allowed for some muscular flexion in life. The distal section, as in solutes, is long and narrow. Again as in solutes, each segment of the distal section comprises a thick hemicylindrical ossicle, although this is surmounted by a left and a right plate (rather than just a single plate, as in solutes), leaving a central interior canal. Occasionally, the distalmost end of the stele could be flexed to form a hook. In solutes, as we have seen, the middle section comprises a somewhat uneasy intergrading between the proximal and distal sections. In cornutes, by contrast, the middle section is a well-demarcated region in which several of the hemicylindrical ossicles as found in the distal part of the stele appear to be fused to form a single unit, the stylocone, surmounted by several pairs of plates.

The shorter 'feeding' appendage characteristic of solutes is absent in cornutes. Indeed, cornutes have no trace of an obvious ambulacrum, either borne on an arm as in solutes or starfish, or prostrate on the theca as in cinctans and sea urchins.

This has been the source of great debate. Some, such as Bather, suggested that a series of feeding channels had formed on one face of the theca but had been overgrown by calcite plates, as seen in the tegmen of crinoids: the 'mouths' formed the entry points for this system of buried feeding channels. Others, notably Ubaghs, modelled the stele along the lines of an ophiuroid arm: he held that the stele in cornutes and mitrates constituted a feeding arm which he called an 'aulacophore' (Ubaghs, 1961, 1963; Figure 4.3) to distinguish it from the cognate solute stele, which according to Ubaghs' view resembles the cornute–mitrate stele only coincidentally. According to this view, the pair of thin plates in each

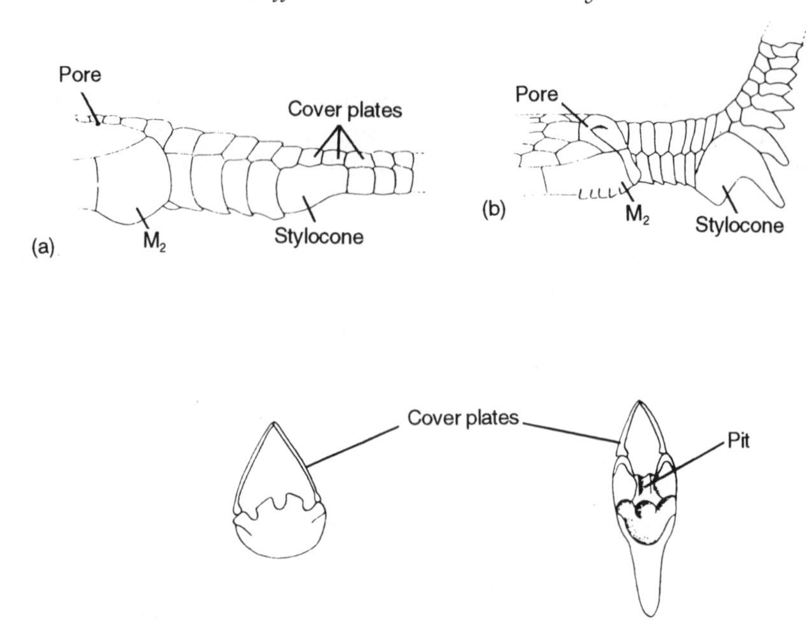

Figure 4.3 Homologies of cornute and mitrate stems according to Ubaghs. On the left (a) is shown a lateral view of the tail of the cornute *Phyllocystis crassimarginata*, with a cross-section below it; on the right (b) are the same views of the mitrate *Mitrocystites mitra*. Ubaghs' interpretation preserves the obvious similarities between the two groups – note how the stylocone faces downwards and the 'cover plates' face upwards in both groups. Jefferies, in contrast, would invert the mitrate with respect to the cornute, turning the mitrate stylocone to face upwards and the 'cover plates' downwards. Jefferies' inference of homology between cornute and mitrate 'tails' is therefore not as straightforward as is Ubaghs': rather than a simple one-to-one correspondence, the entire mitrate tail represents a secondary regionation of a cornute fore-tail (anterior to the stylocone), the rest having been autotomized. (From Jefferies, 1986, based on Ubaghs, 1981.) See also Figure 4.7.

segment protected an ambulacrum running along the internal space of the stele. They hinged on the outer edges of the hemicylindrical ossicle and could open outwards, revealing the ambulacrum. The food was drawn proximally to an internal mouth. This would not have been possible in either solutes or cinctans, which in any case bore their ambulacra elsewhere. Ubagh's interpretation reads cornutes and mitrates as aberrant suspension-feeding echinoderms with a single ambulacrum. This interpretation has been followed by others as in, for example, the description of the mitrate *Diamphidiocystis* (Kolata and Guensberg, 1979).

Others, such as Caster (cited in Philip, 1979) see the solute stele as one in which the primitive feeding function had been lost. But as Philip (1979) points out, this leaves open the question of the origin of the solutes' *other* arm, which does have a feeding function and, in Caster's scheme, would have to have been acquired independently.

Philip (1979) disagrees with Ubaghs' view, pointing out the similarities between the cornute and solute stelae, and the consequent inconsistency that one should have contained an ambulacrum, and the other not. Whatever function it had, the longer, tripartite appendage must have done much the same things in solutes as in cornutes and mitrates.

Again, the model of an ophiuroid arm is inappropriate, given that the latter is in itself a highly evolved structure the features of which are likely to resemble those of a distantly related Palaeozoic form only coincidentally.

Third, all Palaeozoic echinoderms with ambulacra also display the other prominent impedimenta of the water–vascular system, in particular a hydropore or madreporite. Candidate structures are often hard to find in carpoids, especially in mitrates.

Fourth, if as most people agree, the proximal end of the stele was heavily muscled internally, it is difficult to see how a food groove could have passed from the distal region into the theca.

Philip concludes that the best model for the carpoid stele is that of the modern crinoid, in which the stalk is a supporting structure braced internally by the chambered organ.

Although the internal spaces of the cornute stele do indeed contain impressions consistent with the presence of a canal like the radial canal of echinoderms (giving off segmental side-channels to tube-feet), they could equally well have contained, as Jefferies has suggested, segmental muscle blocks, a nerve trunk and an internal supporting structure such as the chambered organ of crinoid stalks. The presence of an ambulacrum seems unlikely, according to Jefferies, given the lack of any 'mouth' at the proximal end of the structure.

It is also inconsistent with the presence of a separate feeding arm in solutes – together with a stele, similar to that in cornutes, which presumably had some other function than feeding. As Paul (1990) pointed out in support of Jefferies' reinterpretation of the solute *Dendrocystoides scoticus* (Jefferies, 1990), it is hard to imagine in evolutionary terms how the feeding function could have been transferred from the shorter (solute) appendage to the longer (cornute) structure, with the loss of the shorter appendage.

Perhaps the best-known cornute is *Cothurnocystis elizae* (Figure 4.4), subject to study by Bather (1913, 1925), Gislén (1930), Jefferies (1968a, b) and Jefferies and Lewis (1978). Ubaghs (1963) describes a close relative, *Cothurnocystis (=Nevadaecystis) americana.*

The theca of this creature consists of a 'frame' of marginal plates supporting, on each side, a more flexible integument tessellated with a mosaic

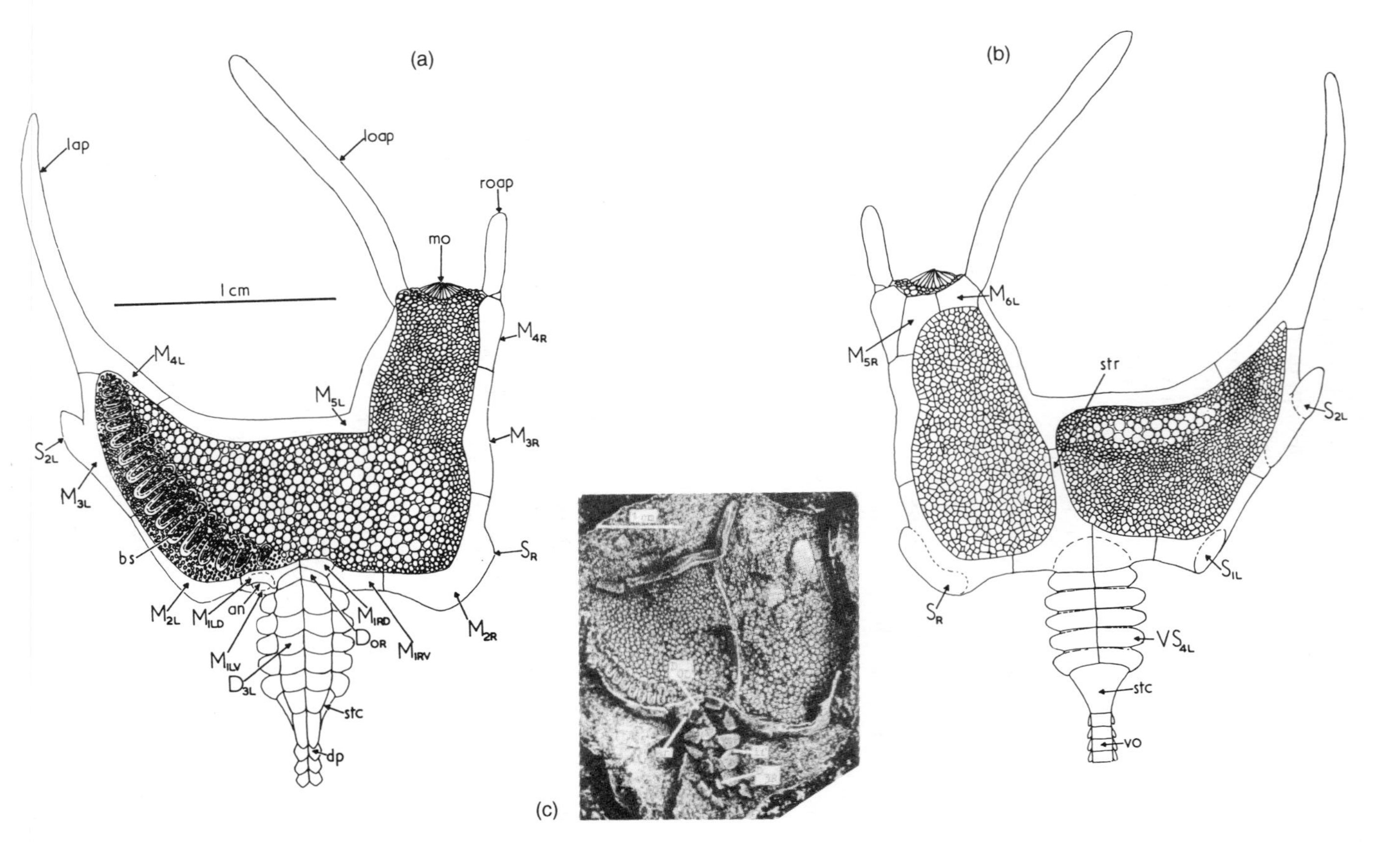

Figure 4.4 Jefferies' interpretation of the cornute *Cothurnocystis elizae* in (a) 'top' and (b) 'bottom' view, with (c) ventral view of the inside surface of the top surface from a fossil specimen. Note the branchial slits (bs) in the reconstruction and specimen, and the position of the mouth (mo). Anterior is top (from Jefferies, 1968b).

of small, irregular ossicles. The 'obverse' side – the one that bears the 'gill' slits – is an almost unbroken field of integument framed by marginal plates. The ossicles are smaller around the slits and near the pyramidal structure variously interpreted as a mouth (Jefferies), anus (Ubaghs) or both (Gislén, 1930), suggesting that it was more flexible in those areas than in mid-field. The stele sprouts from one edge, and the marginal plates are prolonged into several other, smaller appendages, none of which correspond to the solute feeding-arm. In obverse view, the shape of the theca resembles a boot of a medieval jester. Two flexible plates on either side of the mouth/anus resemble the 'tongue' and 'tag' of the boot, and a long, fixed spine near the slits is generally called the 'toe-spine'.

A detailed examination of the 'gill' slits, whatever their function, suggests that they were excurrent openings. Each one was covered by a tiny flap of integument disposed like a one-way valve in a foot-pump. Compression of the body would have forced these slits to open outwards, but pressure from the outside would have forced them shut.

In his description of *Cothurnocystis americana*, Ubaghs (1963) describes the 'gill slits' as 'elliptical structures' and reaches no conclusion about their function, except that they were probably not involved in feeding, that function being taken over by the aulacophore. However, his illustration of that appendage (Figure 4.5) shows the supposed ambulacral cover plates as resolutely shut.

On the other ('reverse') face of the theca, the integument is supported by a bridging 'strut' and the mouth/anus is supported by strap-like extensions of the marginal plates, but there are no slits or other openings. The marginal plates are augmented by small knob-like protrusions like the feet of a small footstool.

Cothurnocystis and carpoids in general are sometimes reconstructed as resting on the edge bearing the mouth/anus, the theca oriented vertically anchored into the sediment with the stele free to wave in the current, fulfilling its function as a suspension-feeding organ. Jefferies and others, including Bather, reconstruct *Cothurnocystis* as lying on the seabed, with the 'reverse' side facing downwards, supported clear of the surface by the knob-like 'feet'. In this orientation, the hook at the end of the large stele is curved downwards into the substratum, and Jefferies thus interprets the stele as an organ of locomotion whereby the creature could anchor itself by way of the hook, and pull itself along, crab-like, by side-to-side flexions of the proximal part of the stele. This conception was elaborated with the use of working models by Woods and Jefferies (1992).

In this orientation, obverse becomes 'dorsal', reverse becomes 'ventral', the hemicylindrical ossicles and the stylocone in the stele become ventral, directed towards the substratum.

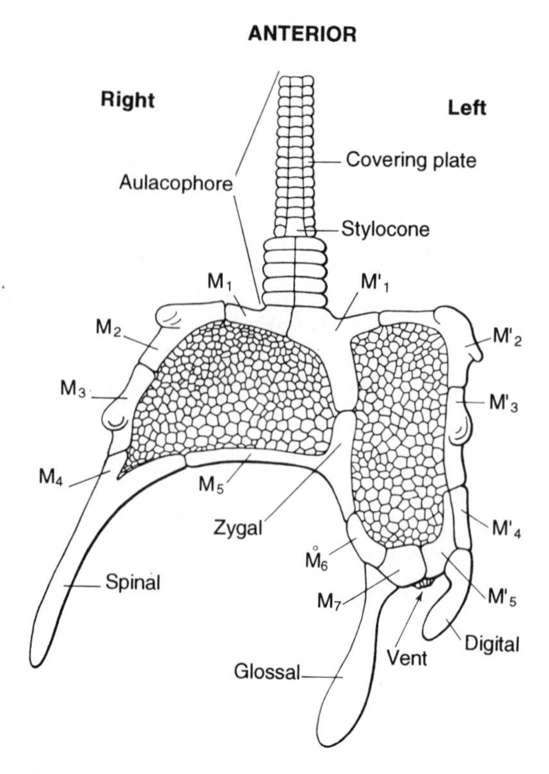

Figure 4.5 Ubaghs' contrasting interpretation of *Cothurnocystis elizae*. Unlike Jefferies' view, the stem (or aulacophore) is anterior, the mouth is a vent, and the branchial slits are not shown (being on the other side of the animal) (from Ubaghs, 1963).

Jefferies compares the obverse (slit-bearing) face of *Cothurnocystis* with the left side of the pterobranch hemichordate *Cephalodiscus* and the dorsal surface of the larval amphioxus. As we have seen in the amphioxus, the left gill slits develop before the right side slits. In *Cothurnocystis*, though, they only *ever* appear on one side. It is as if the entire body of *Cothurnocystis* constitutes the left side of a creature which has, as Gislén supposed, lain down on a right-hand side which subsequently became much reduced.

The embryology of the amphioxus shows that the right side takes some time to catch up with the precocious left, so that the adult ends up more or less symmetrical. Of course, the embryology of *Cothurnocystis* and the other cornutes is totally unknown, but it is tempting to suppose that the adult creature resembles, in morphological terms, an amphioxus larva frozen at that moment when the left-hand gills have formed but the right-hand ones have not. Gislén and Jefferies see in the asymmetry of cornutes a consequence of this suppression. Other authors attempt no explanation for this asymmetry save that it represents a peculiarity of an odd, obscure and extinct group.

4.1.4 Mitrates

The record of the mitrates extends from the Lower Ordovician to the Pennsylvanian (Upper Carboniferous) (Kolata *et al.*, 1991). They are without exception small creatures, the theca no more than two or three centimetres across. Mitrates are the most symmetrical of carpoids, with a clear, streamlined near-bilaterality uncluttered by appendages, except one which resembles that of the stele of cornutes and solutes and with which it is presumably homologous. Some mitrates also have one or a pair of long plates attached to the theca opposite the stele.

A typical representative of the group is *Mitrocystella incipiens miloni* (Figure 4.6). As in cornutes, the theca is flattened and there is a clear difference between the two faces. One face is somewhat flat and clothed with large plates that fold down over the edges to form a flange visible from the other surface: the effect is that of the carapace of a crab. The other surface, in contrast, is convex, largely covered with smaller plates, and may have been flexible in life.

A difference immediately visible between cornutes and mitrates is the lack of obvious 'gill' slits in mitrates. Jefferies (1968b) interprets a pair of minute spaces between plates on the convex face, one on each side, as 'branchial openings'. Whereas Gislén (1930) suggested that mitrates had but one pair of gill slits, Jefferies suggested that mitrate branchial openings were merely atrial openings, and that the 'real' gill slits were concealed within the body cavity. However, there is no doubt that the large, slit-like opening at the end of the theca opposite the stele looks very much like the mouth of a filter-feeder, rather than an anal pyramid.

The stele of mitrates is very similar to that of cornutes (although the distal region is often very much shorter, with fewer segments), with a thickened proximal region, a middle section similar to a stylocone (called the 'styloid' in mitrates) surmounted by several pairs of plates, and a distal section with segments in which paired plates surmounted a hemicylindrical ossicle.

Mitrates are usually reconstructed with the flat, plated face downwards, corresponding to cornute 'reverse', and the convex, flexible face upwards, to cornute 'obverse' (see for example Derstler, 1979). This orientation preserves detailed correspondences between the mitrate and cornute stele: in the mitrates, to match the stylocone of cornutes, the hemicylindrical ossicles and the styloid are directed downwards. This creates problems for Jefferies. For reasons I shall discuss below (con-

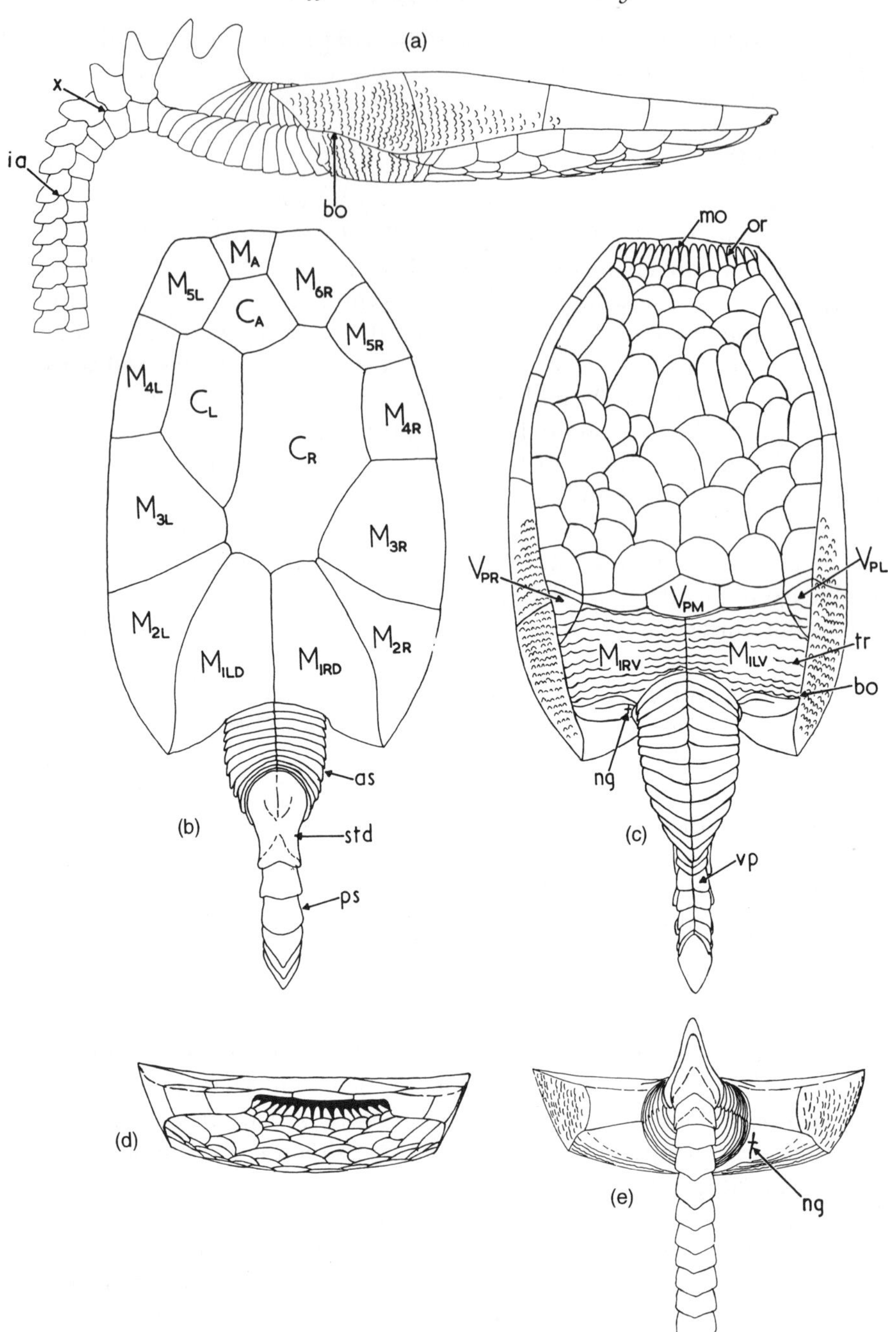

Figure 4.6 Jefferies' reconstructions of the mitrate *Mitrocystella incipiens miloni* in (a) right lateral, (b) dorsal, (c) ventral, (d) anterior and (e) posterior views. Note the inferred presence of a branchial opening (bo) and the 'narrow groove' (ng), inferred as a lateral line canal. The Ubaghsian view would turn the creature upside-down and back-to-front, making the mouth (mo) with lower fringe of oral plates (or) into a curtained anus.

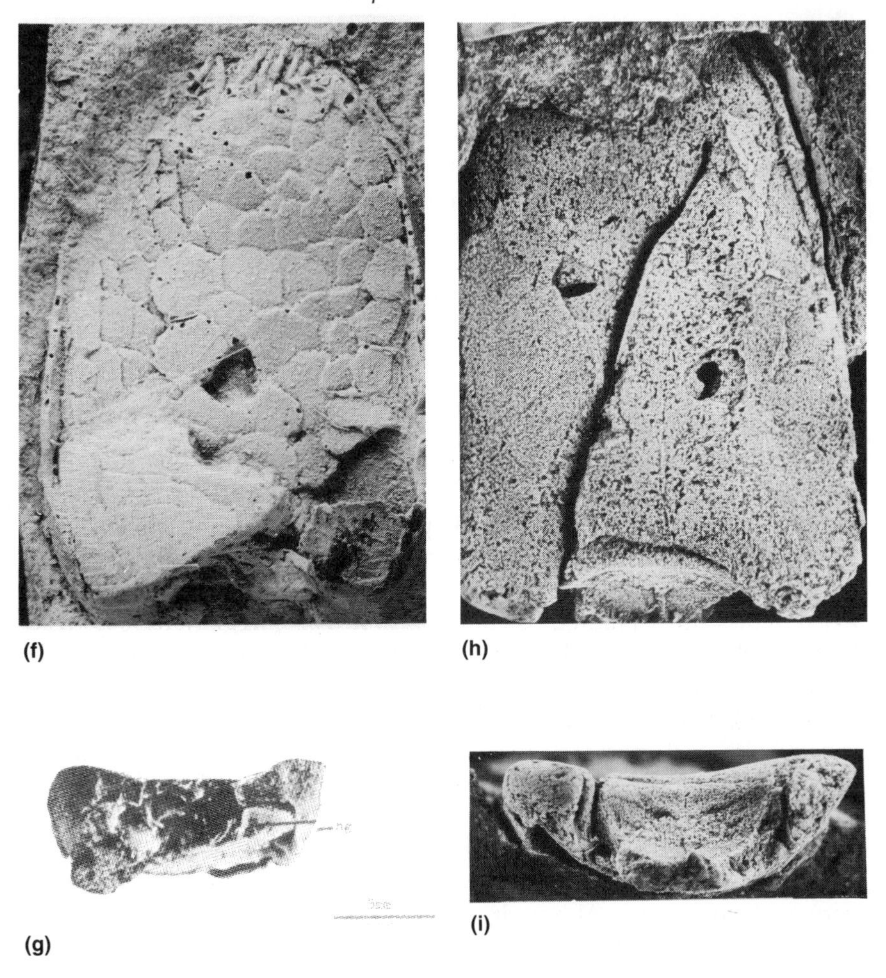

Figure 4.6 (*cont.*) (f) Latex cast of ventral surface (compare with c) and (g) posterior view of body (compare with e, and note the 'narrow groove' feature), (h) natural mould in dorsal and (i) posterior views. (From Jefferies, 1968b.)

nected with the comparative anatomy of the coeloms), his reconstruction of the mitrates demands that the flat, plated face corresponds with cornute 'obverse' or 'dorsal', the convex face with 'reverse' or 'ventral'. This turns the mitrate stele upside-down with respect to that of cornutes, so that the hemicylindrical ossicles and the styloid face upwards, not downwards. Jefferies argues his way out of this by proposing that the cornute and mitrate appendages are not fully homologous. Instead, the mitrate stele ('tail' in Jefferies' terminology) is homologous with the proximal part of the cornute tail only. The mitrate tail represents a regionation of what is effectively the proximal part of the cornute tail, in such a way that the final appearance is 'upside-down' with respect to the cornute organ (Figure 4.7).

Jefferies' detailed argument (best presented in Jefferies, 1981a) in favour of this inversion comes from three sources.

In many cornutes[100] and all mitrates (but not solutes) the tail is not preserved complete – rather than coming to a point at the distal end, it always ends abruptly. Jefferies argues that this is more than an artefact of preservation, because the distal part of the tail is consistently shorter in many mitrates than cornutes. From this, Jefferies argues for a trend towards increasingly drastic autotomy as a normal part of advanced cornute and mitrate ontogeny. Eventually, the distal and middle parts of the tail would have disappeared completely, forcing functional re-specialization of the remaining stump.

Second, Jefferies demonstrates a transition between the two states in a comparison of the stele in the advanced cornute *Reticulocarpos hanusi* (Jefferies and Prokop, 1972) with the primitive mitrate *Chinianocarpos thorali* (Figure 4.7). In the former, the major plates covering the proximal part of the stele are very loose and hardly articulate with one another. Instead, they are separated by chevrons of flexible integument reinforced by small plates. Nevertheless, the mid- and distal sections are very tightly structured as is usual in cornutes. Turning to *Chinianocarpos*, the ossicles of the *whole* stele are interspersed with flexible integument very much as in the *proximal* part of the stele in *Reticulocarpos*. The stele of *Chinianocarpos* looks like an elaboration of the proximal section of the stele in *Reticulocarpos*, with the loss of the old mid- and distal sections.

Third, Jefferies' reconstruction of the mitrates with the flat surface on top, the convex surface beneath, demonstrates an adaptation to life on very soft substrates. Although the average mitrate is conventionally reconstructed as a snowshoe, it may have been too heavy to sit on a muddy bottom without sinking. So rather than prop themselves up on top, like cornutes, mitrates (upside-down) would 'float' in the mud like tiny boats, sustained by Archimedes' Principle.

4.1.5 Dexiothetism: symmetry lost and regained

Jefferies regards the mitrates as having evolved from cornute ancestors, during which the distal part of the old appendage 'autotomized' and the new mitrate appendage evolved, and the mitrates regained a semblance of symmetry.

But features lost in evolution never reappear: like the Panda's new thumb, recruited from a wrist-bone, different organs substitute, functionally, for those that have been lost (Gould, 1980). In Jefferies' scheme, the ancestors of chordates were bilaterally symmetrical, in the same sense as

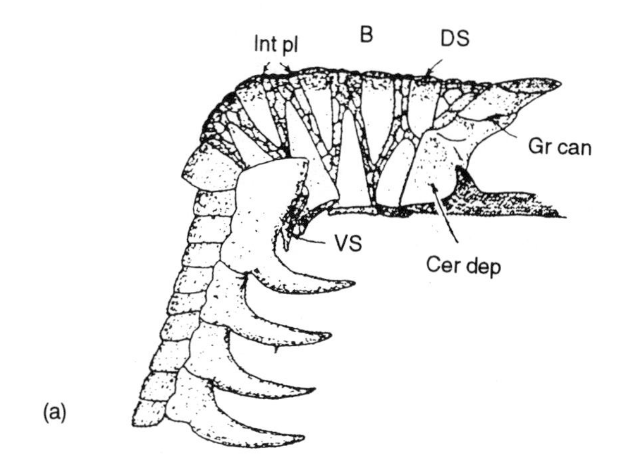

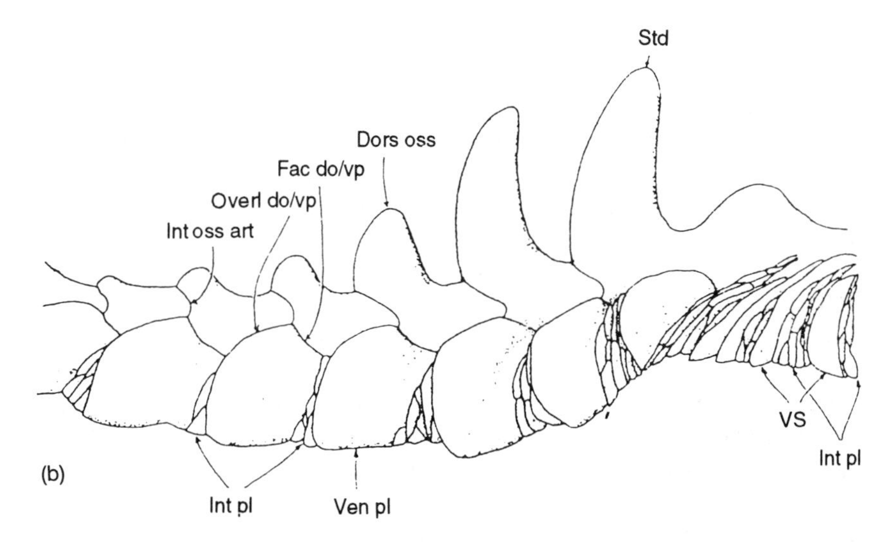

Figure 4.7 (see also Figure 4.3) Jefferies uses comparative evidence from the cornute *Reticulocarpos hanusi* and the mitrate *Chinianocarpos thorali* to resolve the problems of tail orientation posed by his scheme of the cornute–mitrate transition. Briefly, Jefferies proposes that during this transition, the mid- and hind-tail of the cornute tail were lost, the entire mitrate tail developing from the autotomized fore-tail. Therefore there is no homology between particular parts of the cornute and mitrate tails strict enough to dictate orientation. The tail of *Reticulocarpos*, shown in right lateral view in (a), is clearly demarcated into a stiff hind- and mid-tail with protruding ventral spikes, and a flexible fore-tail in which the major plates are separated by smaller intercalary plates (Int pl). The tail of *Chinianocarpos* (b), again shown in right lateral view, seems inverted with respect to that of *Reticulocarpos*, as seen from the spikes on the dorsal rather than ventral edge: but the presence of intercalaray plates (Int pl) along the entire tail – not just the fore-tail – may be evidence for Jefferies' autotomy-plus-regionation scheme (from Jefferies and Prokop, 1972).

any other invertebrate. 'Left' in the chordate ancestor corresponds to 'left' in hemichordates such as *Cephalodiscus* and protostomes such as the roundworm *Caenorhabditis elegans* and the fly *Drosophila melanogaster*. By the same token, 'right' means 'right' in all these organisms. But when this chordate ancestor lay down on its right side and became, in Jefferies' terms, 'dexiothetic', 'left' was no longer 'left', but 'dorsal', and 'right' became ventral (Figure 4.8). Most of the lateral structures down the right side were suppressed, explaining why cornutes are so asymmetrical[101].

When mitrates regained their symmetry, they did not revive the right side, but sculpted a 'new' right side from left-hand structures. According to Jefferies, they did this from the inside, forming a new 'right side' of the pharynx by an outpocketing from the left, and so on (Figure 4.9).

This hypothesized formation of the 'new' right side from the pre-existing left side in mitrates resonates with what is known about the embryology of amphioxus, with the otherwise inexplicable precocity of the left side compared with the right.

It is important to remember that, in this scheme, whereas the left gill slits of fish can be thought of as homologues of the left gill slits of enteropneusts, the right gill slits cannot. The right gill slits of fish are not 'right' in the bilaterian sense, but simply mirror-images of the left series, reflected about a vertical mirror plane.

As shown in Figure 4.8, mitrate left *and* right correspond with 'old' left. Both correspond with cornute dorsal, and 'ventral' in both cornutes and mitrates corresponds with 'old' right. And as Jefferies contends that all chordates evolved from mitrates, modern chordates retain this symmetry. In this conception, chordates retain the sense of the old bilaterian longitudinal axis, but their dorsoventral axis has been shifted by 90 degrees. Jefferies *et al.* (1995) present a slight modification of this view, namely that this quarter-turn in the dorsoventral axis affects organs derived from the mitrate head, whereas the orientation in tail-derived organs retains (or has returned to) its primitive bilaterian orientation.

The development of modern echinoderms, as shown in Chapter 1, takes its cue from the 'old' left side, with the suppression of many right-hand structures. From this, Jefferies regards echinoderms as dexiothetes and thus (in contrast to most other views) more closely related to chordates than to hemichordates.

Genetic discoveries in recent years have stimulated discussion about questions of symmetry, in particular the genetic determination of axes of symmetry, and the causes of phenomena such as 'handedness' (see Morgan, 1992). As I showed in Chapter 3, the genetic determination of the dorsoventral axis is just starting to yield to investigation (De Robertis and Sasai, 1996). From such discussions it becomes possible to discuss

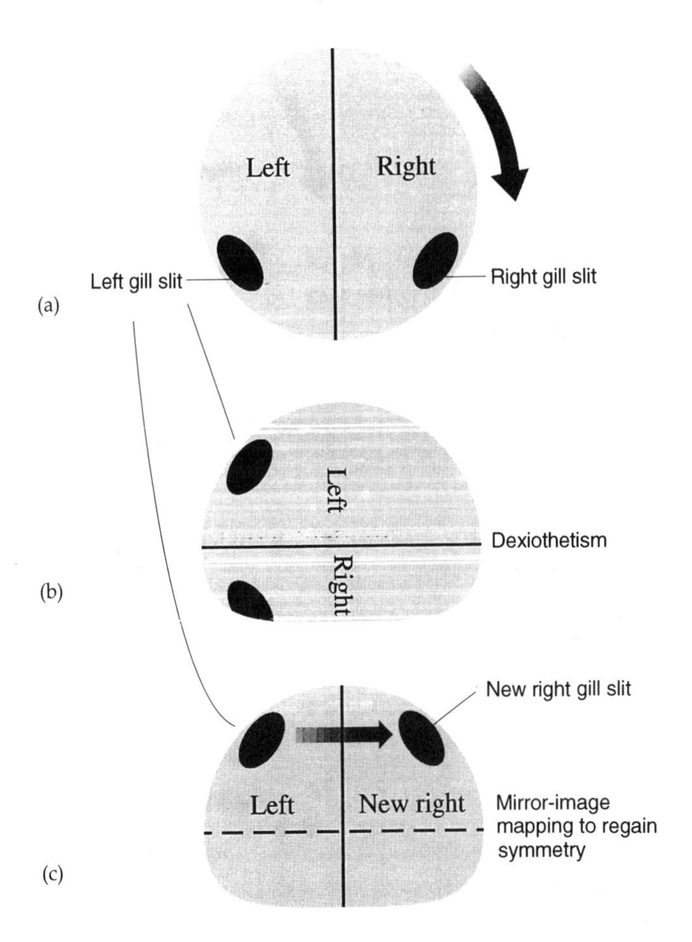

Figure 4.8 Diagram to illustrate the concept of dexiothetism. A bilaterian ancestor, shown here (a) in transverse section, falls over (b), occluding structures on the right side. Bilateral symmetry is regained (c) when the descendant stock carves a 'new' left and right from the 'old' left side. The end result is a bilateral animal, but its dorsoventral axis has been rotated a quarter turn compared with bilaterian animals. (Sketch by the author, redrawn by Sue Fox.)

dexiothetism in terms of hypotheses framed in terms not of comparative anatomy or palaeontology, but of developmental genetics.

4.1.6 The evolution of modern chordates from carpoids

Jefferies supposes that each of the three stocks of modern chordates – cephalochordates, urochordates and vertebrates – evolved from its own lineage of mitrate. This supposition has raised three serious problems worth mentioning here, but which I shall discuss further later on.

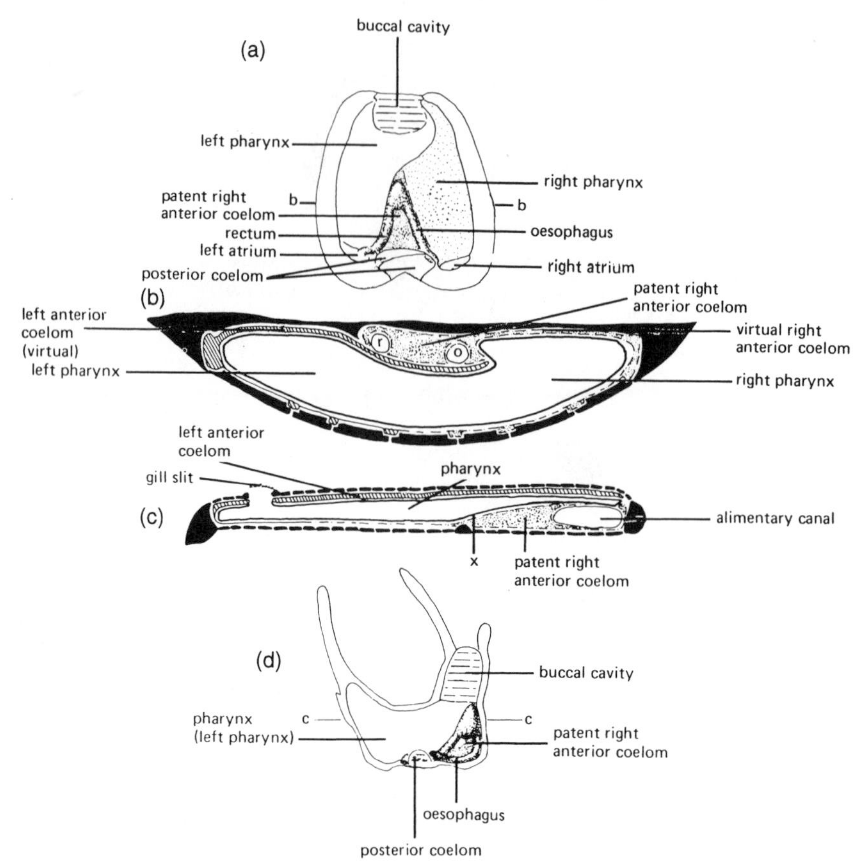

Figure 4.9 Jefferies' interpretation of the thecal chambers in cornutes and mitrates. (a) The mitrate *Mitrocystites* in dorsal aspect, with the dorsal integument removed. (b) Transverse section through b–b, (c) transverse section through c–c in (d) dorsal aspect of the cornute *Cothurnocystis*. In the transition from cornutes to mitrates, a 'new' right pharynx was formed by an outpocketing of the 'old' left, a process which drove the viscera from the floor of the body cavity (as in c), round the inside of the right-hand edge, so that they came to rest hanging from the dorsal midline. r = rectum, o = oesophagus. (From Jefferies, 1986).

(a) Calcite loss and head–tail intergrowth

First, it supposes that the ancestral chordates lost their calcitic skeleton and underwent a pronounced remodelling of the body plan. In the original conception of the calcichordate theory, this loss was postulated to have happened only once, the modern chordates having diversified from a soft-bodied intermediate. However, later elaborations of the theory placed each mitrate on a lineage to one or other of the modern chordate groups, necessitating three independent episodes of loss. In vertebrates, the calcitic skeleton was later replaced with a phosphatic skeleton built on entirely different lines.

I feel that one's conception of the seriousness of this problem depends very much on one's upbringing. For Jefferies, who subscribes to a mild form of recapitulation, the loss of the mitrate skeleton can be seen as the addition of a type of metamorphosis to the life cycle with the subsequent elimination of the calcitic stage. Jefferies notes the apparent 'resorption' of calcite in certain advanced mitrates. Prolongation of the life cycle might have resulted in a more complete loss of the skeleton and resculpting of the body.

The remodelling of the mitrate body to produce that of a chordate seems fantastic, though, to those who hold a more Goodrichian view of chordate organization, in which branchiomery, myomery and neuromery are primitively linked. But as shown by Gislén (1930), this neat segmentation does not bear too close scrutiny, especially in the case of primitive craniates such as hagfishes.

Instead, Gislén interpreted embryology as indicating that neuromery and myomery originate in a posterior region of the animal, which grows up and over an anterior region in which branchiomery has already been established. The apparently neat alternation of branchial arches and somites seen in creatures such as the dogfish results from the intergrowth of these two originally separate metameric systems.

In Jefferies' conception, the mitrate body represents a stage just before this intergrowth takes place. As I shall describe below, the 'brain' of the mitrate is found at the base of a stele in which neuromery and myomery can be demonstrated. The theca represents the 'old' anterior region in which branchiomery is evident, but not myomery or neuromery. During the 'remodelling' process, the neuromeric and myomeric parts of the stele, the 'brain' in the van, would grow up and over the theca, so that the visceral organs would hang below the axial skeleton rather than being held in front of it.

As I have said, this idea seems most odd to biologists schooled in the segmentationist tradition and the Garstangian mistrust of recapitulation. But when the chordate is viewed in terms of Romer's 'somatico-visceral animal', and allowing that the condition of the mitrate adult (and the phenomenon of 'head–tail intergrowth') might have left its relics in the larval stages of modern creatures, the process seems like the logical outcome. As Jefferies elsewhere notes, the mitrate looks like a tunicate tadpole in armour, essentially identical to Romer's somatico-visceral animal.

As shown in Chapter 3, modern genetic studies on the expression of homeobox genes in the developing craniofacial neural crest tend to support the idea of head–tail intergrowth. In mice, homeobox genes of the *Hox* family are expressed in particular rhombomeres of the developing hind-brain. As neural crest cells migrate from the axis down the sides of the head, they take the expression patterns with them (notwithstanding the reservations about this scheme detailed in the last chapter), these then dictate the formation of certain elements of a pre-existing branchial skeleton. The overlapping patterns of *Hox* expression suggest that the combinatorial *Hox* 'code' is framed with reference to the posterior end of the animal (likewise the phenomenon of 'posterior prevalence'). Furthermore, the greatest spatial density of changes in *Hox* combination occur in the hind-brain/branchial region, which is just as one would expect given that this was the 'interface' between the old head and tail. This concentration is most evident in vertebrates, where it is associated with neural crest tissue: it is also associated with the presence of four paralogous *Hox* clusters, which would have allowed for a greater degree of combinatorial 'fine tuning' and developmental redundancy than the single cluster present in the amphioxus (Garcia-Fernàndez and Holland, 1994) and (presumably) tunicates. This would not have been necessary in the amphioxus, in which cephalization is far less marked. The anatomy of the tadpole larva suggests that tunicates retain the mitrate distinction between head and tail.

The posterior origins of the *Hox* system are demonstrated by the fact that its constituent genes are not expressed in the anteriormost (prenotochordal) parts of the vertebrate head. Anterior specification is accomplished by other genes (such as *empty spiracles* and *orthodenticle*) which have a long history among metazoa of specification of anterior (or terminal) structures. In *Drosophila*, the *Hox* genes cognate with those involved in the organization of the 'intergrown' region of the vertebrate head are expressed far more extensively along the body axis. The concentration in the middle and hind-brain regions of the vertebrate head is presumably a specialization peculiar to head–tail intergrowth in vertebrates.

(b) Three separate lineages

Second, the later versions of Jefferies' scheme suppose that this loss happened not once but as many as three times, in each independent lineage – urochordates, cephalochordates and vertebrates. This must be seen as highly unparsimonious, given the drastic modifications involved. This is not a problem of evolution, but of parsimony. Maximum parsimony is not a natural law, but a lower bound on what evolution can achieve, adopted by convention for the convenience of systematists. The triple loss of a calcite skeleton is far easier than a threefold independent gain, and one supposes that examination would reveal profound differences between the process of remodelling in each case. After all, only vertebrates among extant animals have neural crest, only tunicates have the tadpole larva, and so on. In any case, strict parsimony is not always a friend to those seeking to reconstruct deuterostome phylogeny. Structures such as gills, lophophores and so on pop up from time to time as deuterostome *motifs*, illustrative more of deuterostome genetic potential than phylogenetic relationship.

(c) Problems of the fossil record

Third, the fossil record of carpoids is no older than that of chordates. If *Pikaia* from the Middle Cambrian Burgess Shales is a chordate, and conodont animals are regarded as vertebrate-like chordates of a grade of organization similar to that of hagfishes, then the records of cephalochordates and chordates are as venerable as that of carpoids. There was no time in which carpoids could have evolved into modern chordates.

This is not as much of a problem as it might seem, for two reasons. First, we know that the Cambrian was a time in which many new body plans originated in a relatively short space of time (Philippe *et al.*, 1994). Given the patchy, stochastic nature of fossil preservation, resolving the order in which all these forms appeared is difficult. Therefore it does not necessarily follow that the antiquity of the chordate record, of itself, militates against their origin from carpoids. After all, it is unlikely that the common ancestor of metazoans, let alone deuterostomes, lived much before 800 million years ago at the outside (Knoll, 1992; Knoll and Walter, 1992).

Second, the creatures Jefferies uses in his descriptions of carpoids are not meant to be actual ancestors, but only representatives of their groups, and of particular stages in evolution within those groups. *Cothurnocystis* is held up simply as a representative cornute, not as the ancestor of all mitrates. *Reticulocarpos hanusi* is held to represent the condition of a cornute in transition to the mitrate state, not a cornute actually undergoing this transition. Evolution is a busy hedgerow full of cousins, not a *scala naturae* of ancestors and descendants. That many of these exemplars are younger than the fossils of creatures supposedly descended from their groups may not be relevant: nevertheless, such discrepancies are amenable to statistical test (Marshall, 1990).

It has taken the adoption of cladistic methodology to expunge the conceits of ancestry and descent from evolutionary argument. After all, few are concerned nowadays that the fossil bird *Archaeopteryx* is millions of years older than the advanced theropod dinosaurs from which it is supposed to have 'descended'. Of course, the discovery of Precambrian solutes or cornutes would help, but would be immaterial to the argument itself.

4.2 CALCICHORDATES, BALL BY BALL

The summary of Jefferies' work as presented in the last section does not represent a conception arrived at all at once, but the end product of more than thirty years of work. Most ideas concerning the origins of vertebrates are presented as single, speculative papers, meant more as challenges or talking-points than finished items of work. Others are presented in book form and are often tangential to the principal research interests of the author. Jefferies' work, in contrast, is unusual for its detail and depth, the volume of which can be daunting.

To make things yet more difficult, Jefferies' papers are less about discrete projects than *communiqués*, accounts of work in progress, which makes them hard to understand in isolation. To be sure, Jefferies has published occasional review-length summaries, but he tends to use these to introduce important new concepts, so they are more like theoretical discussion papers than reviews.

One exception is his paper 'In defence of calcichordates' (1981a), published as a response to a critical review by Philip (1979). In this paper he discusses many aspects of his theory to that date, which many people had found problematic or (from Jefferies' point of view) misunderstood.

The other is his book, *The Ancestry of the Vertebrates* (1986), which contains exhaustive summaries of his work to that date together with reviews of the anatomy of Recent deuterostomes. However, *The Ancestry of the Vertebrates* says very little about the phylogenetic position of solutes, and nothing at all about how molecular developmental genetics might contribute to our understanding of vertebrate ancestry. This is hardly surprising, as very little relevant research had been done on either topic at that date.

I devote the remainder of this chapter to a breakdown of Jefferies' most important papers, in the order in which they appeared. Much of the discussion goes over the ground presented in Section 4.1, but the chronological approach will allow for an appreciation of how his ideas have changed and developed.

Jefferies has often adapted his ideas to accommodate theoretical innovations (notably cladistics) and has modified them – even to the extent of complete changes of mind – in response to published criticism and new experimental data. This would not be apparent from anything other than a 'ball-by-ball' commentary.

4.2.1 Calcichordates: beginnings

Jefferies joined the British Museum (Natural History) on 1 April 1960. In February 1964, he had the chance to examine some mitrate material from Shropshire, England, that had been brought into the Museum.

This prompted him to re-read Gislén (1930), and, over the next few years, a mass of disparate facts and speculations fell into place. The result was 'Some fossil chordates with echinoderm affinities' (Jefferies, 1967) in which the cornutes (*Cothurnocystis elizae* and *Cothurnocystis curvata* from Scotland) and mitrates (*Mitrocystella incipiens miloni* from Brittany and *Mitrocystites mitra* from Bohemia) were re-appraised and interpreted not as echinoderms, but as chordates 'with echinoderm affinities'. A more monographic treatment with a few amendments (Jefferies, 1968b) was accompanied by a presentation at a meeting of the Geological Society on 21 February 1968 (Jefferies, 1968a). These works were followed soon afterwards by a re-examination of the oldest and most primitive-known cornute, *Ceratocystis perneri* (Jefferies, 1969).

These papers are innovative in their suggestion that mitrates evolved from cornutes, and that the class Stylophora, comprising the orders Cornuta and Mitrata, be removed from the Homalozoa and incorporated as a new group within the chordates. Thus, the echinoderm class Stylophora became the chordate subphylum Calcichordata. Gislén (1930) did not go that far: he regarded carpoids as echinoderms (at least provisionally), noting only their chordate 'affinities'.

These papers are likewise innovative in their demonstration that it is possible to reconstruct the anatomy of cornutes and mitrates using chordates as the 'model'. This is quite different from traditional lines of echinoderm research, in which carpoids are reconstructed very much with reference to living (or at least pentamerous) forms of echinoderm. For example, in a rebuttal of Jefferies' view that mitrates are chordates, Kolata and Guensberg (1979) emphasize the characteristically echinoderm-like features of mitrates including the anal pyramid, calcite plates, spines, sutural pores, and

> ... most probably the most convincing echinoderm feature of all, the aulacophore. The median furrow and movable cover plates of the aulacophore are very similar to the food grooves of some crinozoans. (1979, p.1132)

and, referring to the internal spaces of the aulacophore, that

> lateral depressions mark the site of head-bulbs of tube feet like those in living ophiuroids. (1979, p.1133)

But the use of chordates as a model should prompt researchers to wonder whether such features really do define echinoderms uniquely. If independent evidence shows that they do, then mitrates are echinoderms. But as soon as one suggests that carpoids might, in fact, be chordates (even ones with 'echinoderm affinities'), one is inclined to question the validity of characters traditionally advanced as definitive of echinoderms *sensu stricto*, such as the presence of a water–vascular system and a calcite skeleton. If carpoids are chordates, then these features are no longer definitive of echinoderms, but primitive features of chordate–echinoderm ancestry which the chordates have since lost.

In that case, one cannot simply dismiss the idea of chordate affinity by claiming – as Kolata and Guensberg do – that mitrates are echinoderms on the grounds that they have a broad, general resemblance to echinoderms of more familiar stripe. Instead, the notion must be tackled with reference to chordates, and in terms of cladistics, something that echinoderm specialists (with notable exceptions such as Smith, 1984) seem disinclined to do.

Traditional research on fossil echinoderms generally concentrates on their external features. Gislén's (1930) reconstruction of the internal soft-part anatomy of carpoids is very much the weakest part of his paper. In contrast, Jefferies presents us with a fully worked-out examination of carpoid internal structure. The reconstructions are based on latex casts from natural moulds left in fossils, in which the calcite skeleton had dissolved away. Occasionally (as in the case of *Placocystites*, Jefferies and Lewis, 1978) they are based on ground serial sections re-drawn and enlarged with a pantograph onto thin sheets of polystyrene, which are then cut out and glued together. Patterns of lines, furrows and ridges could be used to infer the edges of internal organs and body cavities. Although the identification of a small blob of rock with a particular internal organ might seem questionable in isolation, the reconstruction when viewed as a whole makes for a complete picture, consistent with both chordate and echinoderm anatomy[102].

Jefferies deduced the presence of four chambers in the theca of cornutes: he interpreted these as a buccal cavity (nearest the 'mouth'), a pharynx (opening to the exterior through the 'gills'), a right anterior coelom (underlying and to the right of the pharynx when the animal is seen from the 'obverse' side) and a posterior coelom (around the point at which the appendage attaches to the theca). The right anterior coelom would contain the heart, non-pharyngeal gut, gonad and other viscera.

The right anterior coelom was equated with the right half of the rearmost coelom (metacoel/somatocoel) in the standard deuterostome tricoelomate body plan; the posterior coelom was equated with the epicardium of tunicates. From a comparison with extant deuterostomes, Jefferies later inferred the presence of a 'virtual' left anterior coelom (that is, without a cavity) covering the obverse face of the pharynx and buccal cavity.

On functional grounds and on a reassessment of their skeleton he reaffirmed Gislén's view that the gill slits of cornutes really were gill slits (excurrent openings, as opposed to incurrent 'mouths'), as they debouched from a left pharynx. The absence of a right pharynx was linked to the lack of a right-hand set of gill slits.

The internal layout of mitrates was different. Due to the presence of a right pharynx, which presumably grew as an outpocketing of the left pharynx during ontogeny, the right anterior coelom was displaced from its position in the lower right-hand corner into a median position, to some extent above the pharynges and hanging down between them.

The presence of gills in mitrates remained inferential. The pharynges would have met the outside through paired atriopores which were possibly present, although hard to discern, in the forms of gaps between moveable plates.

Crinoids provided the modern model for Jefferies' 1967–1968 discussion of the structure of the cornute and mitrate appendage. The crinoid stalk is built around the 'chambered organ', five sausage-shaped compartments sheathed as a bundle in a skin of nerve fibres, the peduncular nerve. A blood vessel called the haemal strand runs through the space in the middle.

The chambered organ derives ultimately from the right-hand half of the rearmost coelom (somatocoel), and therefore the cavity in cornutes that Jefferies interpreted as the posterior coelom. In living crinoids it is an anti-compressional device, working against the calcite skeleton to keep the ossicles in alignment and preventing the structure as a whole from telescoping and collapsing. The peduncular nerve runs forwards to the aboral nerve centre, an important ganglion at the base of the theca. The aboral nerve centre is essential for regenerating arms or parts of the calyx lost through injury.

Now, Jefferies proposed that all these structures – haemal strand, chambered organ and peduncular nerve – were homologous with those he supposed (from reconstructions) to have existed in the cornute or mitrate appendage, except that in cornutes and mitrates the peduncular nerve was restricted to one edge of the chambered organ, rather than surrounding it. The segments of the appendage (as deduced from the calcite segments) were each furnished with branches from the peduncular nerve and the haemal vessel, breaking through the walls of the chambered organ.

Then, by setting crinoids (and by implication cornutes and mitrates) next to chordates, Jefferies compared the chambered organ with the notochord, the peduncular nerve with the dorsal nerve cord and the crinoid aboral nerve centre with at least some parts of the chordate brain. To be sure, these structures show the same mutual relations in crinoids as they do in chordates, but without supporting evidence from other structures, this does not necessarily imply homology. This evidence came from Jefferies' elaborate reconstructions of the mitrate brain and central nervous system, much more complex than in cornutes, based on internal casts of the marginal ossicles.

Although the cornute brain may have differed little from the crinoid aboral ganglion, the mitrate brain resembled that of fish in many respects (Figure 4.10). It was divided into two parts which Jefferies compared with the prosencephalon and rhombencephalon of a primitive vertebrate such as a lamprey. This comparison is highly detailed, and Jefferies also finds evidence for the presence of trigeminal ganglia, optic nerves, a hypophysis and nerves (traced through the calcite skeleton) to all parts of the theca. Some of these nerves may have terminated with simple eyes, of which there were primitively two pairs in mitrates, called cispharyngeal and transpharyngeal.

In this interpretation, the aboral ganglion cannot be seen as anything other than a homologue of the chordate brain. If so, then by virtue of its position, the peduncular nerve and chambered organ become, respectively, homologues of the dorsal nerve cord and notochord.

For all their revelations, these early papers are hampered by a system of phylogenetic reconstruction unable to manage the detail that Jefferies required to support his argument. This is apparent whenever Jefferies tries to make sense of fine-scale relationships between the various different kinds of cornute and mitrate: first, because the differences involved are not as obvious as those between calcichordates as a whole and, say, tunicates; and second, because it is difficult to make decisions about the soundness of character-states when dealing with a group of extinct creatures that look very different from their supposed Recent relatives among the chordates.

This caused confusion in these early papers. In one place (1967, p.199) Jefferies lists resemblances suggesting affinity between *Mitrocystella incipiens miloni*, *Mitrocystites mitra* and tunicates: 'indeed, a mitrocystitid resembled a giant, calcite-plated tunicate tadpole'. (1967, p.199) However, on the very next page he states that:

> some features, indeed, connect mitrocystitids, in particular, to craniates, and separate them from tunicates and Amphioxus. (1967, p.200)

But later on he writes:

> further it no longer seems likely, though it is still possible, that the Cephalochordata derived from a different group of mitrates to the Urochordata and Craniata. (1968b, p.247)

Confusion deepens with the implication (1968b, p.332) that tunicates split from the calcichordate lineage before the amphioxus and the vertebrates parted company.

The implication of the quote from p.199 is that there may be a particular relationship between mitrocystitid mitrates and tunicates (as distinct from the amphioxus and vertebrates). But on the next page he implies that the tunicates and the amphioxus are more closely related to each other than is either one to the vertebrates.

These discussions can be reduced to two of four possible solutions to a 'three-taxon problem', and can be summarized either in the form of a

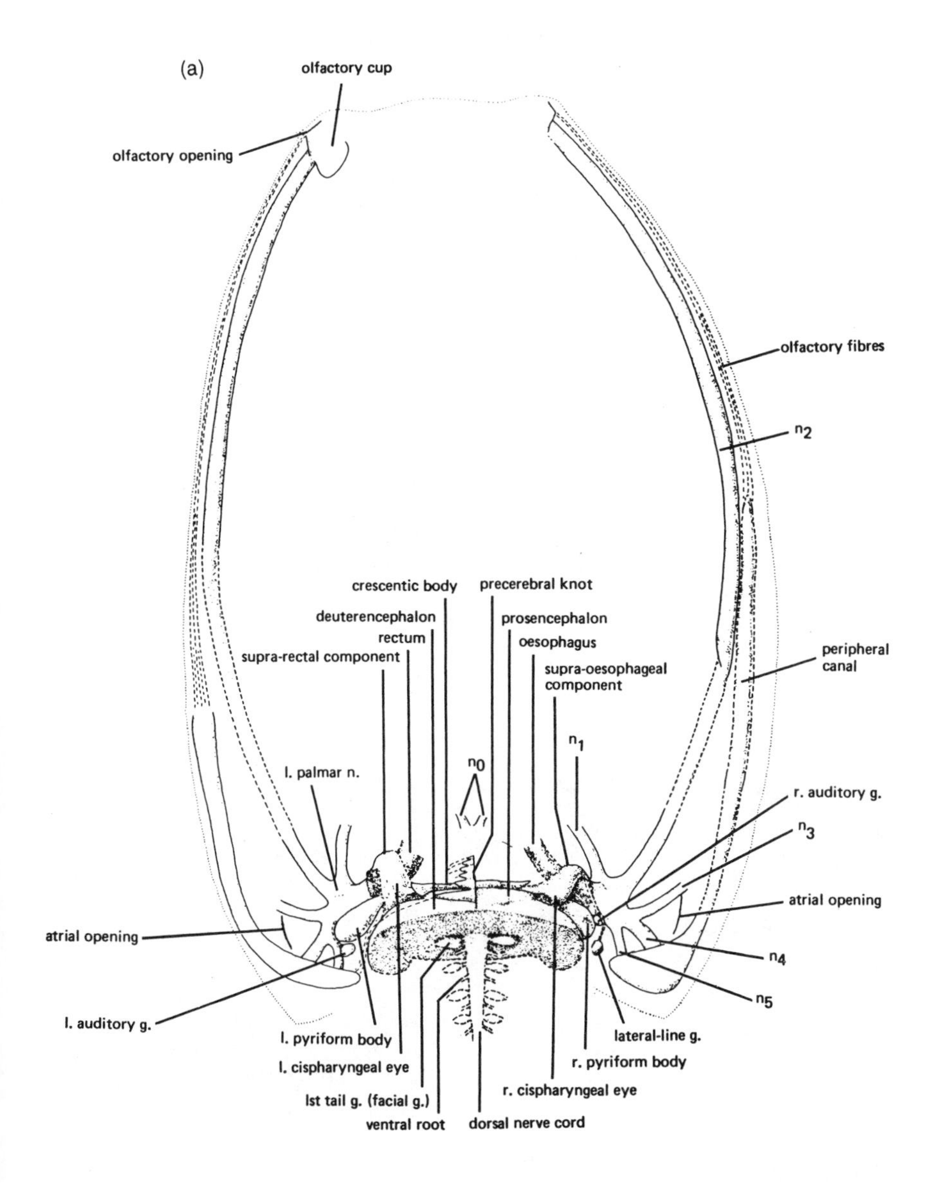

Figure 4.10 Jefferies' interpretation of the nervous system of the mitrate *Mitrocystella incipiens miloni*. Solid outlines are based on direct evidence, dashed lines are reconstructional. (a) Dorsal aspect, (b) ventral aspect and (c) anterior aspect of posterior part of the head (from Jefferies, 1986). (d) Natural mould representing the brain, here in anterodorsal aspect, mp = medial part of brain, ng = narrow groove, pb = pyriform body, pp = posterior part of brain; (e) natural mould of the cast of the plate M_{IRV} (see Figure 4.6c) illustrating the palmar nerve complex (from Jefferies, 1968b).

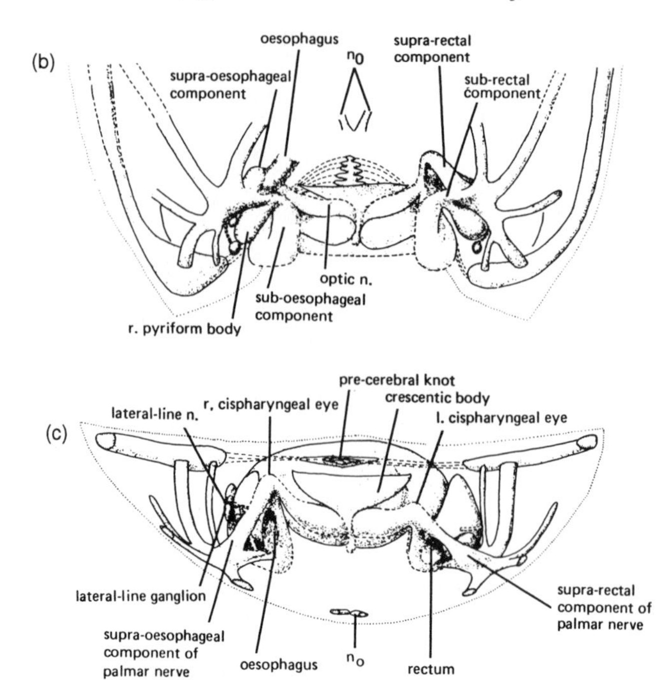

Figure 4.10 (b and c)

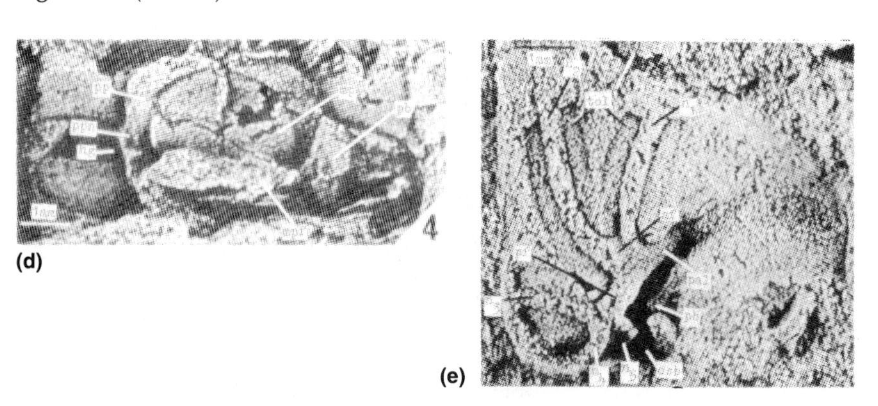

Figure 4.10 (d and e)

tree-diagram or, more conveniently, by the following typographic convention. The quote on p.199 of the 1967 paper reduces to

(amphioxus, vertebrates), tunicates

in which the parentheses express the distinction between tunicates on the one hand, and vertebrates and the amphioxus on the other. The position of the parentheses implies that the amphioxus and the vertebrates are more closely related to each other than either is to tunicates – that the tunicate lineage went its own way before those of the amphioxus and vertebrates had diverged.

On the other hand, this statement says nothing specific about the relationship between vertebrates and the amphioxus, only that the tunicate lineage may be distinct from that of either of the other two groups. It could be that the amphioxus and the vertebrates are just as remotely related to *each other* as either is to tunicates. The solution in which all three groups could be related equally closely is called an 'unresolved trichotomy'

amphioxus, tunicates, vertebrates.

The quote on p.200 of the 1967 paper becomes

(amphioxus, tunicates), vertebrates

but, again, the statement is worded sufficiently ambiguously to allow for an unresolved trichotomy. The next quote (1968b, p.247) also supports this trichotomy although it also raises a fourth possibility, the special nature of the amphioxus lineage,

amphioxus, (tunicates, vertebrates)

before dismissing it as unlikely. The argument on p. 332 (1968b), though, can be boiled down to the statement

(amphioxus, vertebrates), tunicates

identical with one of the interpretations of the statement on p.199 of the 1967 paper.

So, Jefferies presents us with all four possible solutions to the three-taxon problem (including that of the unresolved trichotomy) and seems unable to support any one unequivocally. Interestingly, the one he initially dismissed as unlikely (1968b, p.247) was the solution he eventually adopted, in a study on the mitrate *Lagynocystis* (Jefferies, 1973).

This ambiguity is understandable given that the papers were written in the days before Jefferies applied cladistics to the problem. Until he did so, there was no satisfactory way of resolving these issues. Therefore Jefferies is reduced to the traditional method of explaining the evolution of structure and function as a kind of Just-So Story.

The nebulous tone in which these stories are told does not sit well with Jefferies' clear, precise style, but he rounds it up with an excuse that nonetheless has a defiant ring:

> The foregoing story is admittedly speculative, but differs from previous speculations on the origin of vertebrates in that it starts from a known beginning. (1968b, p.333)

4.2.2 The monsters and the critics I

Jefferies, in 1967 and 1968 stood, like Gislén, against the segmentationist school as regards the 'head problem'. He saw vertebrates very much as divisible into a 'head', unsegmented but for branchiomery, and a 'tail' of much more segmental character, the whole similar to Romer's somatico-visceral animal (Romer, 1972) as described in Chapter 3. Much of the early criticism of Jefferies' views can be interpreted in this light, the critics having assumed that branchiomery is primitively associated with neuromery and myomery.

The 1967 paper was presented at the Zoological Society of London (Jefferies, 1967), and one of the 1968 papers as a demonstration at a meeting of the Geological Society of London (Jefferies, 1968a). The responses of the discussants are included in both cases. Tarlo (the late L. Beverly Halstead) commented with characteristic gusto that

> the whole idea of having the entire nervous system [of mitrates] squashed up in a great heap at the back end of the animal is something that verges on the ludicrous. (1967, p.206)

Jefferies' counter was that Tarlo's 'sweeping rebuttal' came from one schooled in the segmentationist tradition, in which segmentation of the muscles and nerves is physically and ontogenetically related to the positions of the gill arches. This view, as Jefferies explained at some length (Jefferies, 1968b) has a number of difficulties, to which I have already alluded.

Second, although Jefferies is quick to note the concordance between asymmetries in cornutes, mitrates and the larval amphioxus, he does not explain why right-hand gill slits in cornutes are totally absent, rather than merely reduced. If cornutes are ancestral to mitrates, the new set of right gill slits presumed to have been present in mitrates must be innovations, and not recapitulations of right-hand gill slits lost, when, as Gislén supposed, the deuterostome ancestor lay down on its right side, occluding them. Symmetry as an issue is very much in the background in these early papers, but became more prominent as the years progressed.

Third, none of the discussants at the 1967 symposium drew attention to what came to be seen as a serious problem with Jefferies' thesis, even when the author went to some pains to mention them in the paper. He points out that many cornutes and mitrates appear in the stratigraphic record at about the same time as, or rather later than the earliest-known vertebrates, so the cornutes and mitrates under review could not have been vertebrate ancestors, only cousins:

> As regards chronology the earliest known mitrates are about contemporary with the earliest known vertebrates (lowest Ordovician).

> The mitrate ancestor of the vertebrates must therefore have been an unknown form of Cambrian, probably Upper Cambrian, age. (1967, p.201)

This problem has only got worse since the 1960s: the fossil record of vertebrates (or at least, chordates) now stretches down to the Cambrian, yet *Ceratocystis perneri* remains the earliest-known cornute, today as it did in the 1960s. However, as discussed above, there are reasons why this problem may not be as great as it first seems.

Furthermore, the transition from calcitic mitrate to phosphatic vertebrate requires a substantial leap of faith, but without cladistic methods it was not evident in the late 1960s that not one but *as many as three* such leaps might be required, from calcitic mitrates to naked amphioxus, naked tunicates and naked (later phosphatic) vertebrates, each independently. In the beginning, though, it was envisaged that the loss of the calcite skeleton need only have occurred once (Eaton, 1970a), irrespective of the particular lineage of mitrate considered ancestral to each chordate group. This changed later, with the idea that the (armoured) mitrate *Lagynocystis pyramidalis* stood on the amphioxus lineage (Jefferies, 1973), suggesting that the calcite skeleton was lost at least twice.

In their written comments on Jefferies' demonstration at the Geological Society meeting (Jefferies, 1968a), Tarlo and A. P. Hill raised the problem of transition from calcite to phosphate skeletons, from both the physiological and the systematic perspectives. Jefferies conceded that the transition presented problems, although less so if the calcite skeleton had disappeared completely in evolution before a phosphatic one had been acquired by vertebrates. Lampreys and hagfishes, for example, have neither, and at least one genus of modern holothurian has secondarily lost its calcite skeleton. This point was discussed in depth in the monograph (Jefferies, 1968b).

Jefferies found no problem with Tarlo's contention (Jefferies 1968a, p.131) that 'the possession of a calcite skeleton of the echinoderm type is strong evidence to exclude the carpoids from the phylum Chordata' in that, for example, 'the egg-laying habits of *Ornithorhynchus* [the duck-billed platypus] are evidence of reptile affinities, but do not preclude its being a mammal' (1968a, p.133).

This, of course, is simply a matter of what features one chooses as definitive of chordates, and is just one step away from saying that the distinctive skeleton of echinoderms is a primitive feature of the chordate–echinoderm pedigree that has been lost by chordates, and cannot therefore be used to characterize the echinoderms uniquely[103]. But once again, the lack of a language in which to articulate this notion with precision made this step a great one indeed, and left some interesting arguments not so much unresolved as unbroached.

Denison (1971) offered detailed criticism of the first two papers (Jefferies, 1967, 1968b) in a review of vertebrate origins. Whereas Gislén (1930) had offered carpoids as allied to chordates without necessarily being ancestral to them, Jefferies makes clear his view that carpoids are properly chordates themselves, and, furthermore, ancestral to all extant chordate subphyla. According to Denison, this assertion (and the evidence on which it is based) leads to a number of problems.

First, although one might imagine that the viscera of a crinoid might leave traces of their shape on the inner surfaces of the dermal skeleton, this is harder to imagine in a vertebrate. The structure of the body wall is entirely different in each case. Second, in order that the arrangement of thecal chambers in mitrates corresponds to that in cornutes, Jefferies must turn mitrates upside-down with respect to the orientation followed by other workers, posing problems of interpretation as regards the organization of the ossicles in the appendage. Third, as regards the appendage: although Ubaghs and others had regarded the appendage in cornutes and mitrates as a feeding arm with an ambulacrum protected by moveable cover plates, Jefferies regards it as a solid crinoid-like stalk that may be compared in detail with a chordate post-anal tail. In a crinoid, the hollow centre of the tail houses the chambered organ, completely surrounded by the peduncular nerve. Were a carpoid a chordate, the same space must accommodate a notochord and overlying dorsal nerve cord, yet the fossil evidence seems to provide space for just one such axial structure. In addition, spaces conventionally reserved for tube feet and their associated vessels are interpreted as housing segmental muscles, nerves and blood vessels. Difficulties rise when comparing these structures with those of vertebrates as structures in the latter would never be immediately encased by dermal armour (the ossicles). This being so, it is hard to compare a carpoid appendage with the tail of a fish.

Fourth, the interpretation of cornute gill slits as such is difficult to credit because they occur in a single series, whereas chordate gill slits occur in paired series; they are physically distant from the mouth, much bigger than the mouth, and could only have been opened by stretching the theca. Most seriously, these slits are absent in mitrates, and the evidence for internal gills and atriopores proposed by Jefferies seems weak.

Fifth, and turning to the internal reconstruction of the nervous system: Jefferies' reconstruction of the carpoid brain is exactly right for a crinoid aboral nerve centre, but not for the brain of a vertebrate. It is too weakly developed in its middle and anterior portions; and situated behind the gut and at the base of the appendage, is in entirely the 'wrong' place. The course of the cranial nerves, traced through the dermal skeleton, is entirely appropriate for a crinoid but not for a vertebrate. In which case, it is hard to make homologies between carpoid nerves (if that is what they are) and vertebrate cranial nerves stand – the problem is exactly the same as that of the dermal ossicles of the tail and their relationships with the underlying structures. Sixth, and unavoidably, carpoids have a calcite skeleton that resembles that of echinoderms down to the most microscopic scale.

Denison concludes that carpoids are best considered as echinoderms, and any resemblances as might exist between carpoids and vertebrates must be convergences.

Jefferies dismisses several of these points explicitly in a paper (Jefferies, 1981a) largely concerned with criticisms raised by another (Philip, 1979). Jefferies accuses Denison of basing his conclusions on a misreading of interpretative drawings, rather than a study of the photographs or the specimens themselves:

> Denison (1971, p.1134), on the curious basis of three of my diagrammatic drawings [Jefferies, 1968b: Figures 6, 12, 17] ... has asserted that only one longitudinal organ is recorded in the groove [within the carpoid appendage], and seems to imply that I have imagined the evidence (1981, pp.363–364)

Jefferies is similarly dismissive of Denison's implication (1971, p.1134) that the ossicles in the carpoid appendage are homologues of the vertebrae of vertebrates.

With regard to Denison's fourth point, that chordate gill slits occur in paired series, Jefferies notes (1981, p.359) the presence of a single series in the ontogeny of amphioxus.

The remainder of Denison's points are either solved elsewhere (the cornute–mitrate orientation question) or are simply reassertions of echinoderm characters (calcite skeleton) which of themselves cannot refute the assertion that carpoids are chordates.

However, Denison is right to note the problems of reconciling a chordate internal structure with an echinoderm-style skeleton. The integuments of modern echinoderms and chordates are rather different from one another, and nowhere does Jefferies speculate on how these differences might influence the reconstruction of the calcichordate skin and body wall (Nichols, personal communication).

4.2.3 Calcichordates: *Ceratocystis* and *Cephalodiscus*

An explicit statement of the closeness of echinoderms to chordates, closer even than hemichordates, remains likewise elusive in the 1969 paper on

Ceratocystis (Jefferies, 1969), although it is edging nearer. In addition to gill slits, chordate-like tail and the other features described in other cornutes, boot-shaped *Ceratocystis* (Figure 4.11) has what Jefferies interprets as a hydropore: a remnant of a pterobranch-like internal organization and an echinoderm-style water–vascular system otherwise absent from cornutes and mitrates.

If this interpretation is correct, it has important implications for phylogeny, for it places *Ceratocystis* between hemichordates such as *Cephalodiscus* on the one hand, and echinoderms and chordates on the other. Jefferies orders a large number of character states to produce a Just-So Story that has a more confident tone than the one presented in the 1968 monograph (Jefferies, 1968b). The difference comes from an argument about symmetry.

In the 1969 paper, Jefferies shows that the arrangement of coeloms and openings in *Ceratocystis* in dorsal view is topologically equivalent to that of *Cephalodiscus* as seen from the left side (a comparison that adds validity to the interpretation of the hydropore in *Ceratocystis*, through its positioning relative to other body openings). If this arrangement is homologous (as Gislén thought), it explains the sinistral tendency evident in echinoderms and chordates.

It also demands that echinoderms are more closely related to chordates than are hemichordates, because of the sinistral tendency of echinoderms and chordates which hemichordates such as *Cephalodiscus* do not share. Further, it suggests that the water–vascular system is not a definitive character of echinoderms, but a primitive holdover of the ancestors of both echinoderms and chordates, which chordates have since lost.

In the accompanying scenario, Jefferies notes that a *Cephalodiscus* zooid, although it usually lives in its tube, can venture out of doors and crawl about. Perhaps a population of wandering Cambrian pterobranchs living on the exposed sea floor were (by some mischance) prone to falling over on their right sides, occluding the tentacles, holes and coeloms on that side. Eventually they stayed like that, and the right-hand structures withered away. Developing a protective calcite skeleton, some of these pterobranchs elaborated their left-hand set of tentacles into the echinoderm water–vascular system, and the pterobranch stalk into a plated, crinoid-like appendage. 'Evolution' writes Jefferies on p.532, 'now proceeded in two directions'. The echinoderms evolved from a line in which the gill slits were lost: the chordates arose from another that specialized in pharyngeal filter-feeding at the expense of the water–vascular system.

4.2.4 The monsters and the critics II

The first constructive criticism of the calcichordate theory came from T. H. Eaton, first on Jefferies (1968b) (Eaton, 1970a); then on all the papers up to and including *Ceratocystis* (Jefferies, 1969) (Eaton, 1970b). Eaton's

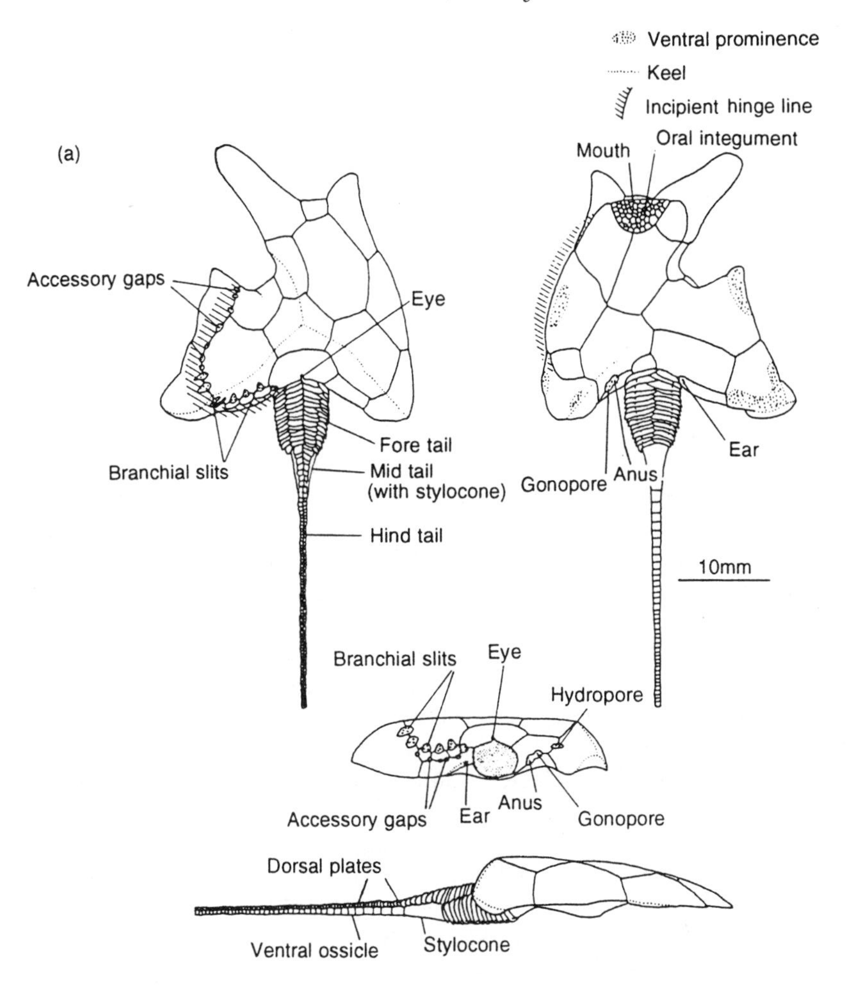

Figure 4.11 (a) Reconstruction of the cornute *Ceratocystis perneri* in dorsal, ventral, posterior and right lateral view. (b) (overleaf) Comparison between the left side of the pterobranch *Cephalodiscus* (left) and the dorsal surface of *Ceratocystis*. The topology of the various orifices is circumstantial evidence for dexiothetism: moving clockwise round both animals we see a mouth (m), a hydropore (h), a gonopore (g), an anus (a), a tail (t) and branchial slits (bs). The letters mp refer to the mesocoel pore of *Cephalodiscus* (from Jefferies, 1990; compare (b) with Gislén's comparison between *Cothurnocystis* and the amphioxus, Figure 2.3).

main criticism (the only one to be articulated in both papers) is that the appendage of crinoids and the tail of chordates have different embryological origins – so if the carpoid appendage is a homologue of one, it cannot be a homologue of the other.

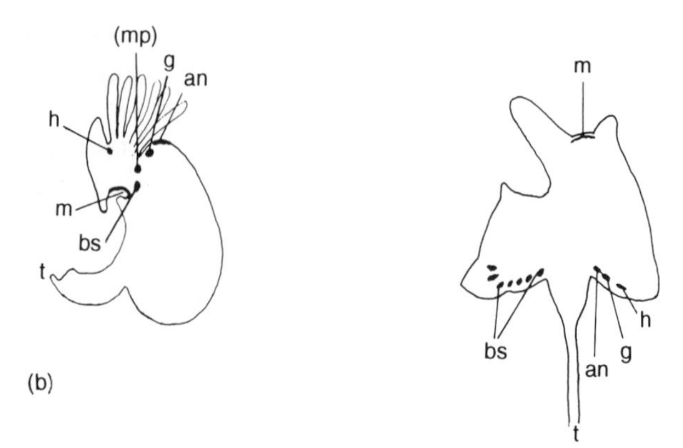

Figure 4.11 (b)

Crinoid larvae are at first free-living, but eventually attach themselves by the preoral lobe to a convenient substrate. This lobe develops as the stalk, the rest of the animal undergoing a rotation in which the animal loses its bilaterality and mouth and anus are brought to face upwards, away from the substratum. So for all that the stalk looks like a posterior structure, it actually develops from the embryonic anterior end. In chordates, in contrast, the tail is post-anal both in embryology and adult position, and thus a truly posterior structure. Furthermore, the crinoid stem is a pentamerous, radially symmetrical structure, quite different from the bilateral chordate tail: the calcichordate appendage is the only truly bilaterally symmetrical feature these creatures possess.

The solution, (Eaton, 1970b) may be sought in hemichordates: pterobranchs have a preoral lobe with an adhesive and suctorial function, as well as a posterior peduncle used in locomotion. Could not the first correspond to the crinoid stem, the second with a chordate tail? In this scheme, the calcichordate appendage is a homologue of the hemichordate peduncle and the chordate tail, but not the crinoid stem. Indeed, as Eaton says

> the lack of homology with the echinoderm stem is, to me, a convincing reason for accepting the bulk of Jefferies' interpretations. (1970b, p.972)

Eaton's other main criticism (1970b) concerns Jefferies' supposition that mitrates evolved from cornutes, noting that

> it seems more probably the opposite, because asymmetry can scarcely be an ancestral feature of anmals [*sic*] whose anatomy bears strong evidence of bilaterality, and whose larvae and embryos must have been bilateral (1970b, p.973]

He goes on to suggest

> that mitrates were evolved from small, unarmored [*sic*], free pterobranchs, that they paralleled crinozoan echinoderms in developing a calcite skeleton but remained essentially bilateral, never completely sessile, and that they gave rise early to the cornutes by loss of symmetry ... (1970b)

From this, it seems clear that although Eaton is aware of the asymmetrical features of modern chordates (in particular the amphioxus) he does not tie them in with Jefferies' (and, ultimately, Gislén's) idea of symmetry lost and regained, expressed in the suppression of right-sidedness in cornutes and the evolution of a 'new' right in mitrates. Instead, Eaton sees the chordates evolving from (bilateral) mitrates, with cornutes an aberrant offshoot. This leaves the modern asymmetries of chordates unexplained. So although Eaton, like Jefferies, is unusual in tying chordates more closely to echinoderms than to hemichordates, he does so for different reasons. For Eaton, it reconciles a difficulty concerning the homology of the calcichordate appendage. For Jefferies, it is a natural outcome of dexiothetism.

Jefferies' formal response to Eaton (Jefferies, 1971) aims to clear up the confused issue of stem–tail homologies. The trouble is that the problem is more involved than it first appears. Of the two kinds of modern echinoderm that habitually attach themselves as larvae, the temporary stalk of embryonic asteroids (starfish) derives from the axocoel (the anteriormost coelom). It is thus a truly anterior structure and can have nothing to do with the hemichordate peduncle or the chordate tail.

Crinoids, though, are different. Although the stem is deployed at the anterior end during attachment, the chambered organ within actually derives from the right somatocoel (one of the posterior coeloms) so is embryologically equivalent to the hemichordate right metacoel, which forms (the right half of) the peduncle.

Jefferies owns to the possibility that an extension of the crinoid somatocoel might have been used as an attachment structure in a way quite unconnected with the origin of the calcichordate appendage or chordate tail and, therefore, that the one has nothing in particular to with the other. However, as a coelomic structure, the chambered organ originates as an evagination of the archenteron, just like the notochord. 'Having said all this', notes Jefferies,

> I agree that the calcichordate stem is more certainly homologous with the tail of other chordates than with the stem of crinoids. Stems enclosing an extension of a somatocoel (= metacoel) may have arisen more than once in the deuterostome stock. (1971, p.910)

Jefferies also makes clear, for the first time, his views on the phylogenetic position of echinoderms, since 'the echinoderm affinities of calcichordates seem more important to me than Eaton'. (1971, p.910)

> In my view these common features [between echinoderms and *Ceratocystis*] indicate a period of common descent for echinoderms and calcichordates which hemichordates did not share. (1971, p.911)

For the first time, Jefferies makes his views on the left-handedness of chordates explicit, in a passage worth quoting at length:

> Eaton's view [that cornutes evolved from mitrates, rather than *vice versa*], however, fails to explain why primitive cornutes are such a strange shape. I have suggested that this shape indicates descent from a bilateral ancestor lying on its right side. Furthermore, such an ancestor would probably be a hemichordate resembling *Cephalodiscus*. It follows that the plane of bilateral symmetry of vertebrates is perpendicular to that of hemichordates. The right side of a hemichordate corresponds to the ventral side of a fish. An analogy can be drawn with flatfishes which have a plane of near bilateral symmetry perpendicular to the usual one for vertebrates, and also perpendicular to the sea bottom. This resulted when the bilateral ancestor of the flat fishes took to lying down on one side. Something similar probably happened in the early evolution of the echinoderm–chordate stock. The internal asymmetries of mitrates are best explained as awkward inheritances from a cornute condition. They were gradually reduced by an increasing tendency towards bilateral symmetry. (1971, p.911)

This discussion answers one of the questions posed above, that right-side gills in mitrates were new inventions, whereas the left gills are homologous with the 'old' left gills of cornutes.

Jefferies' hitherto rather confused position on the amphioxus–tunicate–vertebrate 'three-taxon problem' is also nearing a resolution. Referring to unpublished work that would eventually surface as the paper on the mitrate *Lagynocystis* (Jefferies, 1973), Jefferies suggests that the amphioxus arose independently from mitrate stock, and, more tentatively, that tunicates and vertebrates had a more recent common ancestry – a reversal of the hypothesis more usually adopted (such as that of Garstang) that the amphioxus is more closely related to vertebrates than are tunicates.

Regnéll (1975) comments on Jefferies' ideas in a review on pelmatozoan echinoderms. Without citing any particular evidence to support his dismissal, he concludes

> Jefferies has to be saluted as a very clever advocate of an elegant and seductive hypothesis which has the one drawback not to be compatible with the interpretation of carpoid anatomy that is best in accord with factual evidence, as far as I can judge. (1975, p.535)

4.2.5 The cornute–mitrate transition

Apart from the deficiencies caused by the lack of an adequate phylogenetic method, the biggest hole in Jefferies' ideas up to 1971 relates to the cornute–mitrate transition, in particular the orientation of the appendage. The succeeding pair of papers aimed to address this. Both concern calcichordates from Ordovician deposits near Prague, one of them an advanced cornute, *Reticulocarpos hanusi* (Jefferies and Prokop, 1972; Figure 4.12), the other the mitrate already mentioned, *Lagynocystis pyrimidalis* (Jefferies, 1973; Figure 4.13).

Reticulocarpos is important because it is in many ways transitional between cornutes and mitrates. Unlike many primitive cornutes, the distal section of its appendage is rather short. From a functional analysis of the appendages of cornutes and mitrates, informed by the anatomy of *Reticulocarpos*, Jefferies reasoned that the proximal, middle and distal sections of the cornute tail are not in fact homologous with the apparently corresponding regions in the appendages of mitrates. Instead, mitrates have shed the cornute middle and distal sections, the proximal section evolving into a completely new, tripartite structure. This explains the anomalies between the orientations of the appendages: cornute appendages cannot be compared with the (superficially) upside-down mitrate appendages for their entire length, because the two structures are not homologues. As I noted above, Jefferies finds morphological evidence for this view in a comparison between the appendages of *Reticulocarpos* and that of a primitive mitrate, *Chinianocarpos thorali*.

From this, Jefferies explains that cornute tails were adapted for pulling the ventrally flat-sided animals backwards over the sandy seafloor. But *Reticulocarpos*, unlike its more primitive relatives such as *Cothurnocystis*, may have spent some time on the surface of very soft mud. Its body was flat, small, symmetrical and light, spreading as lightly and evenly over the mud as possible, like a snowshoe on snow. Its appendage was short, for a longer one would have dragged the animal further into the mud with each power stroke. Getting rid of most of it would have carried a selective advantage.

Mitrates, unlike most cornutes, are adapted for life in mud. Rather than perching precariously on the surface like flat-bellied cornutes, their convex ventral surfaces give them the ideal profile for 'floating' in soft mud like boats, using the displacement of mud to compensate for their

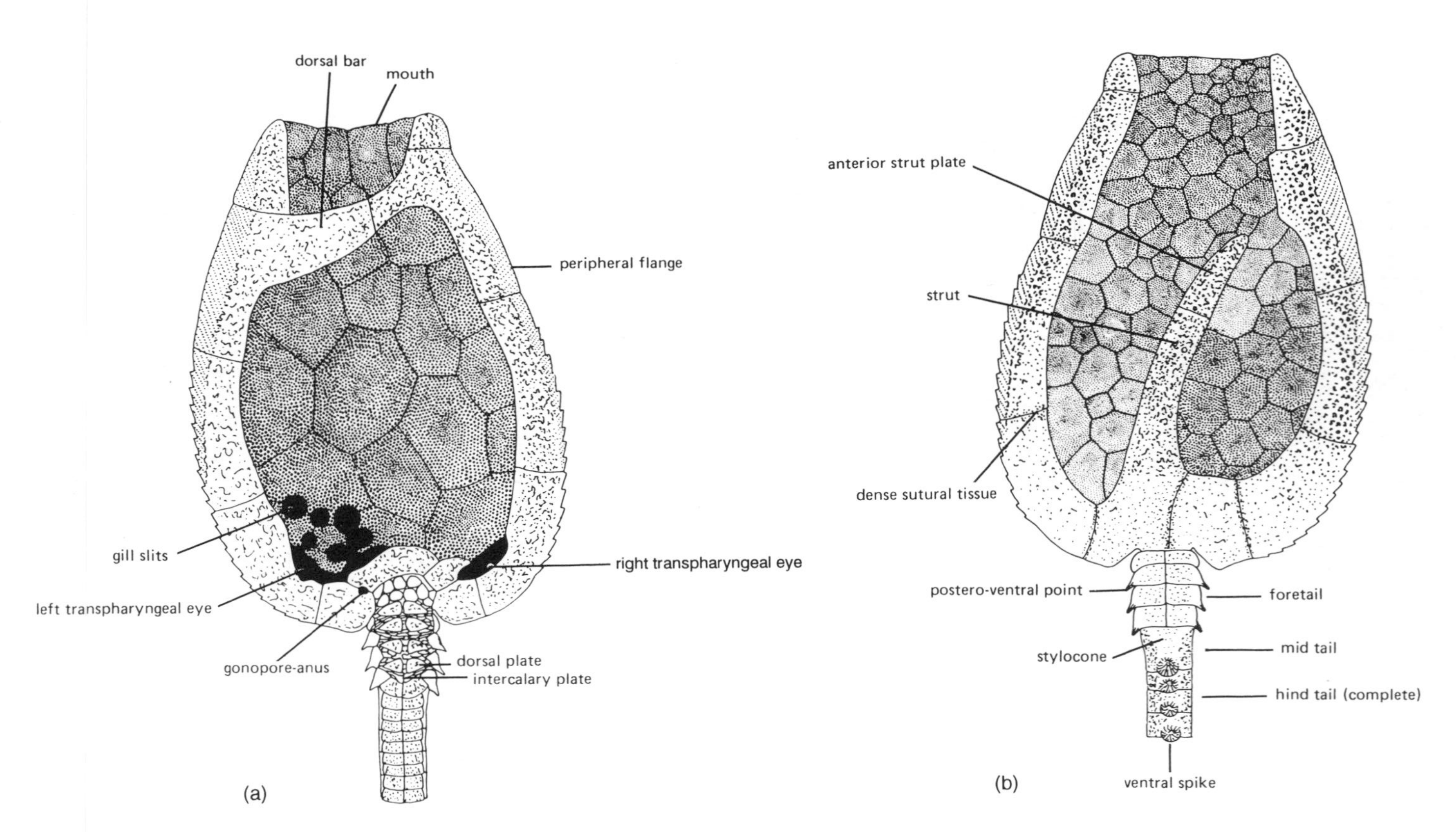

Figure 4.12 The cornute *Reticulocarpos hanusi*, in (a) dorsal and (b) ventral views (from Jefferies, 1986).

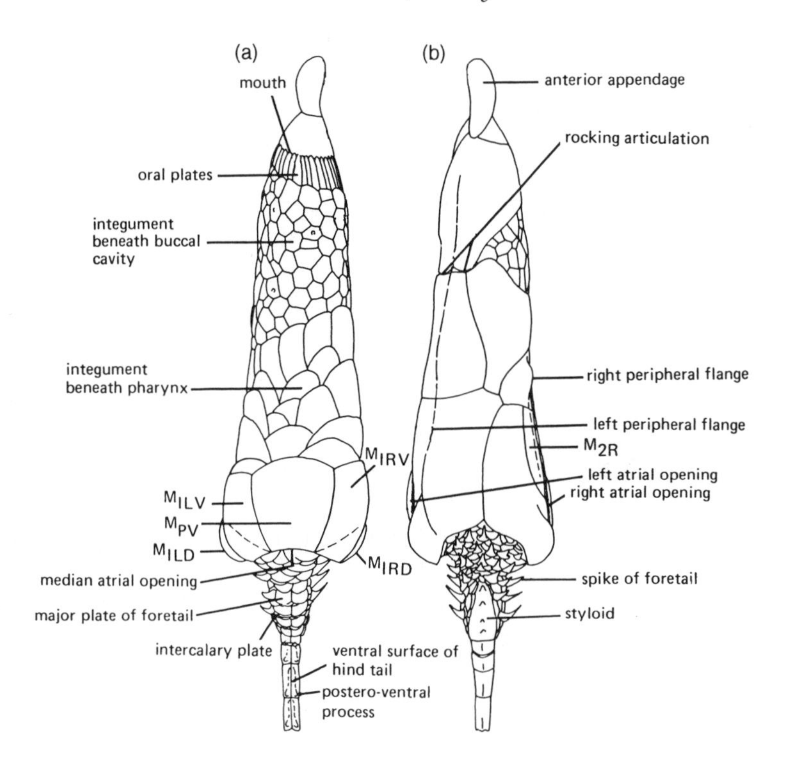

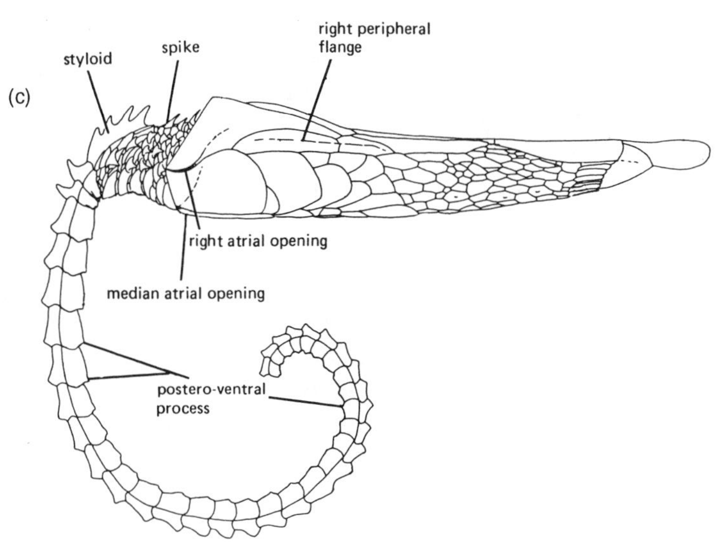

Figure 4.13 The mitrate *Lagynocystis pyramidalis*, in (a) ventral, (b) dorsal and (c) right lateral views (from Jefferies, 1986).

weight and thus avoid sinking. Mitrate appendages are different from those of cornutes functionally as well as structurally, because they were adapted to haul the animals through the mud rather than over it. The symmetrical, streamlined external shapes of mitrates, the ribbed sculpturing of their external skeletons, with the steeper slope of each rib facing the anterior, and comparison with modern bivalve molluscs supports this idea. This argument justifies not only the differences between cornute and mitrate appendages, but the long-standing problem with the orientation of the mitrate body with respect to that of cornutes. Until then, Jefferies' view of this had been justified by an argument based almost entirely on the inferred layout of the internal organs.

The 1973 paper on *Lagynocystis* extends this theme, but is surprising for another reason: uniquely among mitrates, it has not two but three atria. The third lies in the middle of the animal at the posterior end of the head, and presumably meets the world through a median atriopore, quite separate from the (inferred) left and right branchial openings more usual in mitrates. The evidence for this comes from internal anatomy. For once, Jefferies does not have to infer the presence of internal gill slits, for the median atrium of *Lagynocystis* sports a basket of well-preserved, calcified gill bars (Figure 4.14) connecting it with the left and right pharynges[104]. This is reminiscent of the amphioxus, which as a young adult develops tertiary gill slits on either side, behind the left and right gill slits. Jefferies suggests that one of these gill bars belongs to the left atrium, providing, for the first time, direct evidence for mitrate gill slits. In addition, Jefferies traces the lines left by the soft tissue of the gill bars themselves upwards into the posterior coelom. This would have allowed each gill to serve as the outlet for coelomoducts, in the manner of the nephridia of the living amphioxus. This is because each gill bar would have carried a ventral extension of the posterior coelom (epicardium), and each such extension was homologous with a nephridium associated with a gill bar of amphioxus.

What with the general primitiveness of *Lagynocystis* in other respects, Jefferies reasons that the lineage leading to the amphioxus is older and distinct from that leading to tunicates and vertebrates (which share features such as motor end-plates, lacking from the amphioxus). The 1973 paper sheds light on a number of primitive and asymmetrical features of the amphioxus:

> To sum up, *Lagynocystis* shows that amphioxus is far from being the ideal text-book ancestor of vertebrates. (1973, p.465)

The *Lagynocystis* paper received rare approbation in the form of an anonymous note in *Nature* (Anon., 1973) which while noting Jefferies' 'spectacular success' in reconstructing the calcichordates, noted that

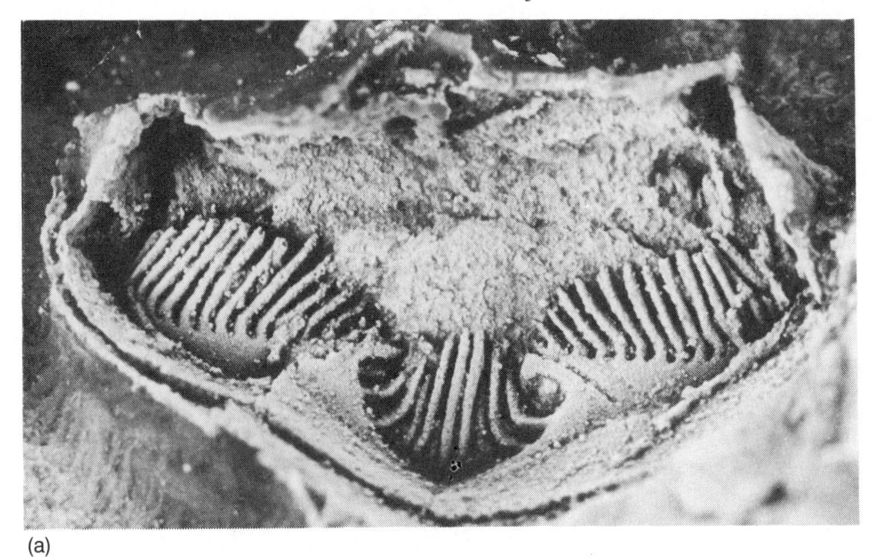

(a)

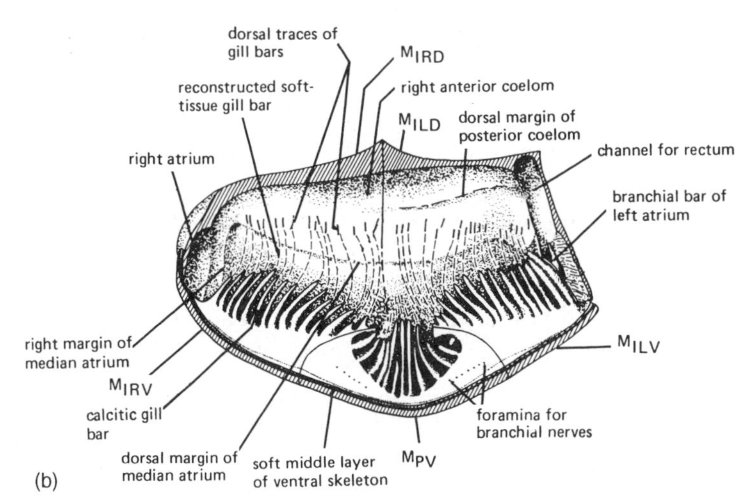

(b)

Figure 4.14 (a) The atrium of *Lagynocystis pyramidalis* (from Jefferies, 1973), with (b) reconstruction (from Jefferies, 1986).

many of Jefferies's colleagues find it difficult to accept such a radical reassessment of the phylogeny of *Amphioxus*. In trying to convince them Jefferies is up against the perennial problem faced by palaeontologists when describing groups, such as the calcichordates, which are very different from any living today. There is no way of knowing whether the soft parts have been accurately identified; this being the case it is difficult to see how hypotheses concerning the evolutionary relationships of such animals can be tested. (Anon., 1973, p.124)

Quite apart from the animals themselves, the 1972 and 1973 papers are curious mixtures of old and new. Both include Just-So-Stories and the old-fashioned family trees in which traditional taxonomic groups such as families and orders are shoehorned into geological time: but it is clear that Jefferies was beginning to organize his ideas along Hennigian lines.

For, just now and again in these two papers, Jefferies uses terms associated with cladistics more than with old-fashioned systematics. The 1972 paper has just one, 'paraphyletic' on page 111, referring to the fact that the cornute family Amygdalothecidae is not a natural group among cornutes because it contains the ancestors of the mitrates, as well as the cornute *Reticulocarpos* and its immediate relatives. The 1973 paper has three: 'parsimony' (p. 460), 'sister-group relationship' (p. 463) and 'Hennigian' (p. 465). Figure 17 (p. 464) shows a series of branching diagrams that are cladograms in all but name, and makes an interesting foil to the family tree in Figure 1 (p. 413). Both figures are reproduced here as Figure 4.15.

A book chapter (Jefferies, 1975) sums up the work to date before Jefferies openly adopted Hennigian principles. It is far from a simple summary, though, and contains a notable reverse, the abandonment of the homologies so carefully drawn between the crinoid stalk and the chordate tail:

> All calcichordates consist of two parts which hitherto I have called theca and stem by supposed homology with crinoids. I have always maintained, and indeed it is now virtually certain, that these parts are homologous with the body and tail of other chordates[105]. I now think, however, that they are probably not homologous with the theca and stem of stemmed echinoderms but represent a parallelism to these structures. (1975, p.254)

This re-appraisal was forced by the description of *Gogia*, an eocrinoid from the Middle Cambrian of Utah (Robison, 1965), a representative of a group possibly ancestral to crinoids, in which the stalk is only weakly developed: suggesting an acquisition of a stalk independently of carpoids. This convinced Jefferies that the suspicion raised five years earlier by Eaton (1970a, b) had been justified: that crinoid stalks and chordate tails must be parallelisms, not necessarily derived from a structure present in the latest common ancestor of echinoderms and chordates. If the ancestor had such a structure, it was lost in the very earliest echinoderms, only to be re-evolved as a convergence.

The implication of this change was serious. If the calcichordate appendage is not homologous with the crinoid stem, but a parallelism, then why should the calcichordate appendage bear any closer relationship with the chordate tail? The inferred presence of essential chordate features such as the notochord and dorsal tubular nerve cord suddenly

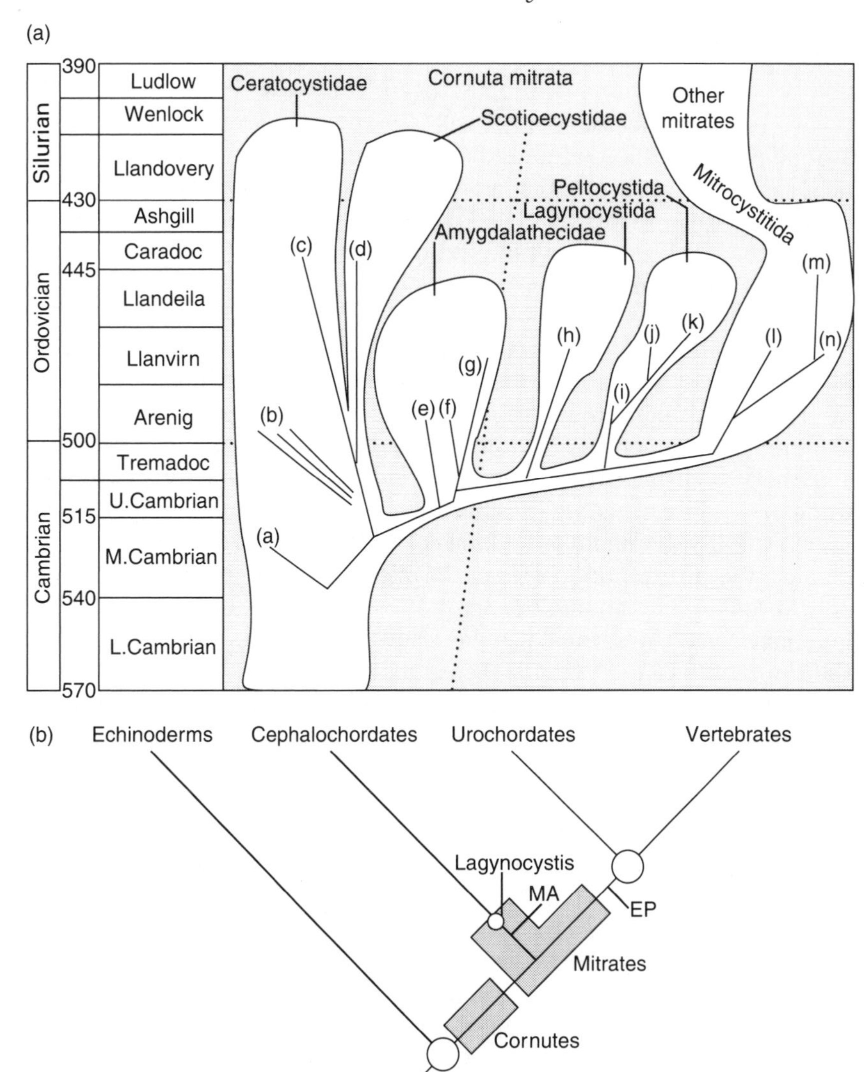

Figure 4.15 Cladistics appears somewhere in the middle of Jefferies' 1973 paper on *Lagynocystis*. Contrast (a) the stratigraphy-pinned family tree at the start of the paper (from Figure 1) with (b) the resolution of a Hennigian three-taxon problem from near the end (Figure 17c). EP = acquisition of motor end plates in urochordates and vertebrates; MA = acquisition of a median atrium in cephalochordates. Note, however, that *Lagynocystis* is still cast as an ancestor rather than a sister group of extant cephalochordates (from Jefferies, 1973).

seemed weaker than before. Jefferies subsequently went to some lengths to find chordate features in the mitrate theca that would support a phyletic link between the calcichordate appendage and the chordate tail, if not the crinoid stalk (Jefferies and Lewis, 1978).

4.2.6 Hennigian interlude

As I showed in Chapter 1, the aim of phylogenetic reconstruction is to resolve relationships into nested sets of monophyletic groups, preferably using data from extant animals. But what is to be done in the case of calcichordates, which are extinct as well as somewhat remote morphologically from their living relatives? Is it permissible to use cladistics to resolve the phylogeny of wholly extinct animals so different from modern creatures that the choices of characters, polarities and outgroups will be contentious?

Although the strict application of cladistics admits only data from Recent animals, data from extinct creatures can be fitted in later. They are incorporated as paraphyletic offshoots from stem lineages called plesions (Patterson and Rosen, 1977). Convention does not permit the inclusion of a plesion to disturb the topology of a cladogram already established using extant taxa. This rule has been the source of much theoretical controversy. Purists contend that fossils can only be included after the fact, because the incorporation of data from animals living in different periods (Recent and fossil) into the analysis would add a note of implicit evolutionary change and thus bias the result (Rosen *et al.*, 1981; Forey, 1982; Gardiner, 1982; Gee, 1988, 1992; Peterson, 1994).

On the other hand, the effect of appending a large amount of fossil data to an analysis based on relatively little neontological data can hardly be ignored: some distortion is inevitable. The topology of cladograms created using data from, for example, fossil mammal-like reptiles (Kemp, 1988; Gauthier *et al.*, 1988; Gee, 1988, Ahlberg, 1993), Devonian tetrapods (Ahlberg and Milner, 1994) and Eocene whales (Novacek, 1994) allows the tracing of important evolutionary transitions inaccessible to neontological data sets, whether morphological or molecular.

Nevertheless, one is entitled to worry that the mixture of Recent and fossil data might be over-welcoming of conclusions reached *a priori*. In the case of mammal-like reptiles, for example, one already knows the character-state distribution of the end product (modern mammals) so it is easy to trace the order of acquisition of these characters as 'transformation series', and ignore character-combinations which might have produced other major groups, but did not (Ahlberg, 1993; Gee, 1993). Faced with this dilemma, some have appealed for methodological purity to be leavened with pragmatic compromise (Fortey and Jefferies, 1982).

However, I suspect that it is possible to produce a phylogenetic analysis in which the inclusion of fossil data is unproblematic. Such an analysis would satisfy some or all the following criteria.

First, *all* of the creatures in the analysis would be extinct.

Second, they were so nearly contemporaneous that evolutionary change would not have introduced significant distortion[106].

Third, the stratigraphic provenance of their fossils is too uncertain to be of value for primitiveness to be related to stratigraphy (Patterson, 1981).

These criteria seem to apply to the calcichordates. For although Jefferies' analysis results in modern chordates as a crown group, much of his work concerns the acquisition of chordate characters in parts of the stem extremely remote from the crown. For all practical purposes, Jefferies' animals are extinct even though their supposed descendants are not.

Again, Jefferies is dealing with creatures that acquired their major features in the 'Cambrian Explosion', in which evolutionary changes happened extremely quickly.

Finally, Jefferies can use the stratigraphic inconsistencies of the carpoid record to his advantage. Rather than worrying that the fossil record of cephalochordates and vertebrates is coëval with that of the cornutes and mitrates from which they are supposed to have evolved, Jefferies can afford to ignore these problems altogether for the purposes of phylogenetic reconstruction. For unlike traditional evolution- and scenario-based methods, cladistics is – as we have seen – essentially timeless.

4.2.7 Calcichordates in the late 1970s

The adoption of Hennigian methods transformed Jefferies' approach to his work. The immediate result was a long paper on the English mitrate *Placocystites forbesianus* (Figure 4.16) co-authored with Museum colleague David N. Lewis, who made a series of models of the animal from very thin sheets of polystyrene, cut from tracings of ground longitudinal and transverse sections using a pantograph (Jefferies and Lewis, 1978).

Placocystites is a difficult fossil, and Jefferies ranges all his forces against it by re-describing many now-familiar cornutes and mitrates in the light of Hennigian methodology learned when the 1978 paper was being written. This paper can be seen as a dry run for the book, *The Ancestry Of The Vertebrates* (Jefferies, 1986).

Gone is the strict subphylum status of the Calcichordata: ancestral to all living chordates, this subphylum is no longer a 'natural' group, because there were calcichordates ancestral to (separately) tunicates, the amphioxus, vertebrates, as well as all three. (Not long afterwards, Jefferies abandoned categorical ranks such as subphyla altogether.)

Cornutes are reassigned as the stem group of chordates, whereas all known mitrates can be accommodated in the chordate crown group, each one being assignable to the stem groups of either tunicates, the amphioxus or vertebrates, within the chordate crown group as a whole. *Placocystites* is a stem vertebrate, in the same way that *Lagynocystis* is a stem cephalochordate.

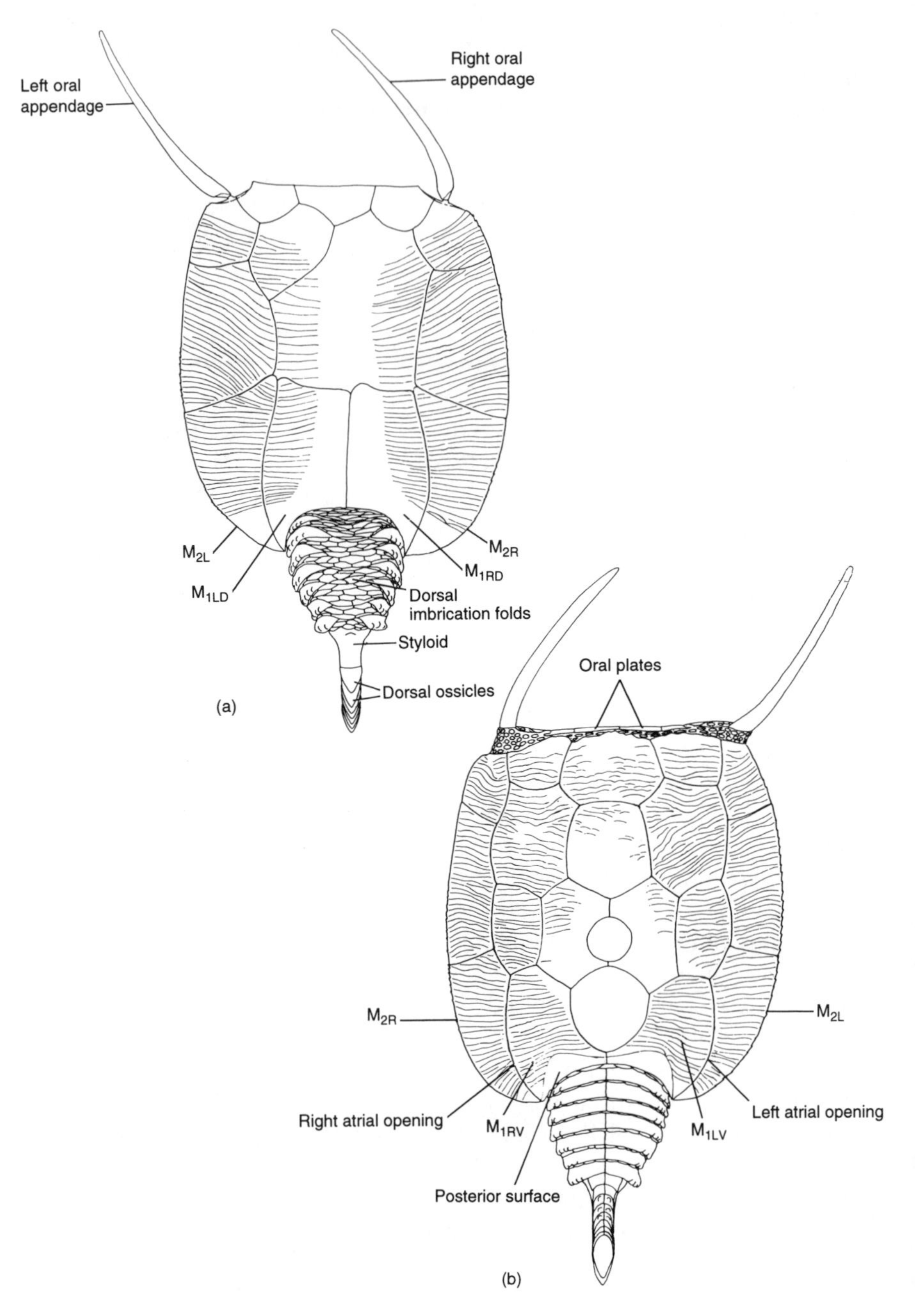

Figure 4.16 The mitrate *Placocystites forbesianus* in (a) dorsal, (b) ventral, (c) right lateral, (d) posterior and (e) anterior views. (f) Polystyrene-sheet reconstruction in ventral view of the inside surface of the dorsal surface (from Jefferies, 1986).

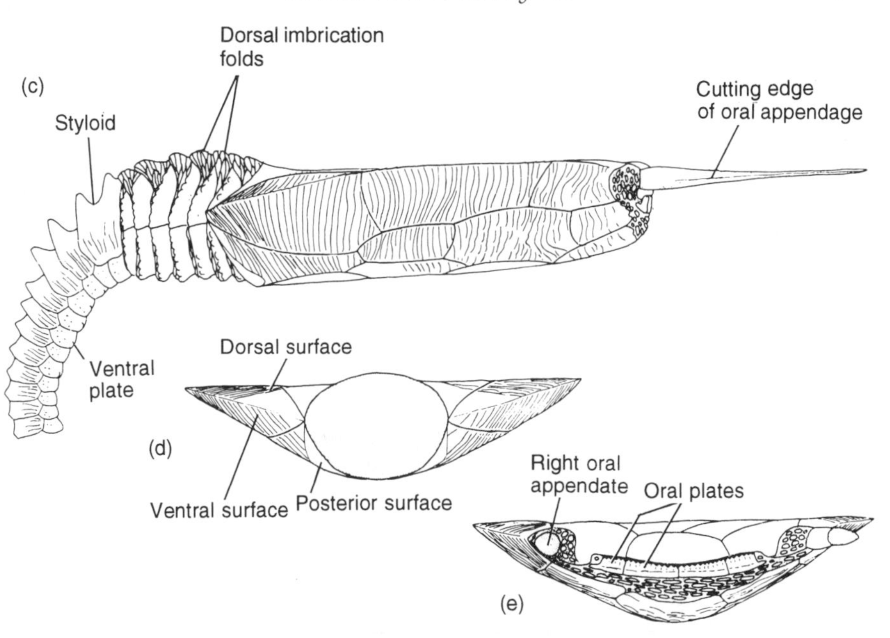
(c)
Dorsal imbrication folds
Styloid
Cutting edge of oral appendage
Ventral plate
Dorsal surface
(d)
Ventral surface
Posterior surface
Right oral appendate
Oral plates
(e)

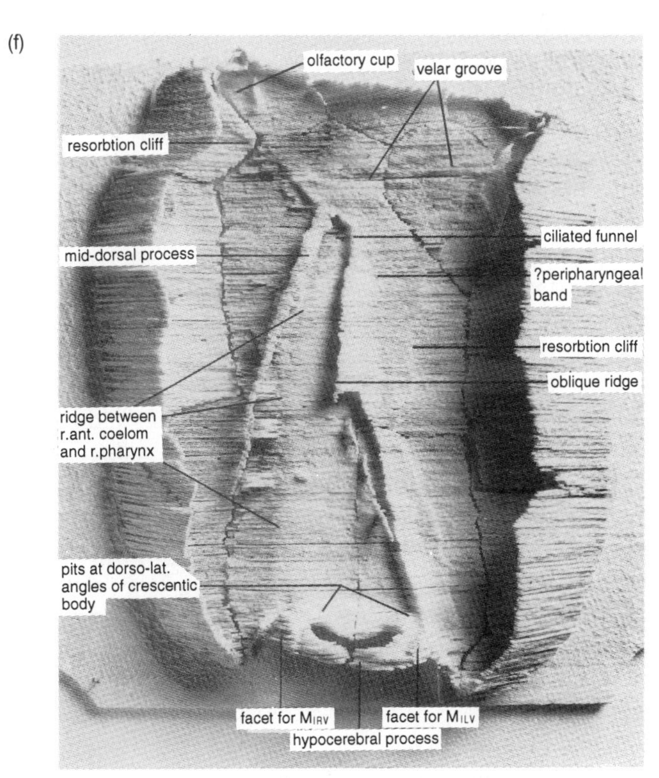
(f)
olfactory cup
velar groove
resorbtion cliff
ciliated funnel
mid-dorsal process
?peripharyngeal band
resorbtion cliff
oblique ridge
ridge between r.ant. coelom and r.pharynx
pits at dorso-lat. angles of crescentic body
facet for M_{IRV}
facet for M_{ILV}
hypocerebral process

Gone also is the homology between crinoid and cornute appendages. As we have seen, this poses a problem, and Jefferies is forced to scour the theca of *Placocystites* for features that would link cornutes and mitrates with chordates.

A particular feature of *Placocystites* is a flange of calcite that he interprets as homologous with the dorsal lamina of the tunicate pharynx, an important part of the filter-feeding apparatus. Much follows from this, and with the illumination of other mitrates and cornutes and comparisons with adult and larval tunicates, the amphioxus and vertebrates, Jefferies is able to infer the innervation and layout of the viscera in *Placocystites* in detail.

To the reader, the difficulty of following several extended anatomical arguments at once makes one suspect sleight-of-hand. But Jefferies is doing his best to be as explicit as possible about complex topics: hence frequent digressions into vertebrate embryology, tunicate anatomy and so on.

Perhaps a little unnecessarily (what with the almost unassimilable wealth of other detail), Jefferies also finds room for head-segmentation in *Placocystites*. He starts by re-naming the calcichordate body (formerly the theca) as a 'head' in the chordate sense of a pre-notochordal head.

Calcichordates had equivalents of a chordate head and tail, but neither trunk, nor notochordal head (these structures would evolve later, as a consequence of mitrate head–tail intergowth). Structures associated with these body regions – the viscera, pharynx and so on – were sited anterior to the pre-notochordal head, so that the head joined the tail directly.

Jefferies cites structures present in the 'head' of *Placocystites* that he equates with the premandibular and mandibular somites which in vertebrate become (in part) the extrinsic eye muscles. Mitrates did not have moveable eyes, but the arrangement of the somite-like structures (the anterior coeloms, and a structure called the crescentic body) is as one would expect, were they eye-muscles-in-waiting. The hyoidean somites in the Goodrichian groundplan were present in mitrates, although in the first section of the tail rather than in the head.

The somites, separated in mitrates, were juxtaposed in the process of head–tail intergrowth, as the axial structures of the tail grew over the top of the pharynges and viscera: the hyoidean and post-hyoidean somites from the back came to lie next to the mandibular and premandibular somites in the front. The gills, formerly completely separate from the somites of the tail, became arranged more or less in series matching the dorsal somites.

Most ingeniously of all, the characteristically vertebrate kidney evolved when somitic mesoderm (once confined to the segmented tail) met lateral plate mesoderm (once confined to the mitrate head). The calcite skeleton was lost along the way, but Jefferies advances the possibility that the calcitic otoliths (ear-stones) of vertebrates might be the remains of a mitrate heritage.

The idea of intergrowth, then, suggests that the components of the modern vertebrate head were originally distributed in various parts of the animal, but came together during a fundamental reorganization of the body parts. The idea that the vertebrate head is a neomorph that formed all-of-a-piece (Gans and Northcutt, 1983) is inconsistent with that view. For one thing, myomery and neuromery cannot have dictated branchiomery, because hemichordates have seriated gill slits but neither somites, neural crest or cranial nerves. Furthermore, as Gislén showed (see Chapters 2 and 3), this 'head–tail intergrowth' is imperfect in some primitive forms such as hagfishes.

As I showed in Chapter 3, experimental embryology echoes this heritage of intergrowth: as neural crest cells migrate downwards on either side of the developing vertebrate head, the expression of homeobox genes in neural-crest tissue in specific parts of the hindbrain can be tracked to specific parts of the neural-crest-derived branchial skeleton (Hunt and Krumlauf, 1991a,b; Hunt *et al.*, 1991a, b).

4.2.8 The monsters and the critics III

In many ways, the calcichordate theory reaches its acme in the 1978 paper, subsequent work serving to consolidate (review articles; Jefferies, 1980, 1981b), fill gaps (an excursion into mitrate locomotion; Jefferies, 1984), refute criticism (for example by Philip, 1979, with replies by Jefferies, 1981a, 1982; discussed further below), or all three (the 1986 book, *The Ancestry Of The Vertebrates*).

A critique of the calcichordate theory (Philip, 1979) gave Jefferies the opportunity to clarify several points that people had found difficult. His rebuttal (Jefferies, 1981a) goes over much the same ground as the *Placocystites* paper (Jefferies and Lewis, 1978), but, as it is a direct response to criticism, is more focused and thus easier to follow. The Philip–Jefferies correspondence is particularly illuminating for this reason, and I shall treat it here in some depth.

Philip's paper actually concerns two theories. In addition to Jefferies' work, Philip criticizes Georges Ubaghs' interpretation of the carpoid stem as a feeding organ or 'aulacophore', in which the 'cover plates' in the distal section could have opened out to reveal an ambulacrum (Section 4.1 for discussion of this idea).

In the 1979 paper, Philip refers to Gislén's point, made explicit in the 1930 paper, that the suggestion of a direct relationship between carpoids and living chordates was neither intended nor should be inferred from his comparisons. (Nevertheless, given the known relationship between modern echinoderms and modern chordates, one should not be surprised to find antique creatures that displayed features reminiscent of both.)

Philip summarizes the calcichordate theory in a long passage (1979, pp.446–447) that is described by Jefferies as 'fair so far as it goes' (1981a, p.354) and which he himself quoted at length (1981a, p.354) with emendations (for example, Jefferies changes Philip's 'theca' for 'head' and 'stele' for 'tail' to accord with his own current usage, as in Jefferies and Lewis, 1978) and annotations reflecting his own changing views. The passage demonstrates how the calcichordate theory was understood by contemporary echinoderm specialists. It is also clearly set out, and so I shall reproduce the passage here exactly as written by Philip, except for the insertion (as in Jefferies, 1981) of numbers in superscripted square brackets that refer to annotations dealing with various points:

> The thecal openings on the left side of *Cothurnocystis* and other cornutes are seen as functional gill slits (cf. the development of amphioxus in which the gill slits appear first on the left-hand side)[1]. The major thecal opening of cornutes is interpreted as a mouth and a pore opening near the stele is seen as an anal opening[2] (in some cornutes, e.g. *Scotiaecystis*, this is internal). The theca of typical cornutes is reconstructed with four (later, in 1975, five) chambers designated the buccal cavity, the left pharynx, the anterior and posterior coeloms (left and right anterior coeloms in 1975)[107]. In mitrates such as *Mitrocystites* right gill slits are also developed and so is a corresponding right pharynx. Also inferred are internal gill slits that open into left and right atria, and a rectum that opens into the left atrium (cf. a tunicate tadpole). Mitrates therefore have paired gill openings with the faeces leaving the theca through the left gill opening. The carpoid stele is interpreted as containing a notochord with a dorsal nerve cord, together with muscle blocks. The various canals running within and along the internal surface of the theca are seen as nerves, which, after reconstruction, are found to lead to an enlarged ganglion at the base of the theca, designated the brain. This is reconstructed in detail, for example, in *Mitrocystella*, and is found to be morphologically similar to that of *Petromyzon* or of a cephalaspid, except for being shorter. As evidence for its fish-like character homologues of the trigeminal complex and trigemino-profundus ganglia of the olfactory, optic and lateralis complexes are all recognized. A small slit in *Mitrocystites* is interpreted as a lateral line; paired pores towards the posterior of the theca of *M. mitra* are held to house eyes; in fact two types of eye are identified, namely transpharyngeal and cispharyngeal; the latter are homologues of vertebrate eyes[3]. A ventral atri-

um is seen in the mitrate *Lagynocystis pyramidalis*. This form is considered to give rise directly to amphioxus and the cephalochordates (with the loss of calcite skeleton, the loss of a fish-like brain and other changes)[4]. Another mitrate group, close to *Peltocystis* and its allies, with an independent loss of skeleton, gave rise to tunicates. A further mitrate group took to continuous swimming, lost its calcite skeleton, acquired a phosphatic one[5] and gave rise to vertebrates. (1979, pp.446–447)

Philip's stated intention at this point is to examine the basic assumptions underlying this scheme and to test their validity. Turning to the specific, numbered points raised by Jefferies (1981a):

1. The morphologically left-hand gill slits in amphioxus do not actually arise on the left, but mid-ventrally, and pursue a rightward excursion before coming to rest on the left in adults (see Chapter 1).
2. More precisely, a combined gonopore and anus.
3. Examination (subsequent to the appearance of Philip's paper) of the presumed stem-tunicate mitrates *Balanocystites* and *Anatifopsis* suggests that the cispharyngeal eyes appeared as way-stations along the nerve trunks connecting the brain with the transpharyngeal eyes. The latter were lost, leaving the cispharyngeal eyes and the (shortened) nerve trunks.
4. Jefferies notes that he once thought *Lagynocystis pyramidalis* directly ancestral to living acraniates 'but never said that it was so' (1981a, p.355), quoting in support a statement on p.410 in his 1973 paper. This latter statement, though, could be read either way. It runs:

   Lagynocystis pyramidalis, from the marine Lower Ordovician of Bohemia, was either directly ancestral to amphioxus, or closely related to such an ancestor (1973, p.410)

5. Hagfishes probably represent the most primitive living sister-group of vertebrates (see Janvier, 1981), and lack bone entirely. Indeed, they need never have had a bony ancestor, in which case the latest common ancestor of living vertebrates need not have possessed bone, either. The implication is that bone is not a shared, derived feature (synapomorphy) of vertebrates, but a character found in certain vertebrate subgroups.

Jefferies further notes that the evidence from *Lagynocystis* shows that gill slits of mitrates are not entirely hypothetical; that his soft-part reconstructions of the mitrate head are supported by skeletal evidence[108]; and complains that Philip hardly mentions the many asymmetries that mitrates share with modern tunicates and the amphioxus, a point that is central to his rebuttal of Philip's critique.

Before addressing Philip's main comments, Jefferies makes a philosophical point addressing the strength of his scheme as a theory:

> Throughout I shall emphasize the interconnectedness of my interpretation and its frequent points of contact with complex skeletal fact. In both these respects the calcichordate theory is greatly superior to its rivals. (1981, p.356)[109]

(a) The Philip–Jefferies debate: general criticisms

Tarlo once commented (Jefferies, 1968a) that it was hard to imagine an animal in which the viscera are in front of the notochord with the central nervous system squashed up in between. Philip echoes this (1979, p.447) but Jefferies responds (1981a, p.357) that tunicate tadpole larvae are just like that, and in turn differ little from Romer's somaticovisceral animal (Chapter 3).

Philip also finds it hard to imagine 'an animal seven-eighths of the body of which is pharynx' (1979, p.447). It hardly needs Jefferies to point out that modern tunicates are just so proportioned. Moreover, brachiopods and bivalves have mantle cavities which are functional analogues of the pharynx and dominate the internal spaces of these animals.

Philip (1979) also objects to the 'enormous plasticity' that Jefferies attributes to calcichordates in their evolution. He notes how Jefferies draws up detailed similarities between *Lagynocystis* and the amphioxus as evidence of their alliance (asymmetry of inferred anus, direction of swimming, etc.) yet glosses over major differences (the loss of the calcite skeleton, head–tail intergrowth with the reorganization of somites, etc.) that indicate their distance. 'The calcichordate theory abounds with similar matters of emphasis' he concludes (1979, p.448). Jefferies replies that such is only to be expected in the early history of a group, in which members of related but distinct taxa should be more like each other than the living members are.

But Jefferies also notes that reorganizations such as head–tail intergrowth should not be regarded as impossible, especially in the early history of a group: and given the metamorphoses undergone by many modern creatures during their lives, they need be no more dramatic than, say, the placing of the anus in the left or right pharynx. For example, Philip (and others) find incredible the consequence of intergrowth in which (in relative terms) the gill slits slip down the theca and come to rest along the sides of the stele. But as Jefferies says,

> Any attempt to derive standard vertebrates from a tunicate tadpole-like ancestor, and there have been many such of which mine is

only the most recent, requires a tailward migration of the gill slits. (1981a, p.357)

As I showed above, Garstang, Berrill and others barely sketch the transition between the tadpole larva and the adult form of the amphioxus or vertebrate, let alone attempt to describe it in the kind of detail routinely offered by Jefferies. And yet a shift in gill slits of this kind would be necessary in Garstang's scheme, for all that he never mentioned it. The implication of Jefferies' comment is that the success of the former corpus of ideas may actually lie in its glossing over of the morphological mechanics involved in the transformation from tadpole larva to vertebrate or amphioxus. Instead, features required for vertebrates are simply dreamed into existence as handmaidens of selective necessity in a free-swimming pelagic descendant of sedentary ancestors, or for marine animals swimming against the tide. The latter approach is popular because it does not require specialist knowledge to comprehend, has a comforting air of cozy fireside storytelling about it – and, of course, cannot be tested or its specifics discussed.

(b) The Philip–Jefferies debate: holes in the head

The traditional way of interpreting the morphology of extinct echinoderms is by examination of external features. Echinoderms and chordates are replete with orifices of every kind, including gill slits (in chordates); hydropores, madreporites, specialized respiratory structures and food grooves (in echinoderms); mouth, gonopore and anus (in both). The interpretation of carpoids as chordates with echinoderm affinities thus poses special problems.

Echinoderm specialists infer the functions of these openings with reference to similar structures of known function in modern echinoderms. For example, many modern echinoderms hold the anus proud of the body surface on a small 'pyramid' of calcite shards arranged to look like a tiny volcano. This makes good functional sense: in the same way that factory chimneys work best if built tall, the stronger currents further from the surface will carry faeces away far more efficiently from a raised anus rather than one flush with the surface. Such things cannot be tested on an extinct echinoderm, though, in which a structure like this is usually interpreted as an anus simply because it looks like one.

When confronted with a structure in an extinct echinoderm for which he can find no modern analogue, an echinoderm specialist will invent a descriptive name that sounds important but actually implies no function. For example, the structures in *Cothurnocystis* which Jefferies interprets as gill slits may be referred to as 'cothurnopores' (bootlace-holes), the comb of gill-like structures in *Lagynocystis* as the 'ctenoid' (i.e. 'comb-like') organ, and so on. This habit illustrates commendable caution, but ultimately acts as a brake on attempts at interpretation.

Jefferies' philosophy of reconstruction is entirely different. He infers the function of structures not on their appearance but on their positions relative to other structures in the animal. It was in such a way that he matched the openings in *Ceratocystis perneri* not only with cystoids, but with openings in the pterobranch *Cephalodiscus* and thus inferred the presence of the hydropore in the former (Jefferies, 1969). It would not occur to an echinoderm specialist to make any such comparison, because *Cephalodiscus* is not an echinoderm; it lacks the distinctive echinoderm integument which (presumably) dictates the form of the openings in the echinoderm body wall, irrespective of their position.

Philip notes (1979, p.448) the many ways in which the thecal openings of *Cothurnocystis* have been interpreted. However, Jefferies and colleagues are alone in interpreting the distal opening (relative to the appendage) as a mouth[110], and a small pore on the left side of the base of the appendage as the anus. The distal opening of *Cothurnocystis* is guarded by a set of plates and looks very like the classic echinoderm anus. But this opening varies a great deal among cornutes and several workers have suggested that the anal pyramid might have been retractable. More certainly, the pore suggested by Jefferies as an anus[111] looks nothing like any other echinoderm anus. It lacks valvular plates, is very small (less than 0.25 mm across in some cases) and passes through solid calcite. It is much more likely to be a hydropore or gonopore.

Jefferies contends that the interpretation of the gonopore-anus as such accords with its position – in the outflow from the gill slits (of which more below). In *Cothurnocystis*, it connects with two grooves, a smaller and a larger, in the internal skeleton which Jefferies interprets as gonoduct and rectum respectively. These grooves can be followed to the pore (which, as noted above, is just to the left of the base of the appendage) from a position just right of the appendage – in which place the gonopore-anus can be seen in the more primitive *Ceratocystis*. The gonopore-anus in *Ceratocystis* can be identified as such with respect to primitive echinoderms: a move in more advanced cornutes from just right to just left of the appendage (and closer to the exhalant gill slits) makes functional sense.

In response to Philip's criticisms, Jefferies notes (1981a, p.362) that anal sphincters in most animals are innocent of valvular plates; that although the pore in *Scotiaecystis* is small, this is the exception rather than the rule (perhaps that form, with its upward-facing mouth, was a suspension feeder producing relatively small amounts of solid waste); and that the pore did not always pass through a single plate of calcite (in *Ceratocystis*, for example, it lies on a suture).

Jefferies' identification of the mouth as such rests on the assumption that all the other thecal openings are outlets; that it is the largest opening in the theca; and that it is at the opposite end of the theca (head) to the appendage (tail). The fact that in some cornutes it looks like an anal pyramid should not deceive: the atrial siphons of tunicates are guarded by pyramids formed of strips of the material tunicin, and the mouth of the holothurian *Psolus* is guarded by spike-shaped plates. And, of course, not every cornute (and no mitrate) has a mouth with the anal-pyramid-like structure.

The interpretation as gill slits of the pores at the posterior left-hand corner of the carpoid theca is, as Philip notes (1979, p.450) a 'keystone' of the calcichordate theory. Yet it may be hard to reconcile this very specific function with the diversity of forms exhibited by these structures in carpoids. *Cothurnocystis* has gills arranged in a linear series, each one covered with an elaborately constructed flap. *Scotiaecystis*, in contrast, has a tightly arranged series of as many as 40 narrow slits pressed between vertical calcite plates. Others, such as *Ceratocystis*, *Phyllocystis* and *Reticulocarpos*, had a much looser arrangement of holes without (as far as is known) any kind of special protective structure (Figure 4.17, based on Philip's Figure 6). It could be that these structures – without being gill slits in the chordate manner – were just some of the many kinds of specialized respiratory organs found in Palaeozoic echinoderms in general (Figure 4.18). If cystoids had their pectinirhombs and diplopores, and eocrinoids their epispires, then why deny carpoids their 'cothurnopores'?

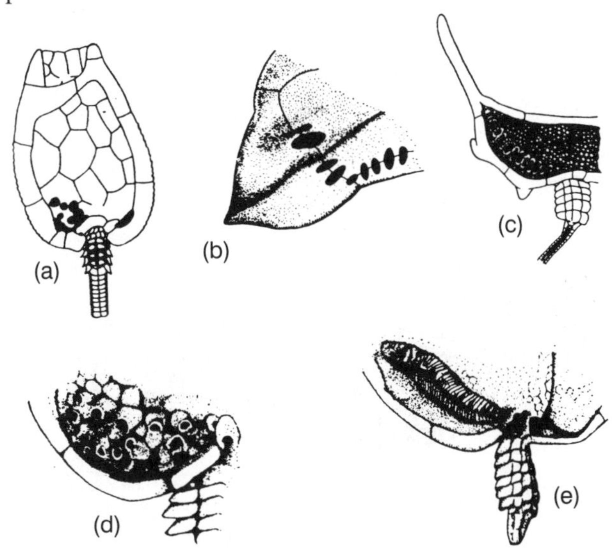

Figure 4.17 Diversity of thecal pores (gill slits) in carpoids. (a) *Reticulocarpos hanusi*, (b) *Ceratocystis perneri*, (c) *Cothurnocystis*, (d) *Phyllocystis crassimarginata*, and (e) *Bohemiaecystis* (from Philip, 1979).

These are alternative interpretations, and as valid as any, but Philip does a poor job of demolishing Jefferies' specific case for gill slits. He cites an objection raised by Ubaghs in which the possibility of a direct connection between the gills and digestive tract (i.e. pharynx) is dismissed, because the digestive tube would have been forced to bend at 'an acute angle at the end of the branchial region, which is unlikely' (p.450). But why is this unlikely? The digestive tubes of all manner of animals (including humans) twist and turn in ways to shame the most ardent contortionist.

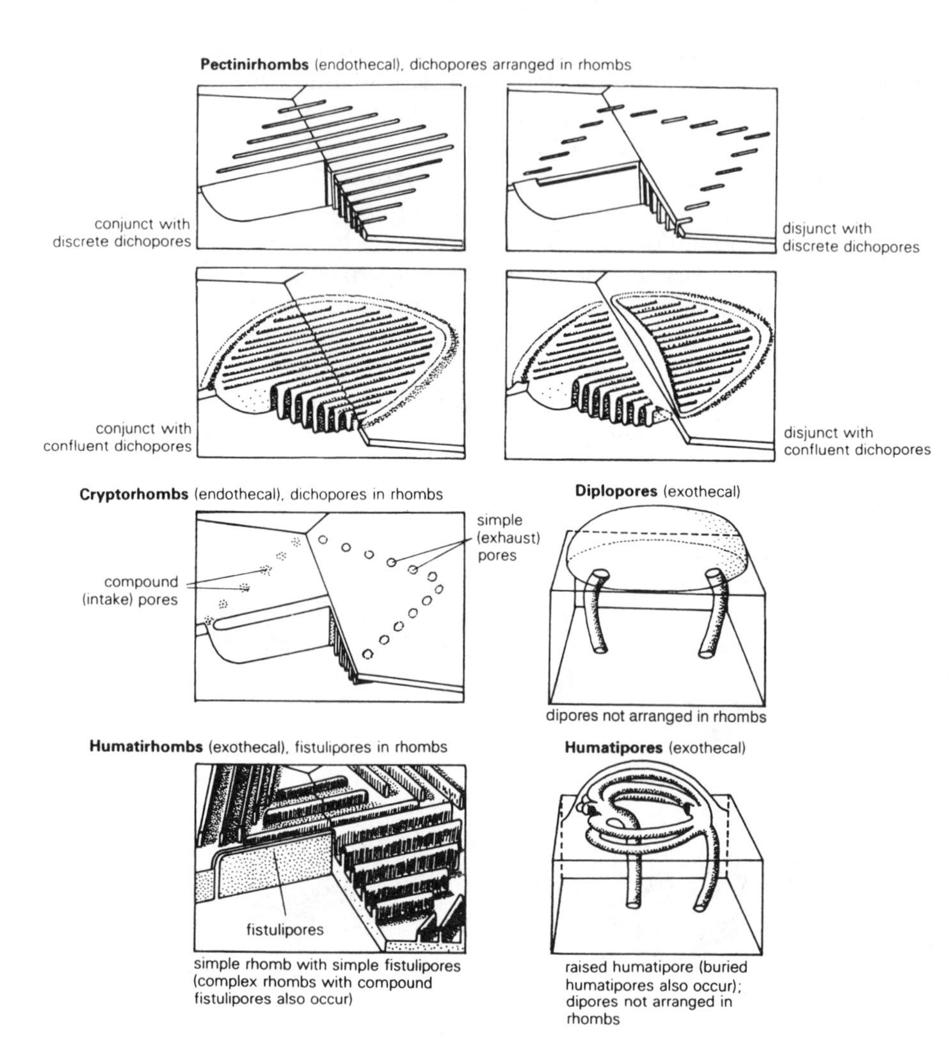

Figure 4.18 Different kinds of respiratory pore structure in cystoids (from Clarkson, 1986, based on Paul).

Apart from recording that cornute gill slits are found in the posterior left-hand corner of the theca, Philip gives no good reason why they should be there rather than anywhere else. However, position is key to Jefferies' detailed explanation (1981, pp.357–361). As is typical of Jefferies' style, the interpretation of these structures as gills depends on their position in relation to other structures.

In *Cothurnocystis* and *Scotiaecystis* at least, the gill slits are plausible excurrent openings. In all cornutes, though, their position on the left side can be compared with the morphologically left-sided gill slits of amphioxus larvae. In *Cothurnocystis*, the position of the gills matches internal sculpturing on the marginal plates, supporting the connection with a pharynx. The gills are also in the right place to serve as exhalant openings for the products of the gonopore. Importantly, the interpretation of gills as such depends on the interpretation of the gonopore, and vice versa.

(c) The Philip–Jefferies debate: the notochord

Philip (1979, p.451) summarizes Jefferies' reconstruction of the carpoid 'tail'. Using *Cothurnocystis elizae* as a model, Jefferies (1967) placed a notochord in the median groove inside the plated stem, overlain with the dorsal nerve cord. Muscles occupied the lateral grooves, the muscle blocks straddling the sutures between successive ossicles. These muscle blocks would have been served each by separate, segmental, ganglionated lateral nerves coming off the main trunk, together with a blood vessel from an axial vessel that would have run along inside the notochord. The placement of a blood vessel inside the notochord may seem strange, raising suspicions that Jefferies placed it there for want of space inside the carpoid stem for putting it anywhere else. However, it was originally reconstructed as such using crinoids as a model – the peduncular vessel runs down the centre of the chambered organ.

In the initial reconstruction, the crinoid stem and chordate tail were proposed as homologous, using the carpoid stem as a kind of intermediary. Later on, the crinoid stem was seen as a parallelism, in accordance with the possibility that such stems had arisen several times in echinoderm evolution. Nevertheless, the homology between carpoid stem and vertebrate tail was maintained, partly as a result of detailed comparisons between the internal anatomy of the mitrate theca and the tunicate pharynx.

But there is a problem. For these thecal homologies to hold, the mitrate tail must be reconstructed 'upside down' relative to that of cornutes. As I showed above, Jefferies copes with this argument by supposing that the mitrate tail is homologous with the proximal part of the cornute tail only. Philip finds this all too hard to take. It is 'almost impossible', he says, 'to envisage animals undergoing such radical morphological changes' (1979, p.452). But that Philip finds such changes impossible to envisage is no good reason for denying the possibility in any general sense. Why spoil it for everyone else?

Although Philip thinks that the 'stele' may have contained musculature and have been an organ of locomotion, he sees no compelling reason why it should have contained a notochord.

Jefferies (1981a, p.363) repeats his arguments in favour of a notochord that would have acted as an anti-compressional structure, particularly in the proximal parts of the appendage in which, in many cases, the ossicles were not securely joined together, but interspersed with areas of softer integument. The notochord, he says 'may have helped to hold the tail together, like the string in a bead necklace'.

The arguments concerning the inversion of the tail during the cornute–mitrate transition are discussed above. The exposition on pp.364-367 of Jefferies (1981a), though forceful, is all the clearer and more succinct. The key passage is on p.366, where Jefferies compares the appendages of the advanced cornute *Reticulocarpos* with the primitive mitrate *Chinianocarpos*. As discussed above, the appendage of *Reticulocarpos* is sharply regionated into a proximal section of loose integument and alternating plates, and a middle and distal section of tightly jointed ossicles. In *Chinianocarpos*, in contrast, the ossicles are interspersed with patches of looser integument throughout the length of the tail:

> The whole tail was ventrally flexible, however, like the fore tail of *Reticulocarpos*, and successive plates of the hind tail were connected by plated imbrication membranes, again like the fore tail of *Reticulocarpos*. These membranes further suggest that the paired hind-tail plates of *Chinianocarpos* are serial homologues of the ventral plates of the fore tail. The total flexibility of the tail and the ubiquity of plated imbrication membranes between the plates indicate that the fore tail of the cornute *Reticulocarpos* is equivalent to the whole tail of the mitrate *Chinianocarpos*. (1981a, p.366)

This is supposed to have arisen through a tendency in cornutes towards an autotomy of the tail which became progressively more severe (and presumably earlier in ontogeny) during cornute evolution. The result was an animal with the merest proximal stump, which then became regionated to form 'new' middle and distal sections. In *Chinianocarpos*, a primitive mitrate, these mid- and distal sections are far more similar in structure to the proximal section of the appendage in the cornute *Reticulocarpos* than are the mid- and distal sections in the appendage of that cornute itself. *Chinianocarpos*, then, shows this process caught, as it were, just before the final regionation of the mitrate tail was complete.

(d) The Philip–Jefferies debate: the brain of a mitrate

Of especial interest in Jefferies' reconstructions of mitrates is the inferred presence of a relatively sophisticated brain and central nervous system. Several authors (including Philip) have suggested that channels running on the insides of the thecal plates once housed nerves. But only Jefferies suggests that they converged on a ganglion of a complexity deserving the status of a brain. This was located at the base of the theca, where it entirely filled a cavity or basin near the joint with the appendage. The brain was similar to that of a lamprey, with cranial nerves and satellite trigeminal ganglia. It also bore nerves leading to two pairs of eyes, the cis- and transpharyngeal.

Philip (1979) is struck by what he sees as a great deal of variation and plasticity between the nervous systems of different forms. 'One can only be impressed' he says (1979, p.452) 'by the differences in detail between [Jefferies' own reconstructions of] *M[itrocystites] mitra* and *M. i. miloni*'. He then gives two examples of this plasticity, the second of which runs:

> A small groove on the underside of *Mitrocystites* is interpreted as the beginning of a lateral line system. The anterior pair of pores in the upper surface are interpreted as 'transpharyngeal' eyes while the posterior ones were thought to be sensory areas for touch ... In the closely related *Mitrocystella*, the eyes are vestigial. Now external eyes and a lateral line system would seem to be eminently sensible developments for organisms experimenting with a motile mode of existence. Yet in later mitrates they are absent. (1979, p.452)

Presumably, the implication is that Jefferies is reading too much structure and order into what was really a less-well-defined, superficial net-like nervous system as seen in modern echinoderms. Philip would prefer to reserve the basin-like hollow in the base of the theca for muscles connecting the theca with the proximal part of the appendage and the stylocone. Rather than a brain, mitrates had a rather more humble aboral nerve centre similar to that of crinoids.

Given that both Philip (1979) and Jefferies (1981a) agree that there was some kind of neural centre in the base of the theca, Jefferies (1981a) contends that their only substantive differences concern its name – 'brain' or 'aboral ganglion'.

Jefferies criticizes Philip's note about plasticity as vague, suggesting (with a scarcely less vague appeal to selection-based story-telling, regrettably) that the answer is sought in adaptation to slightly different modes of life. The later mitrates, he suggests, had adopted the lives of epibenthic burrowers: evidence comes from the all-over ribbed sculpturing of

Placocystites, whereas the plates of *Mitrocystella* are ribbed on the ventral surface only. This lifestyle cost them their eyes and attendant nerves. 'Good eyes, after all, are useful to a shrew but not to a mole, similar though these two animals are' (1981a, p.369).

(e) The Philip–Jefferies debate: calcite and bone

Echinoderms (and carpoids, too) have a dermal skeleton based on single crystals of calcite (calcium carbonate) that form within individual cells. Many vertebrates have a skeleton, both dermal and axial, made from bone (a composite of collagen and calcium phosphate) that is grown and deposited in the intercellular matrix.

Philip objects to vertebrate (and chordate) ancestors having echinoderm-like skeletons. His objections are threefold. First, the two kinds of skeleton are so completely different that it is hard to imagine creatures jumping from one to the other. Second, it is equally hard to imagine how echinoderms could be so careless as to lose their skeleton in the first place, so vital is it to their constitution. Jefferies' citation of calcite loss in some holothuria is compensated by a leathery dermis, in which the remaining calcite spicules are bound within a connective tissue that is different from vertebrate-like cartilage. Third, Jefferies' contention of signs of calcite resorption in later mitrates goes against the grain – these same mitrates show a tendency towards increasingly fused, box-like skeletons.

Jefferies (1981a) answers all three points. First, there is no sense in which he imagines a leap directly from calcite to bone. The two were separated by a gap in which vertebrates had neither, as in modern hagfishes. Second, he cites evidence for the complete loss of calcite in some holothuria, and the presence of cartilage in others. Third, he carefully restates his evidence for calcite resorption in several mitrate genera, and criticizes as selective Philip's supposition of a trend for increasing skeletization (what about the heavily box-like *Ceratocystis*?), as well as the mistaken inherent assumption that the chosen cornutes and mitrates can be treated as a single lineage. Even were such a trend evident in one lineage, why should it occur in another?

I think that Jefferies' rebuttal is altogether too charitable. For my part, Philip's first point reflects poverty of imagination. Why *should* creatures be forbidden from replacing one kind of skeleton with another just because he, Philip, finds the process hard to believe? Disbelief is an unsound basis for argument.

The second objection is illogical: no matter how some holothuria cope with their present nakedness, the fact remains that they *are* naked, and that this nakedness represents the derived condition. It cannot be denied that the ancestors of even the most exiguously dressed holothurians had robust calcite skeletons. In which case, it is pointless to deny that other calcite-clothed creatures could have enjoyed similar *déshabille*. That holothurian connective tissue is made of a substance different from vertebrate collagen is again not relevant, as holothurians and (if Jefferies is correct) vertebrates, tunicates and cephalochordates lost their calcite independently.

Philip's third objection is in any case a *non sequitur*. As a process, the resorption of calcite in any particular plate need have nothing to do with the overall disposition of the plates, the proportion of the animal that is covered with plates, or the relative degree of fusion of the plates.

(f) The Philip–Jefferies debate: the fossil record

As I showed in section 4.2.6, the application of cladistics may salve (if not solve) the problems of stratigraphic inconsistency of the kind encountered by Jefferies. Philip (1979) is quick to appreciate the importance of cladistics as a way of reconstructing phylogeny, but notes that geological occurrence should be taken into account. It is odd, he says (1979, p.455) to read of the mid-Silurian mitrate *Placocystites* as a 'stem vertebrate', when vertebrates had long since appeared.

The problem here, replies Jefferies (1981a, p.371) is that Hennig's concept of the stem group 'is still not widely known'. The difficulty of communicating this concept (without the aid of a diagram) makes such a failure easy to forgive. Jefferies emphasizes (1981a, p.372) that as the stem group includes side-branches from the stem lineage as well as the lineage itself, some of these branches 'would be expected to have survived contemporary with the earliest member of the crown group'.

Failure to appreciate what is meant by a stem group may conceal a larger misunderstanding, that cladistic phylogenies are deliberately framed without reference to the passage of time (Philip's 'geological occurrences'). It is this feature that makes them testable hypotheses as well as graphic summaries of character-state information.

Of course, stratigraphical sequence information can be extremely useful in certain circumstances. (Indeed, ignoring it in the cases of Liassic ammonites or Chalk echinoids would be 'foolish', says Jefferies, who began his career in the Chalk.) In these cases, the sequence is sufficiently complete for there to be little doubt about phylogeny or ancestor–descendant relationships, at least in practice (for an alternative view see Forey, 1982).

But the record of carpoids is so sparse that one can probably afford to ignore it for the purposes of constructing phylogenies. Jefferies notes that only four cornutes (and no mitrates) are known from the Cambrian, during which time the three extant chordate subphyla – in Jefferies' scheme – evolved from unknown mitrates. The Ordovician record is a little better, with seven species of cornute in the key family Amygdalothecidae (including *Reticulocarpos hanusi* and other transitional forms). The cornute *Cothurnocystis* is known from the Middle Cambrian, and the Lowest and Uppermost Ordovician, but not from intervening beds.

> Under these circumstances [Jefferies notes] the Cambrian stratigraphical sequence can forbid nothing, nor provide any useful guide to what features are primitive; and the same is almost as true for the Ordovician. The correct procedure is to reconstruct phylogeny by looking for synapomorphies as Hennig recommended, using outgroup comparison to decide which features are primitive and which advanced ... I advocate ignoring the stratigraphical sequence in this case because its obvious incompleteness makes it more hindrance than help. (1981a, p.373)

Elsewhere in his paper, Philip (1979) recognizes the special problems of poor fossil records. On p.463 he notes (and this is gleefully picked up by Jefferies, 1981a, p.371) that many kinds of echinoderm are known from the Cambrian 'although their sporadic occurrence indicates that their order of appearance is no real measure of evolutionary relationships'.

(g) The Philip–Jefferies debate: cornutes and mitrates

Finally, to address those aspects of the calcichordate theory which Philip feels pose 'insurmountable difficulties'.

The first is the orientation of mitrates with respect to cornutes. As I have shown, the usual mitrate reconstruction has them flat-face downwards, convex face upwards. This preserves the mutual orientation of the appendage, such that (for example) the styloid (or stylocone) always faces downwards, whether the animal is a cornute or a mitrate.

In contrast, Jefferies takes his cue not from the obvious features of external anatomy, but the less obvious internal architecture, interpreted in the light of comparative anatomy: this interpretation produces the left-sided cornutes, and the mitrates striving to regain symmetry. One casualty is the traditional orientation of the theca and thus the appendage, forcing Jefferies to adopt the idea of progressive autotomy in the appendages of the more derived cornutes such as *Reticulocarpos*, as discussed above. Taken at face value, Jefferies' scheme seems unnecessarily complicated and *ad hoc*. 'All such problems are removed' counsels Philip,

> if the traditional and previously accepted view of the relationship between the faces of carpoids is followed. This is the interpretation indicated by the objective morphology of the fossils and so it is the interpretation advocated here. However, it demolishes in its entirety the calcichordate theory. (1979, p.457)

As a reply, Jefferies re-states his views on the layout of the cornute and mitrate theca, with the various internal cavities as deduced from external plating and variations in internal sculpture (1981a, p.379) and asks the rhetorical question 'should we take obvious resemblances of the tails as homologous, or rely on reconstructions of internal head anatomy which lead to an opposite conclusion?' (1981a, p.393).

At first glance this seems lame. But on reflection Jefferies is calling his critics to account: if echinoderm specialists seeking to reconstruct the lives of extinct forms look no further than external anatomy, describing the features simply in terms of their appearance without reference to any larger theoretical framework, one can only expect them to look at the superficially similar cornute and mitrate appendages and conclude, unquestioningly, that they must be homologous. The calcichordate theory is thus demolished only in the minds of those who have failed to grasp its implications, outwith the simple description of externals to which echinoderm palaeobiology is limited.

Jefferies' critics are damned doubly for failing to provide *any* coherent interpretation for the internal anatomy of carpoids, the evidence for which (whether one subscribes to the calcichordate theory or not) can hardly be ignored. Now that Jefferies has interpreted (say) the structure in the posterior part of the theca of *Lagynocystis* as a median atrium with amphioxus-like gill slits (with all that this implies), it is no longer sufficient for a critic to refer to the same structure as a 'ctenoid organ', for no other reason than that is what it looks like.

This apparent inability to interpret carpoid internal anatomy lies at the centre of the final disagreement between Philip and Jefferies. Philip reports (1979, pp.457–8) that there is no evidence for thecal pores in mitrates that would have served as atrial openings, as the calcichordate theory demands. 'The theca of anomalocystitids is rigid' he says (1979, p.458); 'There are simply no openings that could be construed as gill openings in the theca of *Placocystites* (or any anomalocystitid for that matter)'. Jefferies' response is to parade an abundance of evidence based on internal structures (much of which is indirect) to show that such openings must have been present. Most of this evidence is interpreted using modern tunicates as models.

First, as mitrate atrial openings would have been outlet valves (like tunicate atria) one would have expected a tight fit, making them hard to see. Nevertheless, they can be discerned in the suturing of particular plates in forms such as *Mitrocystella*. Rather than normal sutures, these plates articulate by a cylindrical socket-like arrangement that would have allowed a certain degree of flexure, without distorting the overall shape of the theca. More compelling evidence comes from more derived mitrates such as *Balanocystites* and *Anatifopsis* (both alleged stem-tunicates), in which the sutures between the equivalent (i.e. homologous) plates are more easily interpreted as atrial openings (Figure 4.19). In these forms, the ventral surface of the theca is almost entirely covered by

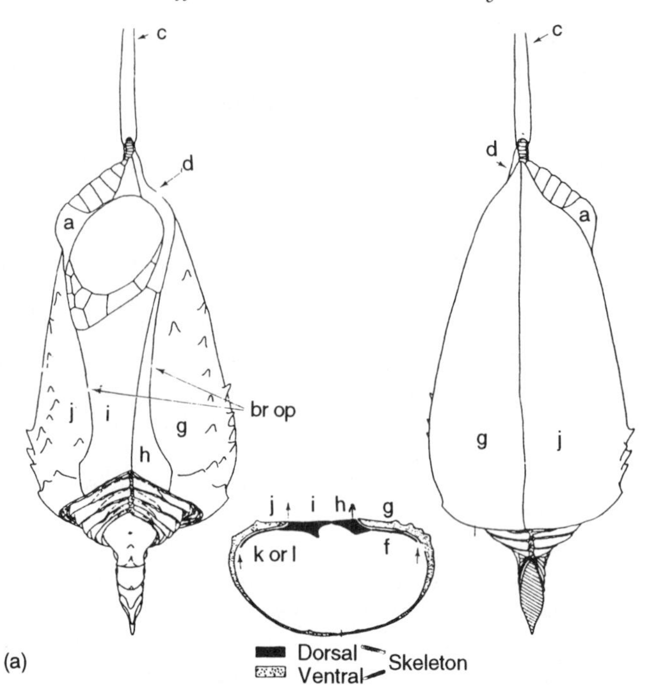

Figure 4.19 Mitrate atrial openings and how they work. (a) The mitrate *Balanocystites* in (left to right) dorsal, cross-sectional and ventral views. Note the expansion of ventral plates *g* and *j* such that they grow up over the dorsal surface, occluding many of the dorsal series, and leaving broad opercular openings (arrows, *br op*) (from Jefferies, 1981a). (b) Right ventrolateral view of *Mitrocystella incipiens miloni*; the expanded cutaway shows the horizontal contact between plates M_{1RV}, and M_{1RD}. This is a cylindrical joint, rotation about which allowing the atrial openings to gape (from Jefferies, 1986).

a pair of very large plates that grow up and over the dorsal surface, occluding many of the dorsal plates but leaving a gap on each side, rather like the operculum that covers the gills of fishes[112]. These same ventral plates can be seen as gross enlargements of the marginal plates in *Mitrocystites* that form the supposed atrial openings. If the plates are homologous, then so (presumably) are the openings. A similar (though less marked) overlap is seen in *Lagynocystis*. Overall, Jefferies agrees with Philip that the thecae of anomalocystitids are rigid, but atrial openings could have worked perfectly well with only minimal flexibility. After all, the operculum of a herring is hardly the most flexible structure.

It is *Lagynocystis*, of course, that displays the most graphic evidence for internal anatomy in any mitrate: the structure in the theca that Ubaghs interprets (or rather describes) as a ctenoid organ is seen by Jefferies as an array of tertiary gill slits along cephalochordate lines (see above).

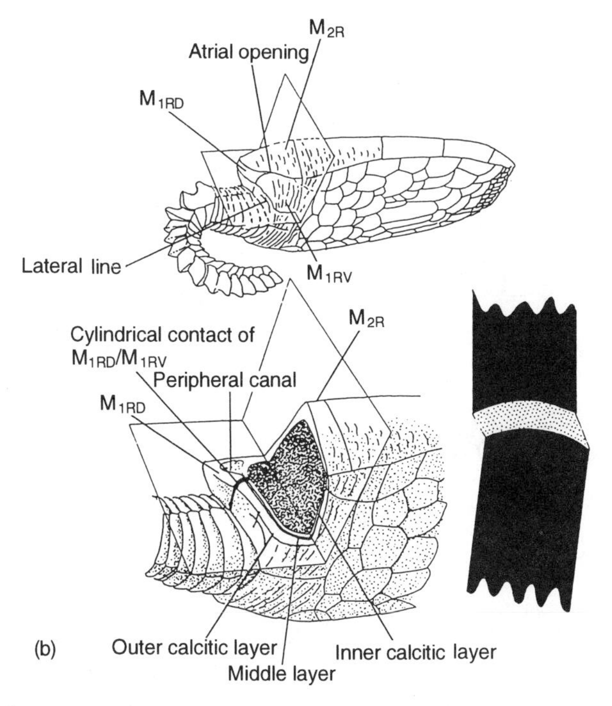

Figure 4.19 (b)

But the core of Jefferies' evidence for the existence of left and right pharynges and atria in mitrates (and thus, by extension, atriopores) comes from comparisons with tunicates. On p.381 he goes so far as to say that 'it is the best evidence that I have correctly oriented the mitrates and for the existence of right and left atrial openings in stem-vertebrate mitrates'. Yet Philip (1979) discusses this comparison not at all – an omission corrected by Jefferies in a long passage (1981, pp.381–394) in which he summarizes much of the evidence described in the *Placocystites* paper (Jefferies and Lewis, 1978).

Jefferies starts with a summary of how tunicates feed (Figure 4.20). Tunicates such as *Ciona* use a number of highly specialized and integrated organs that come together to form the pharynx. Water flows into the animal through the mouth, is strained through the endostylar mucous filter, passes through the gill slits and leaves the animal through the atria. Particles in the water current are trapped by mucous sheets strung across the gill slits. Mucus is secreted by the ventrally situated, gutter-like endostyle, and moved thence by two ciliated strips of epithelium (at left and right of the endostyle) called the marginal bands. At the rear of the animal, the right marginal band (but not the left) passes into the alimentary canal.

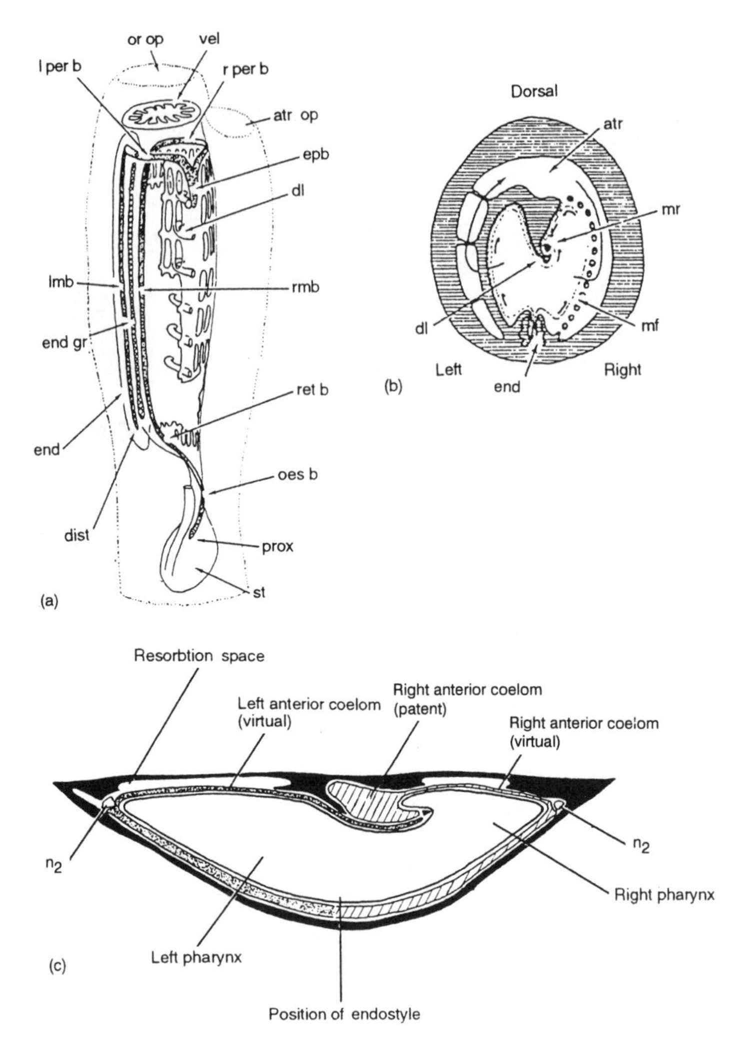

Figure 4.20 (a) The alimentary ciliated loop of a tunicate. Some of the ciliated bands of the pharynx form a continuous series from the distal (dist) to the proximal (prox) ends, running up through the left marginal band (lmb), left peripharyngeal band (l per b), epipharyngeal band (epb), right peripharyngeal band (r per b), right marginal band (rmb), retropharyngeal band (ret b) and oesophageal band (oes b). Other features include dl = dorsal languet; atr op = atrial opening; end = endostyle; end gr = endostylar groove; or op = oral opening; st = stomach; vel = velum. (b) Transverse section through the pharynx of the tunicate *Clavelina*; atr = atrium, dl = dorsal languet; end = endostyle; mf = mucous filter; mr = mucous rope. Compare this with (c) a transverse section of the mitrate *Placocystites forbesianus*. Note how the curvature of the dorsal languet resembles the rightward bend of the viscera in *Placocystites*, (a and b from Jefferies, 1981a, after Werner and Werner; c from Jefferies, 1986).

At the front of the animal, the marginal bands become the peripharyngeal bands which ascend the pharyngeal wall, meeting in the dorsal mid-line as the V-shaped epipharyngeal band. As the mucus ascends the peripharyngeal bands, the water current forces it rearwards, spreading it into sheets which cover the gill slits, and in which particles are trapped. The laden mucus is then moved upwards, along the gill bars, to the dorsal mid-line.

There, along the dorsal mid-line, it meets a row of curved hooks known as the dorsal languets, or a curved fold of flesh known as the dorsal lamina. Whether as languets or as a lamina, this dorsal structure is always found to the left of the epipharyngeal band, and invariably curved ventralwards and to the right when seen in transverse section. This structure is curved to encourage the incoming sheets of mucus to be rolled into a rope of a suitable bore to pass into the oesophagus.

So much for getting the food in. The faeces emerge via an intestine which curves leftwards out of the stomach and (in tunicate tadpoles) debouches into the left atrium.

Why this long digression on tunicates with all the emphasis on left and right? Jefferies can interpret most of the structures inside the theca of *Placocystites* as a tunicate-like pharyngeal filter-feeding mechanism, in which all the geometrical relationships are preserved in detail. All this supports the contention that what Jefferies says is dorsal really is so: that the orientation of the carpoid theca corresponds functionally as well as morphologically with the anatomy of tunicates. And if the internal anatomy of the carpoid theca works, then the inversion of the tail in the cornute–mitrate transition *must* follow. The simple external appearances accepted without question by echinoderm palaeobiology can, after all, be deceptive. No wonder Jefferies finds so little support from that quarter[113].

The evidence for the internal layout of the theca of *Placocystites* (and other mitrates) comes partly from comparisons with cornutes, partly from interpretation of the fossils themselves. This reveals structures – a host of ridges and bumps – which look almost meaningless by themselves, but when taken together can be seen as the vestiges of a pharynx, with everything in its place as one would expect in a tunicate. For example, a ridge known as the mid-dorsal process

> consistently lies right of the low ridge on the surface of the dorsal skeleton that is believed to mark the left boundary of the right pharynx against the dorsal skeleton ... Moreover, the mid-dorsal process is always D-shaped in transverse section, with the vertical stroke of the D facing medially ... If we reconstruct the chambers round the mid-dorsal process, taking these relationships into account, we are almost compelled to assume a fold of flesh extending rearwards from near the presumed ciliated organ and meeting-

> place of the peripharyngeal bands, sloping downwards and rightwards in transverse section, and having the mid-dorsal process along its free edge as a stiffener. In view of its shape in transverse section, the relations of its anterior end and the fact that it grossly separates the right from the left pharynx, this fold of flesh can be interpreted as a dorsal lamina. (1981a, pp.387–388)

It is easy to pluck from this passage words and phrases such as 'we are almost compelled to assume' and use them to suggest that Jefferies is reading too much into the evidence. But such accusations, while easy to make, ignore the parallels that Jefferies draws between the (real) fossil evidence and the (real) relationships between structures in tunicates. The parallels are, indeed, manifold and mutually supportive. On the basis of direct skeletal evidence, Jefferies claims to infer the positions of

> the right and left peripharyngeal bands; the ciliated organ; the dorsal lamina, sloping rightwards and ventralwards as in tunicates; the opening of the oesophagus, right of the midline as in tunicates; the posterior end of the endostyle; the retropharyngeal band running from the right posterior end of the endostyle towards the oesophageal opening as in tunicates; the pharyngo-epicardial openings, right and left of the retropharyngeal band as in *Ciona*; and the rectum opening into the left atrium as in tunicate tadpoles. (1981a, p.393)

This is certainly speculative, but such a sophisticated, integrated scheme, based on real comparative evidence and buttressed by reciprocal illumination, cannot be refuted except by schemes that interpret the same structures in an equally integrated way. As Jefferies himself notes in continuation of the above:

> Anyone who rejects this comparison ought to establish an equally detailed interpretation of his own. It cannot be overthrown, as Philip seeks to overthrow it, by refusing to interpret the evidence in the name of scientific objectivity. The interpretation is hard to expound, partly because it depends on detailed asymmetries in the pharynges of tunicates which interest nobody. But its mere difficulty does not refute it. (1981a, p.393)

Jefferies described his own way of working in a later paper (Jefferies, 1991):

> This argument is like solving a crossword puzzle. If sufficient numbers of things fit with these very complicated fossils, I hope I am getting it right. Any particular argument will only be presumptive, but they all fit together so closely that I hope the ultimate result is correct. (1991, p.124)

Jollie's critique of the calcichordate theory (Jollie, 1982) was written after the appearance of Philip's paper (1979) but before the publication of Jefferies' rebuttal (1981a). It thus falls into very much the same traps that snared Philip, so I shall not discuss it in detail. Jollie's main complaint is that one should beware of any theory positing the transformation of a complex design (a mitrate) to one of similar complexity (a chordate), as such ideas tend to require drastic redesign – a comment that echoes Berrill's (1955) dismissal of Gislén's ideas (1930), and their likening to the Gaskellian scheme of crustacean-like ancestry and transformation (indeed, Jollie made very much the same comparison with Jefferies' work). 'Jefferies' steps in the conversion of a mitrate to a vertebrate are far too complex to be acceptable', notes Jollie (1982, p.172). Jollie's solution is the derivation of chordates from a simple, tricoelomate, and conveniently fictional 'dipleuruloid' ancestor.

But what Jollie puts forward as Jefferies' 'hypothesis' is a highly selective stream of mock-axioms, larded with unsupported commentary: hardly calculated to inspire confidence of impartiality. For example:

> (c) The transition [between mitrate and vertebrate] is assumed to have involved the loss of the 'calcite' skeleton, followed by a naked phase, followed by development of a bony skeleton (resembling the original calcite plate cover!) (1982, p.170)

One might excuse such instances as infelicities of style. More seriously, Jollie (1982) does not attempt any refutation of Jefferies' interpretation of the internal structure of carpoids as seen from the fossils themselves. As I showed above, nothing less is demanded by Jefferies (1981a), and this seems perfectly reasonable given the links proposed between these structures and the anatomies of tunicates. Knocking a straw man is easy – constructing his replacement is much more difficult.

4.2.9 Dexiothetism and recapitulation

Jefferies took the opportunity to explain the principles of Hennigian phylogenetic reconstruction as a 'methodological essay' in a 1979 book, *The Origin Of Major Invertebrate Groups* (Jefferies, 1979). It is one of the more lucid explanations of Hennigian methods, and is worth reading for that aspect alone.

It is doubly interesting in its discussion of the Haeckelian component in Hennig's systematics. Organisms can be arranged in a branching pattern of progressively more inclusive sets because in ontogeny, fundamental features of the *Bauplan* are set out before the more trivial, specific details. This must recapitulate phylogeny, because fundamental change in the *Bauplan* will have a deleterious effect on subsequent embryonic stages and is likely to be lethal, whereas the trivial variations that might one day delineate species are not encumbered by a similar 'burden'. The boles and trunks of the *Bauplan*, as with those of a phylogeny, started out as twigs and acorns, but became buried under stems and crowns of characters as time and evolution wove onward. The idea of a 'burden' in development is similar to that of historical contingency and constraint: organisms cannot develop ideal structures *de novo* in response to the dictates of natural selection, they must make do with the genetic heritage and the structures with which they were born.

It is this concept of burden that allows one to discern fundamental phylum-level homologies without recourse to the stratigraphic criterion of primitiveness, because argument from comparative anatomy can be used to deduce the order in which major features appeared. Stratigraphic arguments become more important as the distinctions between organisms become finer.

This argument, of course, helps Jefferies get round the problem of inconsistency as regards the phylogeny he creates for the cornutes (stem chordates) and mitrates (crown chordates), which does not fit easily with the stratigraphic record.

Chordate phylogeny takes a back seat in the 'methodological essay', but two refinements emerge. The first is the crown group as a re-launching of what Hennig called the ★group: a group of organisms that includes all living members of that group together with their common ancestor and any extinct descendants of that ancestor.

The second is the Dexiothetica, the monophyletic group that descends from a *Cephalodiscus*-like animal that lay down on its right side. The word derives from the Greek *dexios* (right) and *thetikos* ('suitable for laying down') (Jefferies, 1979, p.463). Dexiothetism had appeared in earlier papers of course – both by Jefferies (from 1969) and Gislén – but without a formal name.

4.2.10 Calcichordates in the 1980s

The calcichordate theory up to this point is summarized in book form in *The Ancestry of the Vertebrates* (Jefferies, 1986). Thematically, *Ancestry* is an extension of the large paper on *Placocystites* (Jefferies and Lewis, 1978) and the 'defence' paper (Jefferies, 1981a) in that Jefferies goes to some lengths to set the comparative anatomy of carpoids in context. To this end, he presents complete re-evaluations of the anatomies of echinoderms (crinoid and asteroid), hemichordates, tunicates, the amphioxus, agnathans and gnathostome vertebrates, and a discourse on segmentation and the 'head' problem. It is almost 200 pages before one actually meets a carpoid.

The aim of all this, of course, is to minimize the shock at being presented with the lumps and wrinkles in a series of unusual fossils, and being told that these lumps and wrinkles represent the trigeminal ganglia, dorsal laminae, coeloms and all the other fleshly impedimenta required by the theory. The effect on the reader should be that, once reached, all these identifications are quite natural: given what we have learned about deuterostome comparative anatomy in the preceding pages, it should all fall into place.

Unfortunately, the book generally failed in its aim – reviewers welcomed the first part of the book, and rejected the second. As regards the exposition of carpoid anatomy, Keith Stewart Thomson, writing in *Nature* (Thomson, 1987b) finds fault with the practice of

> ... piling on interpretations, each of which is contingent on some previous part of the argument and fails therefore to test the original premise, is particularly weak when the group in question has no living representative. (1987b, p.197)

Thomson does not comment on the possibility that these very interpretations might not be based on one another, but refer back to the comparative anatomy of extant forms as laid out in the first half of the book. The two halves are intended to make a whole. Yet Thomson seeks to divorce one half from the other: we read that

> ... all is far from vain, especially if the 'calcichordate' argument is isolated from Jefferies's other contributions... the half of the book that reviews and analyses chordate structure and relationships contains many useful new insights. (1987b, p.197)

Carl Gans, reviewing *Ancestry* for *American Scientist* in 1988, is more sympathetic, but like Thomson homes in on the matter of interpretation of the fossils themselves, once removed from their comparative-anatomical context.

> What is presented here is an enormous edifice based upon the reinterpretation of the scallopings in an array of relatively small calcite fossils. (1988, p.189)

Of course, such 'scallopings' will always remain thus unless interpreted in the light of the anatomies of animals still extant. Nevertheless, Gans is right to close his review with the following cautious words:

> Until the fossils are reexamined and the interpretations tested and confirmed, this carefully crafted scheme deserves only a watching brief rather than an uncritical acceptance. (1988, p.189)

By this time, Jefferies had moved on, with an essay on the characterization of the echinoderms as a group (Jefferies, 1988a). The calcichordate theory invalidates a number of characters which, one might have thought, would be useful for the unique definition of echinoderms as a group. One such is the presence of a skeleton made out of plates of calcite with a distinct 'stereom mesh' ultrastructure. To be sure, the only creatures alive today with this attribute are echinoderms, but cornutes and mitrates also had skeletons of this type. But if cornutes and mitrates are considered as chordates, the presence of a calcite skeleton is no longer a feature unique to echinoderms. Instead, it is a feature present in the common ancestry of echinoderms and chordates, which chordates subsequently lost.

Another echinoderm feature to come under the microscope is the water–vascular system of ambulacra, tube-feet and so on. In echinoderm development, this entire organ-system is elaborated from the left hydrocoel (left mesocoel). In hemichordates, the left hydrocoel opens to the outside via a pore, but in echinoderms it is connected by the 'stone canal' to the axocoel (protocoel), only thence to the outside by a pore, the hydropore. This indirect connection between the left hydrocoel and the outside is found in echinoderms only of all extant tricoelomate animals, but inference of a hydropore in the cornute *Ceratocystis perneri* (see above) suggests that it, like the calcite skeleton, is a holdover from the common ancestry of echinoderms and chordates, which the chordates have since lost. Only pentameral symmetry, loss of gill slits and a form of torsion are genuine echinoderm hallmarks, that is, features which can be used to characterize echinoderms uniquely, as fossils as well as in the modern fauna.

In 1987, Jefferies returned to description with a re-evaluation of the primitive and difficult cornute *Protocystites menevensis* (Jefferies *et al.*, 1987), mentioned in a popular article the next year (Jefferies, 1988b). Description of *Barrandeocarpus norvegicus*, a mitrate from Norway, followed in 1989 (Craske and Jefferies, 1989). In these descriptive papers, Jefferies refined his ideas on systematics to a high degree, until they were as different (at first sight) from those of other 'cladists' as they were from traditional systematics (Gee, 1989).

But this preoccupation with systematics turns up a difficult problem. That is, how to establish the 'primitiveness' of a character in a series of extinct creatures that lack obvious sister groups. *Protocystites* is the most primitive known cornute apart from *Ceratocystis perneri*. But there is no easy way of telling whether the features we use to determine the primitiveness of *Ceratocystis* (and so orient our ideas about cornute evolution in general) are not simply quirks of that particular species that say nothing, in fact, about cornute evolution.

The problem is rooted in Ax's concept of the 'stem lineage' (Ax, 1985), the chain of actual ancestors and descendants running like a thread through the stem group. Stem lineages must have existed, but it is in principle impossible to prove whether any particular fossil under study actually represents a segment of that lineage. If *Ceratocystis perneri* were on the cornute stem lineage, all features we see would have a bearing on cornute evolution: they would be more 'derived' or 'crownward' than those of forms lower down the stem, and more 'primitive' or 'anti-crownward' than those of forms further up.

But as it is formally impossible to know whether any particular specimen of *Ceratocystis perneri* belongs on the chordate stem lineage or not, it is better placed on a side-branch, in accordance with cladistic practice. So characters found in *Ceratocystis perneri* and no other animal could have evolved either in the stem lineage itself, or in the side-branch leading uniquely to that species.

The only way out of the dilemma (and even then it is deferral rather than escape) is to find fossil cornutes, or forms that can reasonably be posited as ancestors of cornutes (anti-crownward on the same stem), but either way even more primitive than *Ceratocystis perneri*. It would be even better to find fossils that represent the 'nodal group' of the Dexiothetica, thus rooting the chordates and the echinoderms together: and allowing the detailed reconstruction of the common ancestor of these two phyla.

4.2.11 The solute, *Dendrocystoides scoticus*

An answer came in 1990, not from a cornute nor a mitrate, but from a solute, a member of a group of carpoids that Jefferies had hitherto left alone. *Dendrocystoides* looks superficially like a cornute, with a long, segmented tail and the irregular plated head. Jefferies infers the presence of a gill slit, but there was more certainly a hydropore – a structure absent in all chordates with the possible exception of *Ceratocystis*. But there are intriguing differences, mainly that the creature, like solutes generally, had another tail-like structure, protruding from the front right-hand corner of the head (when viewed from calcichordate-dorsal). Again, as with solutes generally, this looks very like the feeding-arm of a crinoid, and there is evidence that it bore a system of crinoid-like tube feet. As discussed at the beginning of this chapter, Ubaghs and others had interpreted the cornute and mitrate tail as a feeding arm, or 'aulacophore'. However, this view may be hard to sustain given the many homologies between the tail of cornutes and the corresponding organ in *Dendrocystoides*, which possesses something rather more obviously a feeding arm in addition.

If the 'aulacophore' idea is right, then the loss of the solute feeding arm in cornutes (should solutes be ancestral to cornutes) must have accompanied a transfer of food-gathering activity from the front of the animal to the other. As one commentator remarked, this would be 'as unlikely as talking from the wrong end of the alimentary canal' (Paul, 1990).

The presence of a water–vascular system, a gill slit and a tail in *Dendrocystoides* resolves the position of *Ceratocystis*, which now lies midway between *Dendrocystoides* and other cornutes. Second, it lets Jefferies off another hook, because the presence of a hydropore in *Dendrocystoides*, more certainly than in *Ceratocystis*, affirms the primitiveness of the latter with respect to the cornutes.

Third, it allows us to glimpse the very beginnings of chordate and echinoderm evolution: for if we were ever able to find the latest common ancestor of starfishes and ourselves, it would (says Jefferies) look like a solute. The newly dexiothetic solutes had primitive features of both phyla – so much so that it is hard to tell whether *Dendrocystoides*, or any other solute for that matter, is a stem dexiothete, stem echinoderm or stem chordate.

Echinoderms lost the tail and the remaining gill slit, acquired a recumbent water–vascular system (borne on the body rather than on a feeding arm) that became triradiate and finally, in crown echinoderms, pentameral. Chordates retained the tail (homologous with the contractile tail of pterobranchs but rotated by 90°), lost the water–vascular system and elaborated the branchial filtering mechanism instead.

Work on calcichordates by Jefferies' students, Anthony Cripps and Paul Daley, has appeared. Cripps has worked on both cornutes and mitrates (Cripps, 1988, 1989a,b, 1990, 1991) and has specialized in the cornute–mitrate transition (Cripps, 1989b). In a cladistic analysis of cornutes (Cripps, 1991), the most parsimonious outcome of a three-taxon problem places mitrates closer to crown chordates than crown echinoderms. Daley has studied cornutes (Daley, 1992a) and latterly solutes (Daley, 1992b). His work suggests that some solutes can be seen as stem echinoderms+chordates, others as stem chordates alone: solutes are the dexiothetic 'nodal' group.

Figure 4.21 shows the phylogeny of the deuterostomes according to Jefferies as presented in the summer of 1994 (which may well not be its final iteration). Most of the phylogeny will by now be familiar, with cornutes as stem chordates giving rise to mitrates, all of which can be assigned to the stem groups of the cephalochordates, the tunicates, or the craniates. But as members of the 'nodal' group at the root of echinoderms and chordates, some solutes are dexiothetes but others are not. The transition is hard to discern, given the extreme asymmetry of solutes.

4.2.12 Fearful asymmetry

Perhaps ironically, it is not the detailed studies of the fossils but the concept of dexiothetism that has brought Jefferies' work to the attention of a general scientific audience, for his elaboration of the place of dexio-

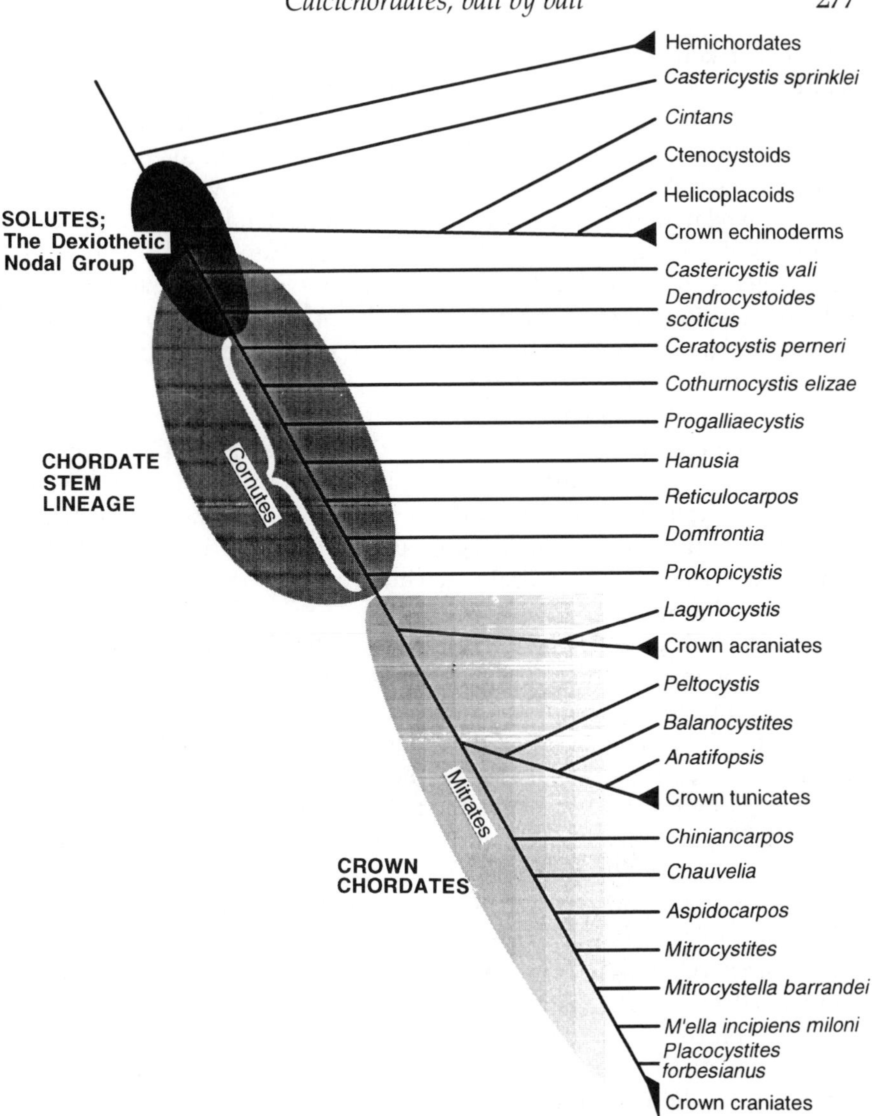

Figure 4.21 A deuterostome phylogeny according to the calcichordate theory. Note the origin of echinoderms from within the solutes (redrawn by Sue Fox from Jefferies, personal communication).

thetism in the *Baupläne* of the deuterostome phyla came at a time when the specification of axes in embryology was beginning to yield to genetic investigation.

It started when Jefferies gave a paper at a CIBA Foundation Symposium entitled 'Biological asymmetry and handedness' (Jefferies, 1991). In his paper, he gave an account of dexiothetism, within the necessary context of the calcichordate theory: all very familiar to palaeontologists who had taken an interest since the 1960s, but doubtless very new to many of the experimental biologists in the audience.

In the CIBA paper, Jefferies suggests that fundamental changes in symmetry have occurred twice in metazoan evolution. In the first, a radially symmetrical animal of the coelenterate grade 'fell over' on its side to produce a bilaterally symmetrical creature, with a definite front and back end. This idea is supported by the evidence from Hox genes, as I showed in Chapter 3: in the hydra, Hox genes are expressed in what seems to be collinear fashion up and down the oral–aboral axis, corresponding with Hox expression along the anterior–posterior axis in bilateral, triploblastic metazoans (Schummer *et al.*, 1992).

The second fundamental shift in symmetry was dexiothetism, in which a pterobranch-like animal similar in its topology to *Cephalodiscus* flipped onto the right-hand side. As explained above (and in Figure 4.8), many structures homologous with the pre-dexiothete right side were reduced in the novice dexiothete, exemplified by cornutes. More advanced creatures, the mitrates, created a 'new' right-hand side by duplicating left-hand structures.

To follow this to its logical conclusion, the chordate dorsal mid-line is not homologous with that of other bilateral animals, but with a line running laterally along the left-hand side. If this is true, then one should be able to test it: genetic systems that determine dorsoventral polarity in animals such as *Drosophila* should be homologous with those which, in vertebrates, determine lateral–medial polarity – and/or vice versa.

The 1991 paper as printed comes with the views of the discussants, many of whom were molecular and developmental biologists. Many were interested – and also cautious. Slack (1992) comments that the discussion on the proposed homologies between protostome and deuterostome axes of symmetry seems 'blissfully unencumbered by data'. Morgan (1992), is more sympathetic, noting that it would be intriguing to speculate on the link between (say) human handedness and 'the dexiothetism of a distant ancestor of the Chordates and Echinoderms, but it is difficult to see how it could be tested. The first step would seem to be the elucidation of the genetic mechanisms of handedness, and their molecular basis'. Morgan (1992) reprints a picture of *Cothurnocystis*.

In his book, Jefferies (1986) draws attention to the prevalent left-sidedness in the development of both echinoderms and chordates. In echinoderms such as starfish, development of the water–vascular system from the left hydrocoel is the rule: but there are occasional examples of starfish in which development proceeds from the right hydrocoel instead, producing an animal whose internal organs are arranged as a mirror-image of the usual condition. This is called *situs inversus viscerum*, and is well known from vertebrates, too, including humans. Jefferies interprets this as evidence for dexiothetism, and explains why he groups echinoderms with chordates, excluding hemichordates (or, at least, pterobranchs).

The genetic basis for *situs inversus* is now under study, but is far from being fully understood. Mutations to a gene in mice (designated *iv*, short for *inversus viscerum*), produce litters in which around 50% of the pups have *situs inversus*. This suggests a mechanism for imposing a polarity, in the absence of which polarity is assigned at random (see Galloway, 1990 for a review). However, Yokoyama *et al.* (1993) isolated a different mouse gene, *inv*, recessive mutations of which produced *situs inversus* in 100% of double-recessive pups. This gene appears to be necessary for the proper direction of 'turning' the embryo immediately after implantation, a process which is vital for the orientation of the viscera (see Ewing, 1993, for a review). Precisely how, physically, such mutations lead to *situs inversus* is unknown, although it may be related to the efficiency of cell adhesion and cell-to-cell contacts in early embryogeny. Gastrulation, for example, is essentially a series of cell movements in which the correct choreography is vital: achieved, fundamentally, by the forging and breaking of contacts between individual cells. Certainly, *situs inversus* can be induced in *Xenopus* by manipulation of the extracellular matrix (Yost, 1992).

Since the publication of the CIBA paper, work on solutes, largely by Paul Daley, has exposed a problem connected with dexiothetism. If some solutes are dexiothetic and others are not, one would expect dexiothetic solutes to be tipped over on their right-hand sides with respect to non-dexiothetic solutes. This would be hard to make out in the blob-like heads in any case, but certainly does not seem to be the case in the tail. In dexiothetic and non-dexiothetic solutes alike, the tails seem to adopt the same, primitively bilaterian orientation.

Jefferies explains this by arguing that dexiothetism affected the head and the tail in different ways. Whereas the head tipped over onto the right, the tail either stayed the way it was, or bounced back with a counter-torsion of a quarter-turn to the left, restoring the original orientation (Jefferies *et al.*, 1996). The ancestral chordate head is as dexiothetic as ever: the tail, though, has the same orientation as in other bilaterally symmetrical animals such as *Drosophila*.

This new observation has consequences for the molecular 'test' for dexiothetism as described above, and resolves some difficulties that recent molecular evidence may have put in its way. As shown in Chapter 3, Arendt and Nübler-Jung (1994) point to homologies between the genetic systems that determine dorso-ventral polarity in *Drosophila* and the frog *Xenopus*. But there is a twist – what is dorsal in *Drosophila* is ventral in *Xenopus*, and vice versa. With this evidence in mind, they pause to reflect on the truth behind Geoffroy's old idea that vertebrates are really the worms that turned (De Robertis and Sasai, 1996). Lacalli (1995) suggests instead that this inversion suggests nothing of the sort, and is a simple geometrical consequence of the smooth continuation of gastrulation into neurulation (Gont *et al.*, 1993; De Robertis *et al.*, 1994).

But what implications can this have for dexiothetism, which predicts relationships between bilaterian dorsoventral and dexiothete lateral polarities? Actually, not very much. Most, if not all the genetic systems so far studied in connection with dorsoventral polarity relate to axial structures such as the notochord and the neural tube. In the calcichordate theory, these structures are associated with the 'tail' end of the animal which resumed its ancestral orientation soon after the dexiothetic event – if it ever really lost it (Jefferies *et al.*, 1996). So one would indeed expect to find homologies between the dorsoventral patterning systems of *Drosophila* and the axial region of chordates, and the inversions seen are probably connected with the invention of neurulation in chordates (Lacalli, 1994; Jefferies and Brown, 1995). Therefore, one would not expect to find molecular signs of dexiothetism along the vertebrate axis, for this is largely tail-derived, and retains the original bilaterian orientation. One should look instead at the remnants of the original and dexiothetic 'head', perhaps parts of the brain and viscera.

In connection with this, one might ask three questions. First, do protostome bilaterians such as *Caenorhabditis* and *Drosophila* exhibit lateral asymmetries in their viscera homologous with those which, in vertebrates, are perturbed by conditions such as *situs inversus*? Dexiothetism predicts that they would not. Second, what are the homologues of *iv* and *inv* in *Drosophila*? Dexiothetism predicts that they would be involved in dorsoventral rather than lateral patterning. Third, is it significant that *situs inversus* affects the positioning of the viscera, parts of the old chordate 'head', and not the axial skeleton, the old chordate 'tail'? If so, this could be explained by the observation that dexiothetism is observed in the 'heads' of solutes and cornutes, but not in their tails.

4.2.13 The monsters and the critics IV: a test of pattern

Recently, the calcichordate theory has been addressed from three different angles; crystallographic (Carlson and Fisher, 1981; Fisher, 1982), ecological (Fisher, 1983) and phylogenetic (Peterson, 1995a).

The first two are aspects of the same: analysis of the microstructure of carpoid stereom (Carlson and Fisher, 1981; Fisher, 1982) suggests that mitrates and cornutes had the same orientation, and that mitrates cannot be considered as upside-down relative to cornutes. Further, the discovery of a slab of mitrates preserved in 'traditional' carpoid orientation with respect to the substrate (Fisher, 1983) can be interpreted as a refutation of the calcichordate theory. But as Peterson (1995a) points out, challenge of the ecological implications of the calcichordate theory is no refutation of the theory itself.

To do that, Peterson continues, one must examine either Jefferies' anatomical interpretations, or his phylogenetic predictions. Like Thomson (1987b) and Jollie (1982), Peterson finds Jefferies' anatomical reconstructions essentially untestable and circular. This is because they depend on the assumption of living chordates as models, and the similarities are then used as proof of relationship. Peterson then turns to what he calls the 'test of pattern', the reasoning being that whatever the anatomy of calcichordates might be, any reconstruction should be testable indirectly, through the phylogeny predicted on the basis of that anatomy. If Jefferies' interpretations of carpoid anatomy are correct, then the phylogeny will look like Figure 4.21. If not, then Jefferies' anatomical reconstructions of calcichordates are called into question. So, can the phylogenetic predictions of the calcichordate theory be sustained on the basis of what is known of calcichordate anatomy?

Peterson's answer is 'no'. The essence of the argument is shown in Figure 4.22. Peterson starts by presenting the most parsimonious phylogeny of extant deuterostomes (A) together with a phylogeny of the same taxa ordered according to the predictions of the calcichordate theory (B). Note that the second differs from the first in the placement of echinoderms and urochordates. The closeness of echinoderms and chordates is, of course, a central and prominent feature of the calcichordate theory: the sisterhood between vertebrates and urochordates, excluding cephalochordates, is based on the anatomy of the mitrate *Lagynocystis* (Jefferies, 1973) and some rather more tentative inferences of higher-mitrate anatomy (see Jefferies, 1986). It is not based on character distributions in Recent forms. The shortest tree (42 steps) (Figure 4.22a) conforms to the traditional view of deuterostome phylogeny (Chapter 1). The tree predicted by the calcichordate theory (Figure 4.22b) is less parsimonious, being one of 18 trees at 57 steps long.

Peterson then adds evidence from fossils, to see if their inclusion changes the predictions made using Recent animals only. Fossil data are used in two subsequent analyses. In the first, he admits only those features of carpoids which are uncontentious, for example the presence of a stereom skeleton and left-sided asymmetry, and the presence of a water–vascular system in solutes. He leaves out features for which the identification depends on debatable interpretation, such as details of the structure variously referred to as the aulacophore (Ubaghs) or the tail (Jefferies).

The calcichordates he uses make an even spread, comprising the solute *Dendrocystoides scoticus*, the cornute *Cothurnocystis elizae* and the mitrates

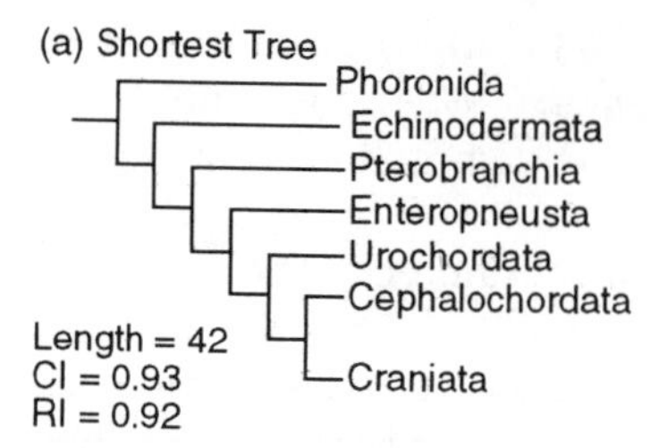
(a) Shortest Tree
Phoronida
Echinodermata
Pterobranchia
Enteropneusta
Urochordata
Cephalochordata
Craniata
Length = 42
CI = 0.93
RI = 0.92

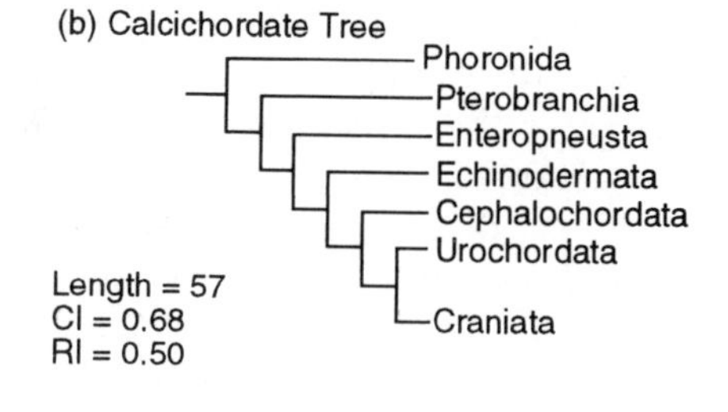
(b) Calcichordate Tree
Phoronida
Pterobranchia
Enteropneusta
Echinodermata
Cephalochordata
Urochordata
Craniata
Length = 57
CI = 0.68
RI = 0.50

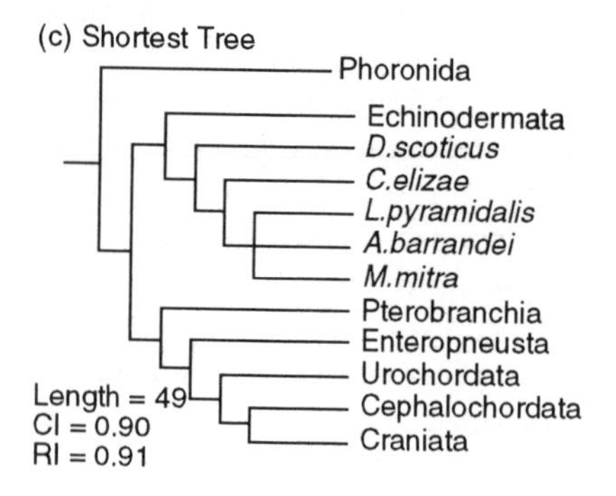
(c) Shortest Tree
Phoronida
Echinodermata
D.scoticus
C.elizae
L.pyramidalis
A.barrandei
M.mitra
Pterobranchia
Enteropneusta
Urochordata
Cephalochordata
Craniata
Length = 49
CI = 0.90
RI = 0.91

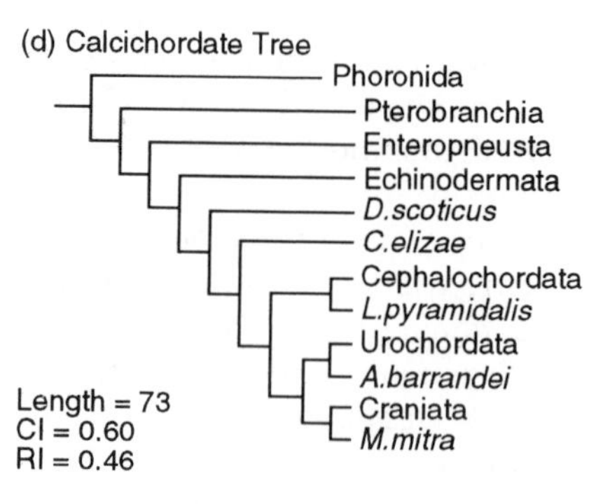
(d) Calcichordate Tree
Phoronida
Pterobranchia
Enteropneusta
Echinodermata
D.scoticus
C.elizae
Cephalochordata
L.pyramidalis
Urochordata
A.barrandei
Craniata
M.mitra
Length = 73
CI = 0.60
RI = 0.46

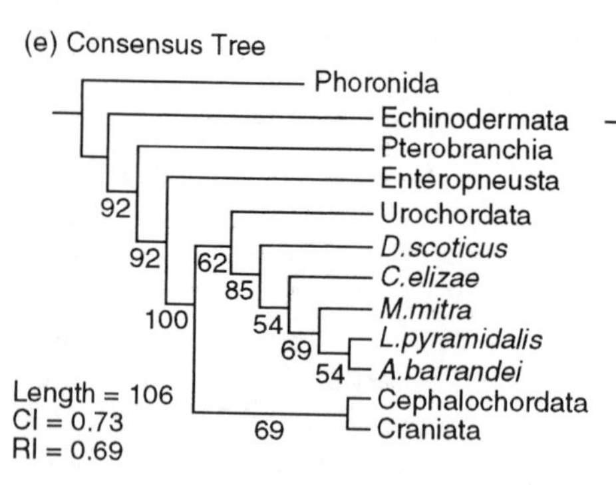
(e) Consensus Tree
Phoronida
Echinodermata
Pterobranchia
Enteropneusta
Urochordata
D.scoticus
C.elizae
M.mitra
L.pyramidalis
A.barrandei
Cephalochordata
Craniata
92
92
62
85
100
54
69
54
69
Length = 106
CI = 0.73
RI = 0.69

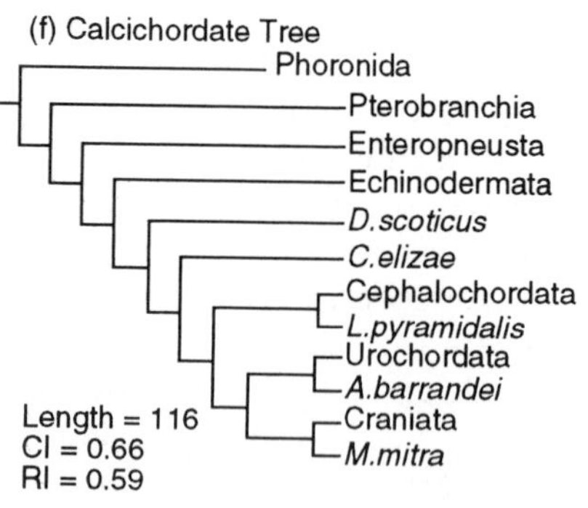
(f) Calcichordate Tree
Phoronida
Pterobranchia
Enteropneusta
Echinodermata
D.scoticus
C.elizae
Cephalochordata
L.pyramidalis
Urochordata
A.barrandei
Craniata
M.mitra
Length = 116
CI = 0.66
RI = 0.59

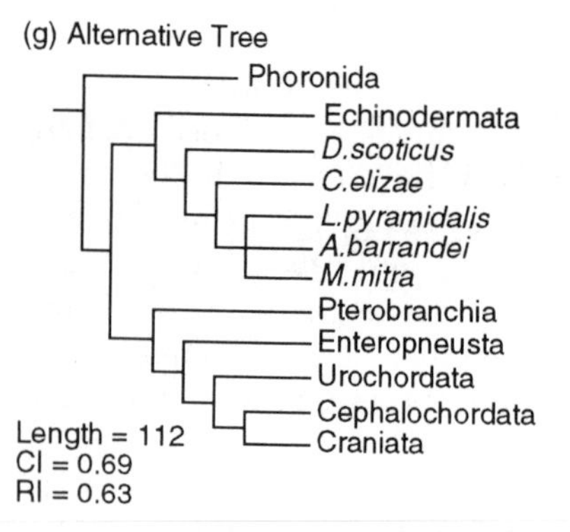
(g) Alternative Tree
Phoronida
Echinodermata
D.scoticus
C.elizae
L.pyramidalis
A.barrandei
M.mitra
Pterobranchia
Enteropneusta
Urochordata
Cephalochordata
Craniata
Length = 112
CI = 0.69
RI = 0.63

Lagynocystis pyramidalis (a stem cephalochordate), *Anatifopsis barrandei* (a stem urochordate) and *Mitrocystites mitra* (a stem vertebrate). The most parsimonious tree (Figure 4.22c), with 49 steps, preserves the deuterostome phylogeny as worked out from Recent animals only, grouping the calcichordates all together as a monophyletic plesion of stem-echinoderms. But the tree arranged according to Jefferies' predictions (Figure 4.22d; compare this with Figure 4.21) has 73 steps, that is 24 steps longer (less parsimonious), and is just one of 9 572 376 different trees of the same length.

This solution should be definitive. It seems that calcichordates are stem-echinoderms with no particular claim on our attention as chordates. But are we not being unfair? Perhaps calcichordates are falling victim to conservative character coding. How would the analysis turn out were one to admit the presence of notochords, optic nerves and so on in calcichordates, just as Jefferies interprets? Peterson thus recasts the characters to reflect Jefferies' interpretations explicitly, and runs the analysis again. The only difference between this and the previous analysis is the inclusion of interpretations peculiar to Jefferies' anatomical reconstructions of calcichordates as well as the interpretations of Recent forms demanded by theory. Unlike the previous analysis, in which a single tree emerged as the most parsimonious, this analysis has no clear winner. There are, in fact, 13 equally most-parsimonious trees, at 106 steps. In ten of these trees, the calcichordates come out as a monophyletic plesion of stem-urochordates. One of these is shown in Figure 4.22(e). Three others place calcichordates as stem-cephalochordates. Again, the topology of Recent taxa is unchanged. These most-parsimonious solutions differ from that predicted by the calcichordate theory (Figure 4.22f), which in this analysis is one of 66 181 trees ten steps longer (116 steps) than the most parsimonious solution. But evolution is not always parsimonious: could Jefferies' phylogeny still be the right one? Peterson suggests not – a phylogeny

Figure 4.22 How the calcichordate theory fails the test of parsimony. (a) The most parsimonious phylogeny of extant deuterostomes. (b) Cladogram of extant groups predicted by the calcichordate theory, which is one of 18 trees 15 steps longer than the most parsimonious solution. (c) The most parsimonious cladogram of Recent phyla and carpoids using a conservative (i.e. non-calcichordate) coding of carpoid characters. (d) The calcichordate tree by the same 'conservative' character set, which is one of 9 572 736 trees 24 steps longer than this. (e) Consensus cladogram of the same taxa, using a character coding based on calcichordate interpretations. Note how the phylogeny of Recent groups is unchanged, and that carpoids are grouped together with urochordates. (f) The calcichordate tree by the same character set, which is one of 66 181 trees 10 steps longer than the most parsimonious solution. (g) An alternative tree just two steps more parsimonious than the calcichordate solution places carpoids among the echinoderms (from Peterson, 1995a).

(Figure 4.22g) just two steps shorter (112 steps) puts calcichordates firmly among the echinoderms. Yet others place calcichordates as stem enteropneusts, stem pterobranchs or stem phoronids. All these trees are more parsimonious than that predicted by the calcichordate theory. Even were one to grant that Jefferies' anatomical interpretations are correct, and that evolution proceeds unparsimoniously, one is entitled to ask why one should place special faith in the calcichordate tree and not in one of the 176 190 trees of equal length or shorter.

If a phylogeny reflects the underlying anatomical reconstruction, one is forced to conclude that there is something amiss with Jefferies' interpretation of calcichordate anatomy. Peterson sees the problem as one of circular reasoning, in which pattern has become mingled with process, one being used to justify the other at different times. Features are interpreted in the way they are not because of inherent likelihood, but because the interpretation fitted the pre-existing scheme. Solutes have gill slits and mitrates have atrial openings because they have to be there to make the scheme work, not because they are there to begin with. Of course, one can always find reasons after the fact why such openings are very hard to make out on the fossils.

Jefferies' argument, though, expounded so well in the 'defence' paper (Jefferies, 1981a) is that although individual pieces of evidence may be weak, they fit together to make a story that is very much stronger. The calcichordate theory explains so much, says Jefferies, that it must, in a sense, be right in most respects. One can see the appeal in the proposal that vertebrate ancestors looked like large tunicate-tadpole larvae, even in a suit of calcite clothes.

But a cladistic phylogenetic reconstruction using parsimony such as Peterson uses sees only individual characters, not how they hang together as a whole. If such an approach is inappropriate, one must answer the question of why Jefferies chooses to frame his hypotheses in a cladistic way such that they are amenable to such testing to begin with – again, as Peterson notes, Jefferies must tell us why we should put special faith in the calcichordate tree rather than any other. Peterson suggests that anatomical reconstruction is untestable, in this sense: were one to propose that calcichordates are stem-enteropneusts, for example, one could come up, in time, with an anatomical reconstruction to support that view. Such an approach gets us nowhere.

But there might be another way. One of the most important features of the calcichordate theory is that it makes predictions that should, in principle, be testable beyond the grooves, lines, bumps and corrugations of the fossils on which they are based.

One of these is dexiothetism. Of course, dexiothetism has in a sense already failed the test – it is included as a character in Peterson's analysis of Jefferies' interpretations. Despite this, Jefferies has performed a valuable service in pointing out asymmetries in the development and structure of extant deuterostomes, the importance of which had escaped the general notice. It took the study of a fossil, *Cothurnocystis*, in which such asymmetries are graphically displayed, for Gislén and later Jefferies to realize that asymmetry is a general feature of the deuterostome condition. It is in part thanks to Jefferies that biologists in general see asymmetry as a general problem that remains unsolved, rather than a collection of miscellaneous oddities such as *situs inversus viscerum*. But if subsequent embryological and genetic work supports this asymmetrical tendency, it does not follow that *Cothurnocystis* or any other calcichordate must necessarily be a chordate. It could just as well be an echinoderm, as Gislén avers. The irony is that even were a living calcichordate discovered tomorrow, its anatomy conforming in all respects with Jefferies' predictions, the phylogeny would *still* be wrong, given that Peterson (1995a) built Jefferies' interpretations into the analysis (Charles Marshall, personal communication). This thought-experiment raises questions about the definition of higher taxa such as echinoderms and chordates, addressed in the next chapter.

Jefferies is still working, and the story of calcichordate theory is not yet at an end. One awaits Jefferies' response to Peterson's paper with interest.

NOTES

99 Judging from the size of isolated tail plates, a giant solute from Russia, *'Dendrocystites' rossicus*, may have been 30 cm long.

100 In some cornutes, the end of tail is rounded off rather than broken abruptly (for example *Scotiaecystis collapsa*, Cripps, 1988, and *Procothurnocystis*, Woods and Jefferies, 1992).

101 The position of solutes is less clear, partly because they are so irregular and their body regions are so indistinct. Jefferies (1990) supposes that solutes are a 'nodal' group in which some (such as *Dendrocystoides)* are dexiothetes, but others may not be. Cinctans are dexiothetes, but closer to echinoderms than chordates.

102 The obvious criticism here is that Jefferies is suffering from some Lowellian 'martian canal' syndrome. Victim of insupportably long chains of inference, he 'reads' the chordate pattern into an otherwise random collection of features. This kind of criticism, although easily made, is almost impossible to refute to the satisfaction of the critic who makes it: and nobody doubts the existence of the rock lumps and ridges Jefferies describes – only the interpretation is in doubt.

103 The brachiopods, members of a single phylum, may have either calcitic or phosphatic hard tissues. This example effectively counters those critics of

Jefferies who seem to think this impossible, and was pointed out by Crowson (1982) in defence of Jefferies' ideas. The phosphatic brachipod *Lingula* has a shell with collagenous, chitinous and phosphatic elements all together (Williams *et al.*, 1994).

104 Ubaghs had also noticed this, but referred to the structure as the 'ctenoid organ', a description which amply describes its comb-like shape but says little about its supposed function.

105 In a footnote added in proof, Jefferies comments 'By strict homology with vertebrates it might be better to speak of the head and tail of mitrates'.

106 That is because there is nothing special about Recent creatures versus those that are extinct. The reason they are accorded such favour is that they demonstrably occur on a single time plane, and so are unlikely to be one another's direct ancestors. The same is not true about animals in the general category 'extinct', but this may not apply if all the animals considered, even though extinct, all come from the same, restricted time plane (e.g. that they all come from, say, the Pleistocene, or the Triassic Period).

107 This seems to be a genuine mistake on Philip's part. On reviewing this passage, Jefferies notes that 'I certainly never held that the posterior coelom was equivalent to the right anterior coelom' (personal communication).

108 '... the head of mitrates is not a "black box" into which soft structures can be imagined at will' (1981a, p.355).

109 This passage is immediately followed by rebuttal of criticism somewhat less considered than that of Philip. 'Its [the calcichordate theory's] very complexity, however, has led Jollie (1973, p.87) to compare it with the arachnid hypothesis of vertebrate origins advanced by Patten (1912). Jollie's objection is irrelevant. The calcichordate theory should be criticized on its own faults and merits since I hold that the echinoderms, not the scorpions, are the closest living relatives of the chordates.' (1981)

110 Gislén (1930) saw it as a combined mouth and anus.

111 As noted above, Jefferies (since 1969) had interpreted it as a gonopore-anus and refers to it as such in Jefferies (1981a).

112 That these gaps open dorsally, like the atria of tunicates, is seen as a synapomorphy and part of the reason why these particular mitrates are regarded as stem tunicates.

113 Caster (1971, cited by Jollie, 1982) dismisses Jefferies' ideas not with a credible alternative but with the simple assertion that 'the "carpoids" as a whole are in every detail of gross morphology and skeletal histology echinodermal. Consequently, for us, all the relevant homologies are with other representatives of that phylum. Such similarities to chordates as "carpoids" possess are in our view wholly convergent and analogous'. This is not an argument.

5

Conclusions

> Geology abounds with creatures of the intermediate class: there are none of its links more numerous than its connecting links; and hence its interest, as a field of speculation, to the assertors of the transmutation of races. But there is a fatal incompleteness in the evidence, that destroys its character as such. It supplies in abundance those links of generic connection which, as it were, marry together dissimilar races; but it furnishes no genealogical link to show that the existences of one race derive their lineage from the existences of another.
>
> Hugh Miller, *The Old Red Sandstone*

5.1 APOLOGIA

The tradition in books with the words 'origin' and 'vertebrates' in the title is for the author to present a new scheme of vertebrate origins which is held to be better than its predecessors for this or that reason. Please, then, indulge me, with your permission to break the mould. The intention of this book was to present contrasting ideas, for the stimulation of new lines of inquiry, not to attempt my own synthesis. Although in its prosecution I have, occasionally, thought to come out with some grand phylogenetic statement, I decided in the end to stay true to my original plan. In the course of this book I have set up – only to knock down again – Gaskell and Patten, Hubrecht and Willmer, Berrill and Jefferies, eminent scholars all. After such cruelty one might understand why I am reluctant to expose my own hostages to fortune.

As a substitute I shall offer brief overviews of the major ideas discussed in this book, pointing out some unsolved problems, and offering some possible solutions. But one should not be surprised if I am occasionally overcome with the temptation to express my own views here and there, for which lapses I offer these apologies in advance.

5.2 THE HIGHER PROTOSTOMES

Several theories of vertebrate origins have held that the solution may be sought among the higher protostomes, whether annelids (St Hilaire), chelicerates (Patten) or crustaceans (Gaskell). Modern molecular sequence evidence indicates that annelids, arthropods, molluscs and some other groups comprise a monophyletic group of coelomate animals (Lake, 1990). If the nemerteans also belong to this large assemblage (Turbeville *et al.*, 1992), then the nemertean schemes of Hubrecht, Willmer and Jensen might also be included.

The alliance of vertebrates and higher protostomes generally demands three things, all of which are closer to satisfaction now than at any time in the past. Even so, there are good reasons for discounting all three.

The first criterion is structural or developmental homology between segmental structures. The remarkable similarities in structure, relative position and expression pattern of homeobox genes reveals a fundamental similarity between arthropods and vertebrates (Manak and Scott, 1994). However, these are manifested only at certain stages in embryogenies which in all other respects are quite different. Moreover, it is clear from work on primitive, non-segmented metazoa that a homeobox 'cluster' is probably a primitive feature of Metazoa in general (Slack *et al.*, 1993), in which case, segmentation arose independently in segmented protostomes and chordates, and cannot be used to unite them.

The second criterion is that there should be some scheme of 'inverting' the body of the vertebrate with respect to the annelid or arthropod, to accommodate the differences in situation between major blood vessels, alimentary tracts, nerve trunks and so on.

It is now evident that many organ systems in arthropods and vertebrates are homologues to the level of gene expression. To give two startling examples, genes that direct the formation of the (ventrally placed) heart in chordates are homologous with those that specify the (dorsally placed) contractile vessels of arthropods. Similarly, genes that direct the formation of eyes in the fruit-fly have homologues involved in eye development in mice and humans (see Manak and Scott, 1994, for a review). It is equally clear that many genes involved in the specification of the dorsoventral axis of arthropods are also involved in laying down the dorsoventral axis in vertebrates – but in the reverse sense.

In other words, there is sound, molecular evidence that a fundamental difference between arthropods and vertebrates is in an inversion of dorsoventral specification (Nübler-Jung and Arendt, 1994; Arendt and Nübler-Jung, 1994). However, the relevant genetic systems come into play very early in embryogeny; unlike Gaskell, one needn't be worried about remodelling entire adult animals. Some have argued that the system seen in arthropods reflects the primitive condition, and that chordates achieved the inversion during neurulation – the ravelling of the aboral cilated band into a nerve cord in Garstangian fashion, internalizing the aboral epidermal field to produce a nerve cord, the consequence of which would be the inversion of patterns of gene expression (Lacalli, 1995). I suspect that this difference between arthropods and vertebrates, though intriguing, says very little specifically about the origins of vertebrates, and more about fundamental differences within metazoa as a whole.

The third criterion is that segmentation in chordates is primitive for that group, in the sense that tunicates have secondarily lost it; tunicates are seen as 'degenerate' vertebrates. This criterion is the weakest of the three, as I have already noted that segmentation was acquired by vertebrates and higher protostomes independently. Furthermore, although it may well be the case that tunicates are secondarily unsegmented, this may have very little to do with the ancestry of chordates and more with present peculiarities of tunicates.

Molecular phylogeny, though, finally nails the vertebrate–protostome link. For although such phylogenies regularly group annelids, arthropods and molluscs together, the deuterostome phyla are invariably set apart. This implies that the ancestry of the vertebrates must be sought elsewhere.

5.3 THE TUNICATE TADPOLE LARVA

As I showed in Chapter 2, Garstang's ideas on the ancestry of vertebrates can be separated into two mutually exclusive tales. The first was the 'auricularia theory', in which the neural tube of the vertebrate ancestor was derived from the fused aboral ciliated bands of an echinoderm-like cilated larva (Garstang, 1894, 1897). The habitus of the adult – sessile or motile – was not stated. I contend that hidden in this idea are the common and strongly recapitulationist assumptions of the time: namely that the vertebrate ancestor was primitively free-living, the sessile habit of tunicates was a secondary stage added on to the end of the life cycle, thus forcing the motile stage back in ontogeny; and that tunicates were secondarily unsegmented. The entire concept was a testament to Haeckelian recapitulation.

The second of Garstang's two theories was based on his idea of paedomorphosis (Garstang, 1922, 1928). In this scheme, the auricularia theory was applied specifically to tunicates, to form a tadpole larva as an interpolation between a ciliated larva and a sessile adult. The process of paedomorphosis in tadpole larvae would elevate them to sexual maturity in that state, abolishing the need for the sessile adult stage.

There are two overt differences between the two theories, and many hidden ones. The overt differences are the invocation in the second theory of the sessile adult as a necessity; and the process of paedomorphosis, to render this sessility obsolete. The hidden differences are more significant, for they are less differences than complete reversals on the original auricularia theory. Namely, that the vertebrate ancestor was primitively sessile, not free-living; that the sessile habit of tunicates was primitive, and that tunicates were primitively unsegmented. Recapitulation was loudly and definitively repudiated.

The first theory is clearly more parsimonious than the second. In the pre-paedomorphic conception, chordates were always seen as primitively free-living. Tunicate sessility was seen as a late, specialized side issue. In the second, a sessile adult has to conjure up a new, motile larval stage, and then abolish itself to produce the same motile result achieved in the first theory with neither interpolation nor paedomorphosis. As did Carter (1957), one must question whether these complexities are really necessary when there is a much simpler alternative.

Another problem with Garstang's second theory is that its application to vertebrates is, actually, somewhat moot. Although one can imagine the advantage to ascidians of inventing a motile, dispersive phase such as the tadpole larva, and that paedomorphosis might explain the motile state of larvaceans, it is harder to conceive of a tadpole larva acquiring the trappings of segmentation necessary to become a cephalochordate or craniate: questions of cephalization and segmentation always find Garstang and Berrill at their weakest.

Others, such as Whitear (1957) and Bone (1960), realized that the application of Garstang's second theory to vertebrates was a metaphorical, not a literal extension from the tunicate condition, in the same way that Darwin regarded the origin of species by natural selection as an extrapolation of the artificial selection applied by fanciers to the variation in domestic pigeons. Once this difference is grasped, it is obvious that for the second theory to apply to vertebrates (rather than ascidians), the sessile adult phase need not be an ascidian, or even a tunicate. The reason is clear: once the tadpole larva is formed as an interpolation, and paedomorphosis has taken place, the identity of the now-abolished adult matters not at all. However, the identity of this lost ancestor was the key to understanding the origin of vertebrates. For Bone (1960), it was a 'proto' hemichordate. Only Gregory (1946) suggested that it could have been something as *outré* as a carpoid echinoderm.

The latest work suggests that the first of Garstang's two theories has much more to recommend itself than the second. Ultrastructural work suggests that the origin of the neural tube in tunicate tadpole larvae, and in chordate neurulation in general, is consistent with the drawing-together of paired ciliated bands (Crowther and Whittaker, 1992; Lacalli *et al.*, 1994).

The second theory, though, is currently in a process of senile and inept collapse. Molecular phylogeny (Wada and Satoh, 1994) and micro-anatomical work (Holland *et al.*, 1988) suggests that larvaceans are primitive tunicates, and so not likely to have evolved by paedomorphosis from an ascidian ancestor. Genetic and ultrastructural studies indicate that the tail of the tunicate tadpole is primitively segmental in character (Crowther and Whittaker, 1994), this segmentation having become muted as a consequence of small size and particularity of habit. It is increasingly likely that tunicates are highly specialized chordates, and that sessility is a derived habit. I suspect that the first tunicates were fully motile, in which case the argument for transposing the whole interpolation-plus-paedomorphosis case from larvaceans to the ancestry of vertebrates becomes meaningless.

After all that, the whole idea of paedomorphosis looks like something of a false trail, an unnecessary source of confusion. On re-reading Garstang's 1922 paper on the subject, I suggest that the central motivation behind paedomorphosis was less to explain the origin of vertebrates than to articulate dissatisfaction with contemporary literal readings of recapitulation by such as McBride. To paraphrase Garstang himself, the theory of larval paedomorphosis is dead, and need no longer limit and warp us in the study of vertebrate phylogeny.

The first tunicates probably looked like larger, more elaborate tadpole larvae, with large, possibly unsegmented heads, and fully segmented tails, rather like Romer's somaticovisceral animals.

5.4 CALCICHORDATES

Given the profound differences between echinoderms and chordates, it is hardly surprising that little effort has been made to investigate how the forms of one might have derived from those of the other, or (more correctly) what their common ancestors looked like. The assumption based on character distributions in the Recent fauna has been that the calcite skeleton, water–vascular system and highly derived mode of development of echinoderms have always been peculiarities of the group, and that the common ancestor of echinoderms and chordates was a bilaterally symmetrical, tricoelomate, soft-bodied creature.

This is what makes Gislén's suggestion (Gislén, 1930) that carpoid echinoderms might have chordate 'affinities', in their possession of pharyngeal gill clefts, so interesting, for it prompts us to look at vertebrate origins in an entirely new way.

Among other things, it raises the problem, hardly discussed elsewhere, of the origin of the marked asymmetries in development and structure which seem peculiar to deuterostomes, and offers the suggestion of a fundamentally bipartite (rather than segmental) character to vertebrate structure. The developmental underpinnings of these phenomena are currently under investigation.

Jefferies (1967, 1968a,b, 1986), of course, took the next step, suggesting that carpoids were not echinoderms with chordate affinities, but chordates with echinoderm affinities. Even though Peterson (1995a) calls this statement into question, his preferred placement of the carpoids in the scheme of deuterostome phylogeny (Figure 4.22c) is consistent with Gislén's view, if not Jefferies'.

Three particular problems emerge from Jefferies' studies. First, the known fossil record of carpoids continues at variance with the proposed calcichordate phylogeny. As I showed above, this is more of an embarrassment than a real problem.

More serious is the difficulty presented by the interpretation of the anatomies of these creatures. The reconstruction of the anatomy of fossil creatures is formally impossible without some contemporary 'model' to serve as a basis. Carpoids are so very strange, even when measured by recent echinoderms, that any proposed anatomical reconstruction will be contentious. Jefferies' detailed reconstructions of the insides of carpoids are thus intriguing, but impossible to test (Thomson, 1987b; Gans, 1988; Peterson, 1995a). It is this problem which, I believe, has hobbled all attempts to refute Jefferies' ideas that are based on anatomical interpretation (for example Philip, 1979).

Hennigian phylogenetic reconstruction is one of the greatest strengths of the calcichordate theory. But, as Peterson (1995a) shows, it is also one of its signal weaknesses. A phylogenetic analysis based on characters used by Jefferies and his associates to support the calcichordate theory does not actually support the phylogeny that is the most succinct expression of the theory. The reason for this is unclear. It could mean that some or all of Jefferies' anatomical interpretations of carpoid anatomy are in error, but Peterson's test cannot tell us anything more.

Nevertheless, I would hazard a guess that the 'cothurnopores' of *Cothurnocystis* really are pharyngeal clefts, homologous with those of chordates and hemichordates. For the moment, I suggest that Gislén's view, that carpoids are echinoderms with chordate (or hemichordate) affinities, is the correct one.

5.5 RECENT AND FOSSIL FORMS

So, carpoids were echinoderms, with the addition of pharyngeal clefts (and possibly other internal structures) nowadays seen only in hemichordates and chordates. This statement conceals a problem of vital importance in systematic biology, that of the relative contributions of Recent and fossil forms to phylogenetic reconstruction. This debate has hardly touched ideas about vertebrate origins, as these are couched almost exclusively in terms of creatures alive today – with the hidden assumptions that distributions of features we see today must have pertained for all time. Calcite skeletons are only found in modern echinoderms, and so must be characteristic of the group. Similarly the notochord and pharyngeal structures must be characteristics of chordates.

The added interest of Gislén's and Jefferies' ideas is that of all the schemes put forward to explain the origins of vertebrates, they have drawn on fossils – specifically, on fossils in which there appear to be combinations of characters which do not occur in any Recent forms. What is one schooled in the neat apportionment of features according to Recent forms supposed to make of an apparition such as *Cothurnocystis*, evidently an echinoderm but with pharyngeal clefts and (possibly) a notochord-braced tail?

If there once existed creatures in which were displayed combinations of features, each one separately diagnostic of different Recent groups, might this not cast doubt on the phylogenetic integrity of these Recent groups? If so, how might this affect phylogenetic reconstruction?

Even more disturbingly, if it turns out that strange, extinct groups are in the majority, and that the Recent fauna is a small, randomly chosen and unrepresentative selection of a large past diversity, is it right to shoehorn ancient forms into groups (such as chordates or echinoderms) defined on the basis of character distributions in Recent forms? Questions like these are being asked increasingly often, and not only in the sphere of vertebrate origins. Further consideration of the problem might help to put the primary subject of this book into context.

5.6 THE CAMBRIAN EXPLOSION

Methods of phylogenetic reconstruction, whether coldly cladistic or romantically hand-waving, always run into trouble when the species under study diverged in a short period of time, a long time ago.

The most problematic such event was the 'Cambrian Explosion', between about 540 and 520 million years ago, in which nearly all major groups of multicellular animals appeared (Lipps and Signor, 1992; Conway Morris, 1993, 1994; Philippe *et al.*, 1994; Gee, 1995b). Most of the extant phyla originated at this time, including the echinoderms (Figure 5.1) in all likelihood with many other phyla that have since become extinct.

There is no doubt that the Cambrian Explosion was a real event (for all that the causes remain obscure) and not an artefact of deposition, or of

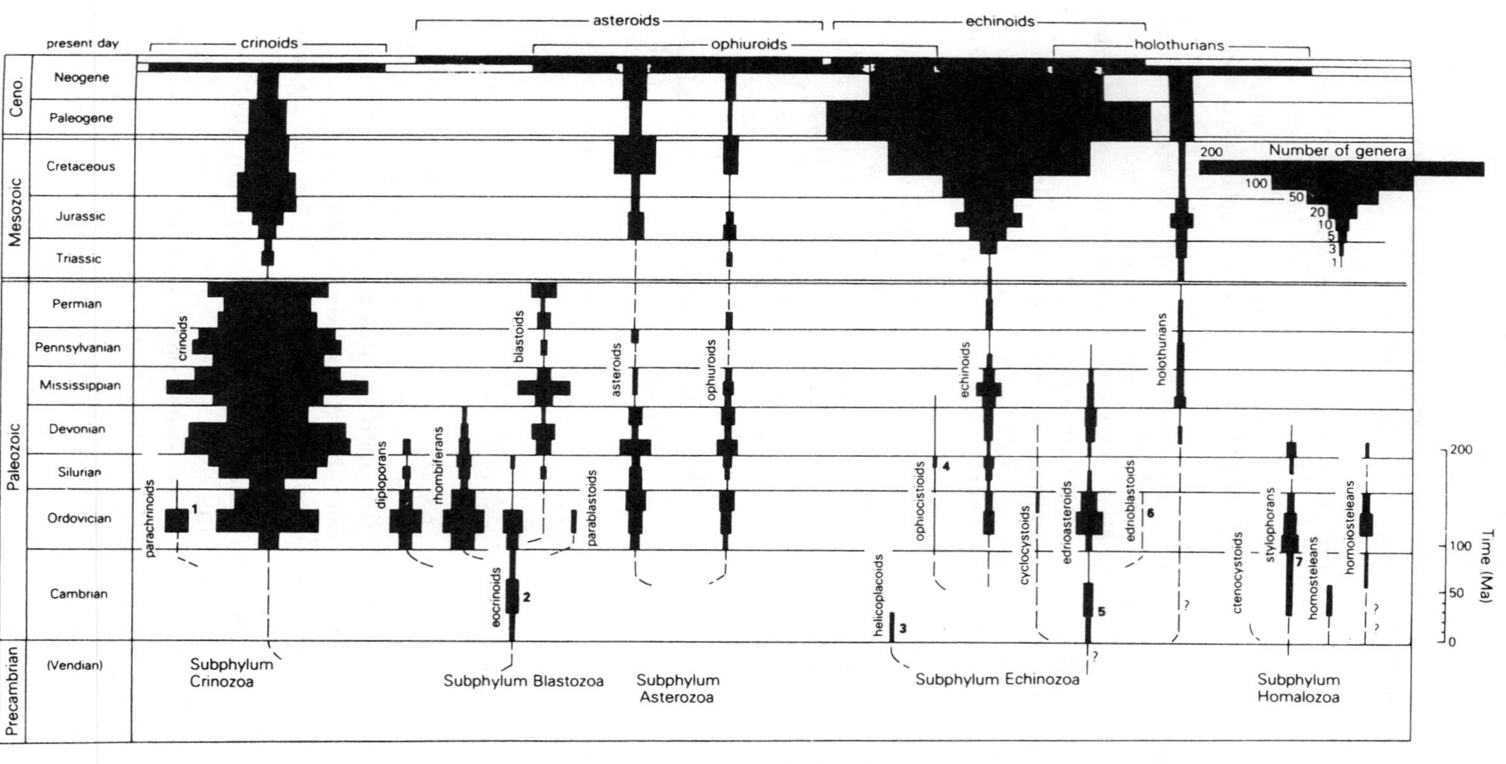

Figure 5.1 Time ranges and relative abundance of various echinoderm groups. The carpoids are relegated to 'Homalozoa' at the bottom right. Note how a large number of disparate groups has been replaced by a smaller number of relatively speciose groups (from Clarkson, 1986, based on Sprinkle).

the sudden acquisition of hard tissues by a variety of long-distinct, soft-bodied forms. Molecular (Knoll, 1992) and palaeontological (Knoll and Walter, 1992) evidence suggest that eukaryotes (that is, organisms with distinct membrane-bound cell nuclei) underwent a dramatic radiation around 1,000 million years ago. These organisms were all of the unicellular-alga grade of organization, so it is highly unlikely that evidence for multicellular animals of this antiquity will be found. Nevertheless, by 600 million years ago, in the Uppermost Precambrian, a variety of large, possibly multicellular forms had appeared: by the middle of the Cambrian period, around 500 million years ago, virtually every familiar group of multicellular animal was represented. (However, one could argue that the origin of a lineage and its preservation in the fossil record are different things. Davidson *et al.* (1995) suggest that multicellular animals could have lived as microscopic rotifer-like creatures for an indeterminate period before the Cambrian Explosion. Fossilization potential was achieved only after the advent of evolutionary innovations such as undifferentiated tissue ('set-aside cells') to serve as adult rudiments in the larval stage.)

So, although it is evident that a great deal of evolution happened between 1000 and 600 million years ago, for which we have little actual evidence in the form of body fossils, it is equally evident that the rate of increase in organic complexity during that period has been unmatched since. Evolution since the Cambrian has been a process of elaboration of existing forms, rather than the creation of radically new body plans. Indeed, one could say that evolution since the Cambrian has been a process of loss (Gould, 1989). Many of the major groups that appeared in the Cambrian Explosion subsequently became extinct, so that the present fauna is highly depauperate by comparison.

This is as true for the deuterostomes as for any other group. Below, I shall argue that the poverty of the modern fauna colours our ideas about the phylogenetic relationships even of animals long extinct, and that one must, as compensation, adopt a more synoptic view of deuterostome evolution if one is to discern therein the ancestry of vertebrates, or indeed of any other deuterostome group.

Above all, one must be prepared to be pragmatic, and accept the possibility that the 'true' story of vertebrate origins lies within a class of several possible schemes – the evidence may yet be insufficient for the unequivocal support of one, to the exclusion of all others.

5.7 DIVERSITY AND DISPARITY

This effect of poverty on phylogeny is best demonstrated not with deuterostomes, but with arthropods.

The phylogeny of arthropods is currently a matter of dispute, but a consensus is emerging from developmental, palaeontological and genetic studies (see Averof and Akam, 1995, for a review). One view of arthropod phylogeny held that the group was polyphyletic, and that any similarities were as a result of convergent 'arthropodization'. Three groups of arthropod were identified: chelicerates (spiders, scorpions and the extinct trilobites), crustaceans and 'uniramians'. The latter group consisted of insects, myriapods (centipedes and millipedes) and the lobopods or onychophora (the velvet-worm *Peripatus* and its allies). It was based on the observation that members of this group exhibited simple, unbranched limbs, whether the long legs of grasshoppers or the short, stumpy legs of lobopods.

The emerging consensus is rather different. Within the monophyletic Arthropoda, insects are closely related to crustaceans, with chelicerates and myriapods standing further away (Averof and Akam, 1995). The lobopods stand outside the Arthropoda altogether (Monge-Najera, 1995), and have no more in common with insects than (say) crustaceans.

But was this always the case? The exquisite soft-part preservation in Cambrian assemblages such as the Burgess Shales fauna of British Columbia, the Sirius Passet fauna of northern Greenland and the Chengjiang fauna of China reveal an enormous diversity of arthropod design that is no longer apparent in the modern fauna.

Most Burgessian creatures are clearly arthropods, in that they have jointed exoskeletons to which are attached a variety of paired appendages. And yet, in many cases, the arrangements of segments and appendages differ from those considered diagnostic of any of the three major modern arthropod groups. If it is the case that the present criteria for distinguishing the major groups of arthropod define clades, there may have been many more different kinds of arthropod body plan in the Cambrian than there are now.

Many other Burgessian creatures do not, at first sight, look like arthropods at all. Indeed, they seem to differ from any known phyla. Celebrated examples include the pelagic *Anomalocaris* and *Opabinia,* and *Hallucigenia,* a creature so odd that it could inspire the design of science-fictional aliens[114]. The presence of so many strange creatures led Gould to expound (1989) and defend (Gould, 1991) the concept of 'disparity', to express difference of form, as distinct from 'diversity', which should refer to variation within the same basic form (the concept was mooted not by Gould, but in a paper by Runnegar (1987) on molluscs). Among animals generally, disparity has decreased since the Cambrian, but diversity has increased. In other words, there are fewer major groups than there once were, but individual groups tend to contain more species. To put it yet another way, there may be millions of beetles, but they are all beetles nonetheless.

However, recent work suggests that many of these supposedly problematic Burgessian creatures are in fact allied not to arthropods *per se*, but lobopods. Several may represent a kind of armoured lobopod, now extinct, but apparently diverse in the Cambrian (see Hou and Bergström, 1995, and Conway Morris, 1994, for reviews and see Chen *et al.*, 1994, for a slightly divergent view). Even *Hallucigenia* can be accommodated in such a scheme. The interpretation of another Burgessian creature, *Kerygmachela* from Greenland as a kind of pelagic lobopod with external gills (Budd, 1993) opens the way for *Anomalocaris* and *Opabinia* to be brought into the fold. The modern lobopods are banished from the modern Arthropods partly because of their nakedness. But if lobopods were once armoured as a matter of course, the point becomes moot. In what follows I shall refer to members of both groups, informally, as 'arthropods'.

Gould's idea of disparity as a concept would be dented if it were shown that the several enigmatic Burgessian groups postulated as unique phyla can be related to modern forms. Cladistic analysis of Burgessian and modern arthropods (Briggs and Fortey, 1989; Wills *et al.*, 1994) suggests that this might be the case – rather than stand apart from modern arthropod groups, Burgessian forms intermingle with modern ones in a single cladogram (Figure 5.2). The purpose of the study by Wills *et al.* (1994), from which Figure 5.2 is drawn, was to challenge Gould's concept of disparity – to show that Cambrian arthropods are no more disparate than modern ones. One might question whether disparity as a concept is amenable to cladistic analysis, or indeed any method of phylogenetic reconstruction that produces a rooted tree, with the inescapable implication of monophyly, particularly when the analysis mixes fossil and Recent forms. Nevertheless, the central message seems clear. Even if one questions that such oddities as *Anomalocaris*, *Opabinia* and *Hallucigenia* properly belong in the same monophyletic group as the more familiar arthropods, there are still a large number of Burgessian creatures which *do* look like arthropods, *can* be incorporated into a plausible cladistic analysis, and yet *cannot* be classed with any modern, arthropod body plan. The conclusion is that the modern range of arthropod body plans is but a small fraction of that which once existed.

This poverty of body plans may exert an effect on phylogenetic reconstruction. This stems from the creation of monophyletic groups based on associations of character states in the few modern remaining body plans, and the interpretation of extinct body plans in the light of these groups. According to Gould (1989), such misconceptions initially led to many misinterpretions of the affinities of the Burgess Shales fossils. But because modern body plans are so scarce in comparison with what has gone before, their perceived importance (and the character combinations arrayed to support them) may exceed their actual worth.

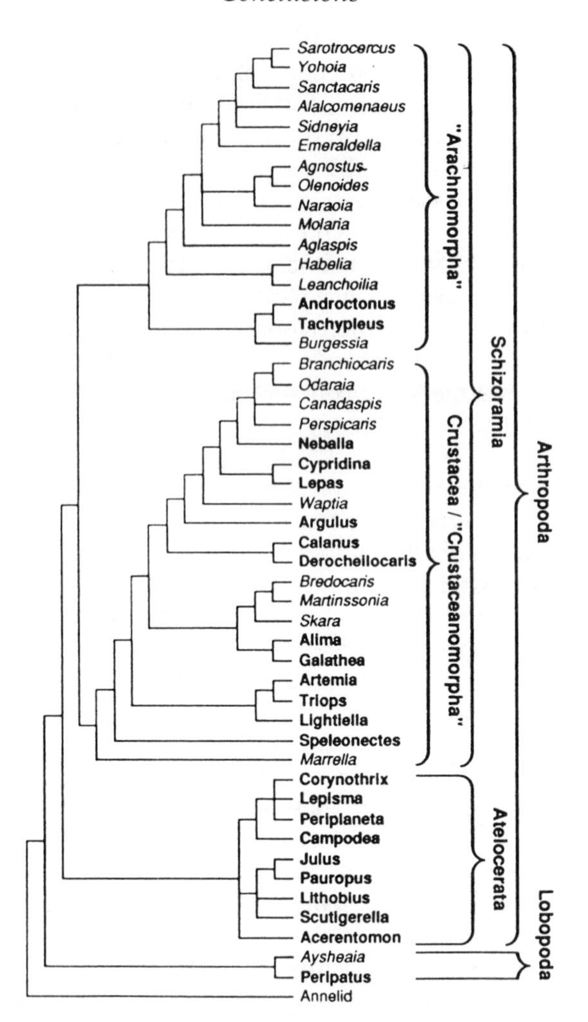

Figure 5.2 Strict consensus of 48 equally parsimonious 353-step cladograms for the data set of Cambrian and Recent arthropods. Recent arthropods are shown in bold type, Cambrian forms are indicated by italics (from Wills *et al.*, 1994).

If, as seems likely, the extinction of high-level taxonomic groups (classes, phyla) is largely governed by stochastic processes unrelated to adaptation, the range of body plans we see today is largely a random assortment – they are not here because they are better, they are here because they are lucky. It is thus all the more inappropriate to organize extinct classes and phyla by the criteria used to define modern monophyletic groups.

For example, one feature taught to generations of students as a characteristic feature of crustaceans concerns the appendages, each of which is forked into two branches (that is, biramous), one branch a gill, the other a 'walking' leg. Lobopods and insects have simple, unbranched legs, a feature once used to unite them as 'uniramians' ('the single-branched ones'). However, there is some palaeontological and developmental evidence to suggest that insects had a biramous ancestry, in which the gills became wings (Carroll *et al.*, 1995). This is part of the reason why insects and crustaceans are nowadays seen as close relatives.

Modern lobopods are terrestrial, with stumpy, single-branched legs, yet are highly derived and specialized compared with their ancient, marine forebears. Yet on these modern creatures rest the defining characteristics of the entire group, past and present. So what, then, are we to make of *Kerygmachela*, an animal with a general resemblance to Cambrian lobopod-like animals, yet with biramous, gill-bearing appendages? Budd (1993) refers to it as a gilled lobopod, but the presence of biramous limbs in a lobopod calls into question the use of biramous limbs as a defining feature of other arthropod groups such as crustaceans.

More worryingly, one begins to ask whether *any* such feature can be of use in defining modern arthropod groups, given the larger disparity of Burgessian arthropods, and the likelihood that a similar feature will turn up among their number. Ultimately, one cannot escape the question of whether it is correct to decree that the criteria used to define extant groups must apply for all time, *simply* because the groups used are extant. Phylogenetic reconstruction that encompasses all known variation in all the constituent groups under study, whether living or extinct, might give a more accurate picture of phylogenetic relationships.

5.8 DEPAUPERATE DEUTEROSTOMES

As a group, the modern deuterostomes are at least as disparate in form as modern arthropods, and as depauperate in relation to their fossil history. Excluding the debatable lophophorates, modern deuterostomes comprise hemichordates, chordates and echinoderms.

Several major subgroups within these major deuterostome groups have come and gone. The hemichordate graptolites, just like the arthropod trilobites, were diverse and speciose, and perished around the Permian period, leaving behind a meagre and motley assortment of pterobranchs. Among the vertebrates, the conodonts disappeared in the Triassic after a long and successful residence: a vast array of species of unguessed anatomical diversity, now all gone. Other important losses among the vertebrates were placoderm fishes and several groups of armoured agnathan, beside which the extinction of the dinosaurs is a minor distraction. Of course, we cannot even begin to estimate the fearful toll among the soft-bodied enteropneusts, urochordates and cephalochordates.

But this carnage is probably as nothing compared with that which fortune has meted the echinoderms – of the approximately twenty classes known to have existed, only five survive. And it would have been four, but for the tenacity of *Miocidaris*, the only genus of echinoid that managed to scrape through the end-Permian extinction event (reviewed in Erwin, 1994).

If the example of *Kerygmachela* and the other Burgessian arthropods is any guide, it should be no surprise if features used to define extant deuterostome groups turn up in fossils in unexpected places, particularly among echinoderms. If *Kerygmachela* is a lobopod with biramous limbs, why should *Cothurnocystis* not be an echinoderm with chordate-like pharyngeal gill slits? The presence of gill slits is now used to define chordates (plus hemichordates), but such definitions are based on the extant fauna only and may not always have been applicable.

Some have questioned whether cladistics (and parsimony) can be applied in the same way to the characters that define major groups as to the much smaller variations within a group whose members share the same overall plan[115]. Different forces govern the fortunes of major groups from those which rule the fates of genera and species (Gould and Eldredge, 1993), and it might be that the methods appropriate for the phylogenetic reconstruction of species cannot be applied to phyla.

Parsimony, for example, may be ill-suited as a guide to untangling the profound developmental shifts that assuredly underlie the origins of major body plans (Ruddle *et al.*, 1994; Peter Holland, personal communication). In a phylogenetic reconstruction designed to express the relationships between many disparate groups (*sensu* Gould), the amount of convergence might well be high, the choice of an appropriate outgroup fraught, and the chosen character polarities questionable.

The choice of outgroup becomes particularly important in such situations. The reason is that most methods of phylogenetic reconstruction produce a connected network, rooted or not, in which relatedness between all the taxa is implicit – common ancestry may be read into the analysis as a prior assumption, even though the chosen taxa may represent a poyphyletic assemblage. This problem becomes more extreme as the range and disparity of the included taxa becomes more inclusive. This makes the choice of outgroup extremely difficult.

The conclusion, I think, is that when dealing with the interrelationships of ever larger, more inclusive groups, such as deuterostomes, protostomes or coelomates, it is progressively harder to justify the assertion that characters tend to be gained or lost the smallest number of times allowed by the analysis.

Even if the calcichordate theory is wrong, it nevertheless raises an interesting possibility – that once one starts to discuss fossils, especially fossils which admit no easy interpretation on the basis of a Recent model, features usually thought of as diagnostic of major groups (gill slits, calcite skeletons, notochords, brains) might pop up all over the place, challenging conventional phylogenetic groupings, and our ideas about their origins.

There is no shame, then, in unresolved polychotomies. In many cases, they may be all we can achieve, and the best we have a right to expect – and they may even be useful (Philippe *et al.*, 1994). The challenge comes in discerning that level of phylogenetic scale, that amount of disparity, wherein we can claim that right.

5.9 TWO PERSISTENT PROBLEMS

Two unresolved polychotomies that have cropped up repeatedly in this book could repay further study. The first is the trichotomy of echinoderms, hemichordates and chordates; the second is that of urochordates, cephalochordates and craniates.

5.9.1 Echinoderms, hemichordates, chordates

The traditional resolution of this trichotomy is

(1) echinoderms, (hemichordates, chordates)

based on such shared features as pharyngeal gill clefts. However, recent molecular evidence (Halanych, 1995), combined with some distinctly old-fashioned embryology, reawakens Metchnikoff's old union of echinoderms and hemichordates, the Ambulacraria

(2) (echinoderms, hemichordates), chordates

Fossils, in the form of calcichordates produce the third of three possible resolved arrangements,

(3) hemichordates, (echinoderms, chordates).

Peterson (1995a) discounts this third option, favouring the traditional phylogeny, option (1) above. However, phylogeny (2) raises an interesting point of character distribution, for it implies that pharyngeal clefts were present in the ancestry of echinoderms, but have since been lost. Conversely, it suggests that calcite skeletons might once have been a feature of hemichordates. Was *Cothurnocystis* an armoured hemichordate?

5.9.2 Urochordates, cephalochordates, craniates

The textbook version of this trichotomy,

urochordates, (cephalochordates, craniates)

is supported by much detailed anatomical and genetic work. Many features of the craniate central nervous system are foreshadowed in cephalochordates, and cephalochordates have a single cluster of *Hox* genes similar to one of the four craniate clusters (Garcia-Fernàndez and Holland, 1994). However, there is much yet to learn about urochordates, and the ongoing genetic work may yet uncover a few surprises. This, together with ultrastructural work revealing segmentation in the tail of the tunicate tadpole (Crowther and Whittaker, 1992) and work on the phylogeny of larvacea (Holland *et al.*, 1988), suggests that tunicates are primitively segmented and motile. Although it is tempting to suggest it, there is as yet insufficient evidence from these lines of inquiry to support one of the two alternative resolved phylogenies,

cephalochordates, (urochordates, craniates).

This is the phylogeny predicted by the calcichordate theory on the basis, mainly, of the primitive mitrate *Lagynocystis* (Jefferies, 1973), held to be a cephalochordate. As far as I am aware, nobody has proposed the third phylogeny,

(cephalochordates, urochordates), craniates.

5.10 THE WAY OUT

It could be that the way to solve problems such as these is not to map the *actual* occurrences of morphological characters, but their *potential* to occur (Peter Holland, personal communication). The potential to express gill slits, for example, presumably occurs in a wider range of animals than those which actually do so. Were such mapping possible, parsimony would be a useful criterion for the judgement of competing phylogenies, for we would be looking at the data in the raw, not filtered through ontogeny and metamorphosis. One would be able to address the problem of telling the difference between primitive absence and convergent loss.

The first step might be to track morphological features by way of gene expression, but this only puts the problem back a stage, as the problems of convergence beset genes as much as morphology. A more interesting strategy might be to map differences in large-scale genetic organization, which are less likely to be convergent, and might shed light on relationships between creatures which look very different, morphologically.

The discovery that metazoa all tend to have collinear clusters of homeobox genes was the first step in this direction. The further finding that the single *Hox* cluster in the amphioxus resembled one of the four mammalian clusters far more than (say) the cognate arrangement in *Drosophila* (Garcia-Fernàndez and Holland, 1994) was another highly significant step, for it allowed one to establish the position of the amphioxus as primitive with respect to the vertebrates, rather than derived and degenerate, as well as solving the problem of the vertebrate head (Chapter 3). Neither should one neglect 'pseudogenes', the relics of genes which once had functions since lost. Such might be useful in tracing the history of highly modified specialists such as urochordates (Ruddle *et al.*, 1994).

Deeper secrets still might be laid bare. Differences in gene-cluster duplication patterns have enormous potential to solve such problems as the phylogenetic position of hemichordates, lophophorates and echinoderm classes.

The same applies to the analysis of large-scale patterns in the order of genes within functionally related arrays, rather than the simple comparisons of sequences of individual genes, such as the 18S ribosomal RNA genes. The analysis of gene order (rather than gene sequence) may provide, for the first time, a sure way of resolving deeply branched phylogenies. This is because changes in the order of genes in a cluster are likely to have occurred much less often than the order of nucleotides within each individual gene in the array. The technique is also far less prone to uncertainty than conventional sequence analysis: an individual gene is easily identified as such, but it is much harder to judge whether a given nucleotide *within* a gene is original, or a replacement that occurred at some unspecified time. Again, molecular phylogenies using nucleotide base-pair sequences must contend with the effects of natural selection on some parts of the sequence and not others, and with relative 'rates' of change. Gene-order analysis suffers far less from such problems.

Wesley Brown and colleagues (Boore *et al.*, 1995) have compared the ordering of genes in mitochondrial DNA in arthropods, producing a phylogeny that confirms the new consensus of arthropod phylogeny (Recent forms, at any rate). Brown's group is currently using this technique to elucidate deuterostome phylogeny, and their results are awaited with interest. They may well be definitive.

NOTES

114 I am convinced that the fearsome Jarts in the science-fiction novel *Eternity* by Greg Bear are modelled on *Hallucigenia*.

115 This is the central concern of Pat Willmer's brave textbook *Invertebrate Relationships* (Willmer, 1990). But rather than try to define the limits of cladistics in any formal way, she starts by rejecting cladistics as of little use in the reconstruction of higher-order phylogenies. The result is a polyphyletic scheme of invertebrate relationships supported by *ad hoc* appeals to adaptation and function.

References

Ahlberg, P. E. (1993) Therapsids and transformation series. *Nature*, **361**, 596

Ahlberg, P. E. and Milner, A. R. (1994) The origin and early diversification of tetrapods. *Nature*, **368**, 507–514

Akam, M. (1984) A common segment in genes for segments of *Drosophila*. *Nature*, **308**, 402–403

Akam, M. (1987) The molecular basis for metameric pattern in the *Drosophila* embryo. *Development*, **101**, 1–22

Akam, M., Dawson, I. and Tear, G. (1988) Homeotic genes and the control of segment diversity. *Development*, **104** (suppl), 161–168

Akimenko, M.-A., Johnson, S. L., Westerfield, M. and Ekker, M. (1995) Differential induction of four *msx* homeobox genes during fin development and regeneration in zebrafish. *Development*, **121**, 347–357

Aldridge, R. J. (1987) Conodont palaeobiology: a historical view, in *The Palaeobiology of Conodonts*, (ed. R. J. Aldridge), The British Micropalaeontological Society, Chichester, pp.11–34

Aldridge, R. J. and Briggs, D. E. G. (1986) Conodonts, *Problematic Fossil Taxa*, (eds A. Hoffman and M. H. Nitecki), Oxford University Press, New York/Clarendon Press, Oxford, pp.227–239

Aldridge, R. J., Briggs, D. E. G., Clarkson, E. N. K. and Smith, M. P. (1986) The affinities of conodonts – new evidence from the Carboniferous of Edinburgh, Scotland. *Lethaia*, **16**, 279–291

Aldridge, R. J., Purnell, M. A., Gabbott, S. E. and Theron, J. N. (1995) The apparatus, architecture and function of *Promissum pulchrum* Kovács–Endrödy (Conodonta, Upper Ordovician) and the prioniodontid plan. *Philosophical Transactions of the Royal Society of London*, B **347**, 275–291

Alldredge, A. (1976) Appendicularians. *Scientific American*, **235(1)**, 94–102

Alldredge, A. (1977) Morphology and mechanisms of feeding in the Oikopleuridae (Tunicata, Appendicularia). *Journal of Zoology*, **181**, 175–188

Anon. (1973) Ancestry of chordates. *Nature*, **245**, 123–124

Arendt, D. and Nübler-Jung, K. (1994) Inversion of dorsoventral axis? *Nature*, **371**, 26

Averof, M. and Akam, M. (1995) Insect–crustacean relationships: insights from comparative developmental and molecular studies. *Philosophical Transactions of the Royal Society of London*, B **347**, 293–303

Awgulewitsch, A. and Jacobs, D. (1992) *Deformed* autoregulatory element from *Drosophila* functions in a conserved manner in transgenic mice. *Nature*, **358**, 341–344

Awgulewitsch, A., Utset, M. F., Hart, C. P., McGinnis, W. and Ruddle, F. (1986) Spatial restriction in expression of a mouse homoeo box locus within the central nervous system. *Nature*, **320**, 328–335

Ax, P. (1985) Stem species and the stem lineage concept. *Cladistics*, **1**, 279–287

Baker, R. A. and Bayliss, R. A. (1984) Walter Garstang (1868–1949): zoological pioneer and poet. *Naturalist*, **109**, 41–53

Bardack, D. (1991) First fossil hagfish (Myxinoidea): a record from the Pennsylvanian of Illinois. *Science*, **254**, 701–703

Bardack, D. and Zangerl, R. (1968) First fossil lamprey: a record from the Pennsylvanian of Illinois. *Science*, **162**, 1265–1267

Barnes, R. D. (1968) *Invertebrate Zoology* (2nd edn), W. B. Saunders, Philadelphia

Barnes, R. D. (1980) *Invertebrate Zoology* (4th edn), Holt Saunders International, Philadelphia and Tokyo

Barnes, R. S. K., Calow, P., Olive, P. J. W. and Golding, D. W. (1993) *The Invertebrates – a New Synthesis* (2nd edn), Blackwell, Oxford

Bateson, W. (1884) The early stages in the development of *Balanoglossus* (*sp. incert.*). *Quarterly Journal of Microscopical Science*, **24**, 207–235

Bateson, W. (1885) The later stages in the development of *Balanoglossus Kovalevskyi*; with a suggestion on the affinities of the Enteropneusta. *Quarterly Journal of Microscopical Science*, **25**, suppl., 81–122

Bateson, W. (1886) The ancestry of the Chordata. *Quarterly Journal of Microscopical Science*, **26**, 535–571

Bateson, W. (1914) Address of the President of the British Association for the Advancement of Science. *Nature*, **40**, 287–302

Bather, F. A. (1900) Part III, The Echinoderma, in *A Treatise On Zoology*, (ed. E.R. Lankester), Adam and Charles Black, London

Bather, F. A. (1913) Caradocian Cystidea from Girvan. *Transactions of the Royal Society of Edinburgh*, **49**, 359–529

Bather, F. A. (1925) *Cothurnocystis*, a study in adaptation, *Palaeont. Zeitschrift*, **7**, H I

Bengtson, S. and Urbanek, A. (1986) *Rhabdotubus*, a middle Cambrian rhabdopleurid hemichordate. *Lethaia*, **19**, 293–308

Benton, M. J. (1987) Conodonts classified at last. *Nature*, **325**, 482–483

Benton, M. J. (ed.) (1993) *The Fossil Record* 2. Chapman & Hall, London

Berrill, N. J. (1955) *The Origin of Vertebrates*, Clarendon, Oxford

Binyon, J. (1979) *Branchiostoma lanceolatum* – a freshwater reject? *Journal of the Marine Biological Association of the United Kingdom*, **59**, 61–67

Bird, A. P. (1995) Gene number, noise reduction and biological complexity. *Trends in Genetics*, **11**, 94–100

Blair, S. (1994) Hedgehog digs up an old friend. *Nature*, **373**, 656–657

Bone, Q. (1957) The problem of the 'amphioxides' larva, *Nature*, **180**, 1462–1464

Bone, Q. (1958) The asymmetry of the larval amphioxus. *Proceedings of the Zoological Society of London*, **130**, 289–293

Bone, Q. (1960) The origin of the chordates. *Zoological Journal of the Linnean Society*, **44**, 252–269

Boore, J. L., Collins, T. M., Stanton, S., Daehier, L. L. and Brown, W. M. (1995) Deducing the pattern of arthropod phylogeny from mitochondrial DNA rearrangements. *Nature*, **376**, 163–165

Bopp, D., Burri, M., Baumgartner, S., Frigerio, G. and Noll, M. (1986) Conservation of a large protein domain in the segmentation gene *paired* and in functionally related genes of *Drosophila*. *Cell*, **47**, 1033–1040

Bopp, D., Jamet, E., Baumgartner, S., Burri, M. and Noll, M. (1989) Isolation of two tissue-specific *Drosophila* paired box genes, *Pox meso* and *Pox neuro*. *EMBO Journal*, **8**, 3447–3457

Bowring, S. A., Grotzinger, J. P., Isachsen, C. E., Knoll, A. H., Pelechaty, S. M. and Kolosov, P. (1993) Calibrating rates of early Cambrian evolution, *Science*, **261**, 1293–1298

Briggs, D. E. G. (1991) Extraordinary fossils. *American Scientist*, **79**, 130–141

Briggs, D. E. G. (1992) Conodonts: a major extinct group added to the vertebrates. *Science*, **256**, 1285–1286

Briggs, D. E. G. and Fortey, F. (1989) The early radiation and relationships of the major arthropod groups. *Science*, **246**, 241–243

Briggs, D. E. G., Clarkson, E. N. K. and Aldridge, R. L. (1983) The conodont animal. *Lethaia*, **16**, 1–14

Budd, G. (1993) A Cambrian gilled lobopod from Greenland. *Nature*, **364**, 709–711

Bürglin, T. R., Finney, M., Coulson, A. and Ruvkun, G. (1989) *Caenorhabditis elegans* has scores of homoeobox–containing genes. *Nature*, **341**, 239–243

Bürglin, T. R., Ruvkun, G., Coulson, A., Hawkins, N. C., McGhee, J. D., Schaller, D., Wittmann, C., Müller, F. and Waterston, R. H. (1991) Nematode homeobox cluster. *Nature*, **351**, 703

Burri, M., Tromvoukis, Y., Bopp, D., Frigerio, G. and Noll, M. (1989) Conservation of the paired domain in metazoans and its structure in three isolated human genes. *EMBO Journal*, **8**, 1183–1190

Carlson, S. J. and Fisher, D. C. (1981) Microstructural and morphologic analysis of a carpoid aulacophore. *Abstracts with programs of the Geological Society of America*, **13**, 422

Carroll, S. B. (1995) Homeotic genes and the evolution of arthropods and chordates. *Nature*, **376**, 479– 485.

Carroll, S. B., Weatherbee, S. D. and Langeland, J. A. (1995) Homeotic genes and the regulation and evolution of insect wing number. *Nature*, **375**, 58–68.

Carter, G. S. (1957) Chordate phylogeny. *Systematic Zoology*, **6**, 187–192

Caster, K. E. (1971) Review of G. Ubaghs: Les Echinodermes carpoides de l'Ordovicien inférieur de la Montaigne Noire (France). *Journal of Paleontology*, **45**, 919–921

Chen, J.-Y., Ramsköld, L. and Zhou, G.-Q. (1994) Evidence for monophyly and arthropod affinity of Cambrian giant predators. *Science*, **264**, 1304

Chen, J.-Y., Dzik, J., Edgecombe, G. D., Ramsköld, L. and Zhou, G.-Q. (1995) A possible Early Cambrian chordate. *Nature*, **377**, 720–722.

Chisaka, O. and Capecchi, M. R. (1991) Regionally restricted developmental defects resulting from targeted disruption of the mouse homeobox gene *hox-1.5*. *Nature*, **350**, 473–479

Cho, K. W. Y., Blumberg, B., Steinbeisser, H. and De Robertis, E. M. (1991) Molecular nature of Spemann's Organizer: the role of the *Xenopus* homeobox gene *goosecoid*. *Cell*, **67**, 1111–1120

Clarkson, E. N. K. (1986) *Invertebrate Palaeontology and Evolution*, 2nd edn, Allen and Unwin, London

Cohen, S. M. (1990) Specification of limb development in the *Drosophila* embryo by positional cues from segmentation genes. *Nature*, **343**, 173–177

Cohen, S. M. and Jürgens, G. (1990) Mediation of *Drosophila* head development by gap-like segmentation genes. *Nature*, **346**, 482–485

Cohen, S. M. and Jürgens, G. (1991) Drosophila headlines. *Trends in Genetics*, **7**, 267–272

Condie, B. G. and Capecchi, M. R. (1994) Mice with targeted disruptions in the paralogous genes *hoxa-3* and *hoxd-3* reveal synergistic interactions. *Nature*, **370**, 304–307

Conklin, E. (1932) The embryology of Amphioxus. *Journal of Morphology*, **54**, 69–151

Conway Morris, S. (1976) A new Cambrian lophophorate from the Burgess Shale of British Columbia. *Palaeontology*, **19**, 199–222

Conway Morris, S. (1989) Conodont palaeobiology: recent progress and unsolved problems. *Terra Nova*, **1**, 135–150

Conway Morris, S. (1990) *Typhloesus wellsi* (Melton and Scott, 1973), a bizarre metazoan from the Carboniferous of Montana, USA. *Philosophical Transactions of the Royal Society of London* B, **327**, 595–624

Conway Morris, S. (1993) The fossil record and the early evolution of the Metazoa. *Nature*, **361**, 219–225

Conway Morris, S. (1994) Why molecular biology needs palaeontology. *Development*, suppl., 1–13

Conway Morris, S. (1995) Nailing the lophophorates. *Nature*, **375**, 365–366

Conway Morris, S. and Peel, J. S. (1995) Articulated halkieriids from the Lower Cambrian of North Greenland and their role in early protostome evolution. *Philosophical Transactions of the Royal Society of London*, B, **347**, 305–358

Corbin, V., Michelson, A. M., Abmayr, S. M., Neel, V., Alcamo, E., Maniatis, T. and Young, M. W. (1991) A role for the *Drosophila* neurogenic genes in mesoderm differentiation. *Cell*, **67**, 311–323

Craske, A. J. and Jefferies, R. P. S. (1989) A new mitrate from the Upper Ordovician of Norway, and a new approach to subdividing a plesion. *Palaeontology*, **32**, 69–99

Cripps, A. P. (1988) A new species of stem–group chordate from the Upper Ordovician of Northern Ireland. *Palaeontology*, **31**, 1053–1077

Cripps, A. P. (1989a) A new genus of stem chordate (Cornuta) from the Lower and Middle Ordovician of Czechoslovakia and the origin of bilateral symmetry in the chordates. *Géobios*, **22**, 215–245

Cripps, A. P. (1989b) A new stem–group chordate (Cornuta) from the Llandeilo of Czechoslovakia and the cornute–mitrate transition. *Zoological Journal of the Linnean Society*, **96**, 49–85

Cripps, A. P. (1990) A new stem craniate from Morocco and the search for the sister–group of the craniata. *Zoological Journal of the Linnean Society*, **100**, 27–71

Cripps, A. P. (1991) A cladistic analysis of the cornutes (stem chordates). *Zoological Journal of the Linnean Society*, **102**, 333–366

Crowson, R. A. (1982) Computers versus imagination in the reconstruction of phylogeny, in *Problems of Phylogenetic Reconstruction*, (eds K. A. Joysey and A. E. Friday), Systematics Association Special Volume no. 21, Academic Press, London, pp.245–255

Crowther, R. J. and Whittaker, J. R. (1992) Structure of the caudal neural tube in an ascidian larva: vestiges of its possible evolutionary origin from a ciliated band. *Journal of Neurobiology*, **23**, 280–292

Crowther, R. J. and Whittaker, J. R. (1994) Serial repetition of cilia pairs along the tail surface of an ascidian larva. *Journal of Experimental Zoology*, **268**, 9–16

Cunningham, J. T. (1886) Dr. Dohrn's inquiries into the evolution of organs in the Chordata, *Quarterly Journal of Microscopical Science*, **27**, 265–284

Daley, P. E. J. (1992a) Two new cornutes from the Lower Ordovician of Shropshire and southern France. *Palaeontology*, **35**, 127–148

Daley, P. E. J. (1992b) The anatomy of the solute *Girvanicystis batheri* (?Chordata) from the Upper Ordovician of Scotland and a new species of *Girvanicystis* from the Upper Ordovician of South Wales. *Zoological Journal of the Linnean Society*, **105**, 353–375

Davidson, E. H., Peterson, K. J. and Cameron, R. A. (1995) Origin of Bilaterian Body Plans: Evolution of Developmental Regulatory Mechanisms, *Science*, **270**, 1319–1325

De Beer, G. R. (1922) The segmentation of the head in *Squalus acanthias*. *Quarterly Journal of Microscopical Science*, **66**, 457–474

De Beer, G. R. (1937) *The Development of the Vertebrate Skull*, Clarendon, Oxford

Delsman, H. C. (1922) *The Ancestry of Vertebrates*, Amersfoort, Valkoff

Denison, R. H. (1956) A review of the habitat of the earliest vertebrates. *Fieldiana, Geology, Chicago*, **11**, 359–457

Denison, R. H. (1971) The origin of vertebrates: a critical evaluation of current theories. *Proceedings of the North American Paleontological Convention, September 1969*, **H**, 1132–1146

De Robertis, E. M. (1995) Dismantling the organizer. *Nature*, **374**, 407–408

De Robertis, E. M., Oliver, G. and Wright, C. V. E. (1990) Homeobox genes and the vertebrate body plan. *Scientific American*, **263**, 46–52

De Robertis, E. M., Fainsod, A., Gont, L. K. and Steinbeisser, H. (1994) The evolution of vertebrate gastrulation. *Development*, suppl., 117–124

De Robertis, E. M. and Sasai, Y. (1996) A common plan for dorsoventral patterning in Bilateria, *Nature*, **380**, 37–40

Derstler, K. (1979) Biogeography of stylophoran carpoids (Echinodermata), in *Historical Biogeography, Plate Tectonics and the Changing Environment*, (eds J. Gray and A. J. Boucot), Oregon State University Press, pp.91–104

Di Gregorio, A., Spagnuolo, A., Ristoratore, F., Pischetola, M., Aniello, F., Branno, M., Cariello, L. and Di Lauro, R. (1995) Cloning of ascidian homeobox genes provides evidence for a primordial chordate cluster. *Gene*, **156**, 253–257

Dillon, L. C. (1965) The hydrocoel and the ancestry of the chordates. *Evolution*, **19**, 436–446

Dilly, P. N. (1993) *Cephalodiscus graptolitoides* sp. nov. a probable extant graptolite, *Journal of Zoology, London*, **229**, 69–78

Doe, C. Q. and Scott, M. P. (1988) Segmentation and homeotic gene formation in the developing nervous system of *Drosophila*. *Trends in Neurosciences*, **11**, 101–106

Duboule, D. (1994) Temporal colinearity and the phylotypic progression: a basis for the stability of a vertebrate Bauplan and the evolution of morphologies through heterochrony. *Development*, suppl., 135–142

Durham, J. W. (1974) Systematic position of *Eldonia ludwigi* Walcott. *Journal of Palaeontology*, **48**, 750–755

Durham, J. W. and Caster, K. E. (1963) Helicoplacoidea: a new class of echinoderms. *Science*, 820–822

Dzik, J. (1986) Chordate affinities of the conodonts, *Problematic Fossil Taxa*, (eds A. Hoffman and M. H. Nitecki), Oxford University Press, New York/Clarendon Press, Oxford, pp.240–254

Dzik, J. (1995) *Yunnanozoon* and the ancestry of chordates. *Acta Palaeontologica Polonica*, **40**, 341–360

Eastham, L. (1949) Prof. Walter Garstang. *Nature*, **163**, 518–519

Eaton, T. H. (1970a) [Review of Jefferies, 1968b]. *Journal of Paleontology*, **44** (1)

Eaton, T. H. (1970b) The stem-tail problem and the ancestry of chordates. *Journal of Palaeontology*, **44** (5), 969–979

Eisen, J. S., Myers, P. Z. and Westerfield, M. (1986) Pathway selection by growth cones of identified motoneurones in live zebra fish embryos. *Nature*, **320**, 269–271

Ennig, C. C. (1982) The biology of Phoronida. *Advances in Marine Biology*, **19**, 1–89

Enquist, M. and Arak, A. (1994) Symmetry, beauty and evolution. *Nature*, **372**, 169–172

Epstein, D. J., Vekemans, M. and Gros, P. (1991) *splotch* (Sp^{2H}), a mutation affecting development of the mouse neural tube, shows a deletion within the paired homeodomain of *Pax–3*. *Cell*, **67**, 767–774

Erwin, D. H. (1994) The Permo-Triassic Extinction. *Nature*, **367**, 231–236

Ewing, T. (1993) Genetic 'master switch' for left–right symmetry found. *Science*, **260**, 624–625

Fainsod, A., Steinbeisser, H. and De Robertis, E. (1994) On the function of *BMP–4* in patterning the marginal zone of the *Xenopus* embryo. *EMBO Journal*, **13**, 5015–5025

Ferguson, E. L. and Anderson, K. V. (1992) Localized enhancement and repression of the activity of the TGF–β family member, *decapentaplegic*, is necessary for dorsal–ventral pattern formation in the *Drosophila* embryo. *Development*, **114**, 583–597

Field, K. G., Olsen, G. J., Lane, D. J., Giovannoni, S. J., Ghiselin, M. T., Raff, E. C., Pace, N. R. and Raff, R. A. (1988) Molecular phylogeny of the animal kingdom. *Science*, **239**, 748–753

Field, K. G., Olsen, G. J., Giovannoni, S. J., Raff, E. C., Pace, N. R. and Raff, R. A. (1989) Phylogeny and molecular data. *Science*, **243**, 550–551

Fietz, M. J., Concordet, J.-P., Barbosa, R., Johnson, R., Krauss, S., McMahon, A. P., Tabin, C. and Ingham, P. W. (1994), The *hedgehog* gene family in *Drosophila* and vertebrate development. *Development*, suppl., 43–51

Finkelstein, R. and Perrimon, N. (1990) The *orthodenticle* gene is regulated by *bicoid* and *torso* and specifies *Drosophila* head development. *Nature*, **346**, 485–488

Finkelstein, R. and Perrimon, N. (1991) The molecular genetics of head development in *Drosophila*. *Development*, **112**, 899–912

Fisher, D. C. (1982) Stylophoran skeletal crystallography: testing the calcichordate theory of vertebrate origins. *Abstracts with Programs of the Geological Society of America*, **14**, 488

Fisher, D. C. (1983) Life orientation of mitrate stylophorans and its implications for the calcichordate theory of vertebrate origins. *Abstracts with Programs of the Geological Society of America*, **25**, A-105

Flood, P. R. (1966) A peculiar mode of muscular innervation in amphioxus. *Journal of Comparative Neurology*, **126**, 181–218

Forey, P. L. (1982) Neontological analysis versus palaeontological stories, in *Problems of Phylogenetic Reconstruction*, (eds K. A. Joysey and A. E. Friday), Systematics Association Special Volume no. 21. Academic Press, London, pp.119–157

Forey, P. and Janvier, P. (1993) Agnathans and the origin of jawed vertebrates. *Nature*, **361**, 129–134

Fortey, R. A. and Jefferies, R. P. S. (1982) Fossils and phylogeny – a compromise approach, in *Problems of Phylogenetic Reconstruction*, (eds K. A. Joysey and A. E. Friday), Academic Press, London and New York, pp.197–234

François, V., Solloway, M., O'Neill, J., Emery, J. and Bier, E. (1994) Dorsal–ventral patterning of the *Drosophila* embryo depends on a putative negative growth factor encoded by the *short gastrulation* gene. *Genes and Development*, **8**, 2602–2616

Fraser, S., Keynes, R. and Lumsden, A. (1990) Segmentation in the chick embryo hindbrain is defined by cell lineage restrictions. *Nature*, **344**, 431–435

Friedrich, W. P. (1993) Systematik and Functionsmorphologie mittel kambrischer Cincta (Carpoidea, Echinodermata). *Beringeria*, **7**, 1–190

Fritzsch, B. and Northcutt, R. G. (1993) Cranial and spinal nerve organization in amphioxus and lampreys: evidence for an ancestral craniate pattern. *Acta Anatomica*, **148**, 96–109

Gabbott, S. E., Aldridge, R. J. and Theron, J. N. (1995) A giant conodont with preserved muscle tissue from the Upper Ordovician of South Africa. *Nature*, **374**, 800–803

Gage, J. D. and Tyler, P. A. (1991) *Deep–Sea Biology: a Natural History of Organisms at the Deep-sea Floor*, Cambridge University Press, Cambridge

Galloway, J. (1990) A handle on handedness. *Nature*, **346**, 223–224

Gans, C. (1987) The neural crest, a spectacular invention, in *Developmental and Evolutionary Aspects of the Neural Crest*, (ed. P. F. A. Maderson), John Wiley, New York

Gans, C. (1988) Review of R. P. S. Jefferies' *The Ancestry of the Vertebrates. American Scientist*, **76**, 188–189

Gans, C. (1989) Stages in the origins of vertebrates: analysis by means of scenarios. *Biological Reviews*, **64**, 221–268

Gans, C. and Northcutt, R. G. (1983) Neural crest and the origin of vertebrates: a new head. *Science*, **220**, 268–274

Gans, C. and Northcutt, R. G. (1985) Neural crest: the implications for comparative anatomy. *Fortschritte Zoologie*, **30**, 507–514

Garcia-Fernàndez, J. and Holland, P. W. H. (1994) Archetypal organization of the amphioxus *Hox* gene cluster. *Nature*, **370**, 563–566.

Garcia-Fernàndez, J., Baguñà, J. and Saló , E. (1991) Planarian homeobox genes: cloning, sequence analysis and expression. *Proceedings of the National Academy of Sciences of the United States of America*, **88**, 7338–7342

Garcia-Fernàndez, J., Baguñà, J. and Saló , E. (1993) Genomic organization and expression of the planarian homeobox genes *Dth-1* and *Dth-2*. *Development*, **118**, 241–253

Gardiner, B. G. (1982) Tetrapod classification. *Zoological Journal of the Linnean Society*, **74**, 207–232

Garstang, W. (1894) Preliminary note on a new theory of the phylogeny of the Chordata. *Zoologische Anzeiger*, **17**, 122–125

Garstang, W. (1897) Physiology and the ancestry of the vertebrates. *Transactions of the Junior Science Club, Oxford*, (NS) **1**, 1–6

Garstang, W. (1922) The theory of recapitulation: a critical restatement of the Biogenetic Law. *Journal of the Linnean Society (Zoology)*, **35**, 81–101

Garstang, W. (1928) The morphology of the Tunicata, and its bearings on the phylogeny of the Chordata. *Quarterly Journal of Microscopical Science*, **72**, 51–187

Garstang, W. (1951) *Larval Forms and Other Zoological Verses*, Blackwell, Oxford

Garstang, S. and Garstang, W. (1926) On the development of *Botrylloides* and the ancestry of vertebrates (preliminary note) *Proceedings of the Leeds Philosophical Society*, **1926**, 81–86

Gaskell, W. H. (1890) On the origin of vertebrates from a crustacean-like ancestor, *Quarterly Journal of the Microscopical Society*, **31**, 379–444

Gaskell, W. H. (1896) Address to the Physiological Section, British Association, Liverpool, *Report of the Liverpool Meeting, British Association for the Advancement of Science*

Gaskell, W. H. (1908) *The Origin of Vertebrates*. Longmans, Green, London and New York

Gaskell, W. H., MacBride, E. W., Starling, E. H., Goodrich, E. S., Gadow, H., Smith Woodward, A., Dendy, A., Lankester, E. R., Mitchell, P. C., Gardiner, J. S., Stebbing, T. R. R. and Scott, D. H. (1910) Discussion on the origin of vertebrates. *Proceedings of the Linnean Society of London*, Session **122** (1909–1910), 9–50

Gaunt, S. J. (1991) Expression patterns of mouse *Hox* genes: clues to an understanding of developmental and evolutionary strategies. *BioEssays*, **13**, 505–513

Gaunt, S. J., Sharpe, P. T. and Duboule, D. (1988) Spatially restricted domains of homeo-gene transcripts in mouse embryos relation to a segmented body plan. *Development*, **104** (suppl), 169–179

Gauthier, J., Kluge, A. G. and Rowe, T. (1988) Amniote phylogeny and the importance of fossils. *Cladistics*, **4**, 104–209

Gee, H. (1988) Friends and relations. *Nature*, **334**, 13–14

Gee, H. (1989) A backbone for the vertebrates. *Nature*, **340**, 596–597

Gee, H. (1992) By their teeth ye shall know them. *Nature*, **360**, 529

Gee, H. (1994) Return of the amphioxus. *Nature*, **370**, 504–505

Gee, H. (1995a) Lophophorates prove likewise variable. *Nature*, **374**, 493

Gee, H. (1995b) The molecular explosion. *Nature*, **373**, 558–559

Gehring, W. J. (1985) The homeobox: a key to understanding development? *Cell*, **40**, 3–5

Gehring, W. J. (1987) Homeo boxes in the study of development. *Science*, **236**, 1245–1252

Gilchrist, J. D. F. (1917) On the development of the Cape *Cephalodiscus* (*C. gilchrist* RIDEWOOD). *Quarterly Journal of Microscopical Science*, NS **62**, 189–211

Gilland, E. H. (1989) Morphogenesis of hindbrain neuromeres in embryos of *Squalus acanthias* and *Alligator mississippiensis*: SEM and LM observations. *American Zoologist*, **29**(4), A143

Gilland, E. and Baker, R. (1993) Conservation of neuroepithelial and mesodermal segments in the embryonic vertebrate head. *Acta Anatomica*, **148**, 110–123

Gilmour, T. H. J. (1978) Ciliation and the function of the food-collecting and waste-rejecting organs of lophophorates. *Canadian Journal of Zoology*, **56**, 2142–2155

Gilmour, T. H. J. (1979) Feeding in pterobranch hemichordates and the evolution of gill slits. *Canadian Journal of Zoology*, **57**, 1136–1142

Gislén, T. (1930) Affinities between the Echinodermata, Enteropneusta and Chordonia. *Zoologiska Bidrag frän Uppsala*, **12**, 199–304

Gont, L. K., Steinbeisser, H., Blumberg, B. and De Robertis, E. M. (1993) Tail formation as a continuation of gastrulation: the multiple cell populations of the *Xenopus* tailbud derive from the late blastopore lip. *Development*, **119**, 991–1104

González-Reyes, A., Urquia, N., Gehring, W. J., Struhl, G. and Morata, G. (1990) Are cross–regulatory interactions between homoeotic genes functionally significant? *Nature*, **344**, 78–80

Goodbody, I. (1974) The physiology of ascidians. *Advances in Marine Biology*, **12**, 1–149

Goodrich, E. S. (1917) 'Proboscis pores' in craniate vertebrates, a suggestion concerning the premandibular somites and hypophysis. *Quarterly Journal of Microscopical Science*, NS, **62**, 539–553

Goodrich, E. S. (1918) On the development of the segments of the head in *Scyllium*. *Quarterly Journal of Microscopical Science*, NS, **63**, 1–30

Goodrich, E. S. (1930) *Studies on the Structure and Development of Vertebrates*, Macmillan, London

Gould, A. P., Brookman, J. J., Strutt, D. I. and White, R. A. H. (1990) Targets of homeotic gene control in *Drosophila*. *Nature*, **348**, 308–312

Gould, S. J. (1977) *Ontogeny and Phylogeny*, Harvard University Press, Cambridge, MA

Gould, S. J. (1980) *The Panda's Thumb: More Reflections in Natural History*, Norton, New York

Gould, S. J. (1989) *Wonderful Life – The Burgess Shale and the Nature of History*, Norton, New York

Gould, S. J. (1991) The disparity of the Burgess Shale arthropod fauna and the limits of cladistic analysis: why we must strive to quantify morphospace. *Paleobiology*, **17**, 411–423

Gould, S. J. and Eldredge, N. (1993) Punctuated equilibrium comes of age. *Nature*, **366**, 223–227

Graff, J. M., Thies, R. S., Song, J. J., Celeste, A. J. and Melton, D. (1994). Studies with a Xenopus BMP receptor suggest that ventral mesoderm-inducing signals override dorsal signals in vivo. *Cell*, **79**, 169–179

Graffin, G. (1992) A new locality of fossiliferous Harding Sandstone: evidence for freshwater Ordovician vertebrates. *Journal of Vertebrate Paleontology*, **12**, 1–10

Graham, A., Heyman, I. and Lumsden, A. (1993) Even-numbered rhombomeres control the apoptotic elimination of neural crest cells from odd-numbered rhombomeres in the chick hindbrain. *Development*, **119**, 233–245

Greenwood, P. H. (1974) Cichlid fishes of Lake Victoria, East Africa. *Bulletin of the British Museum (Natural History), Zoology*, Supplement **6**

Gregory, W. K. (1936) The transformation of organic designs: a review of the origin and deployment of the earlier vertebrates. *Biological Reviews*, **11**, 311–344

Gregory, W. K. (1946) The roles of motile larvae and fixed adults in the origin of vertebrates. *Quarterly Review of Biology*, **21**, 348–364

Gurdon, J. B. (1992) The generation of diversity and pattern in animal development. *Cell*, **68**, 185–199

Gutmann, W. F. (1981) Relationships between invertebrate phyla on functional–mechanical analysis of the hydrostatic skeleton. *American Zoologist*, **21**, 63–81

Halanych, K. M. (1995) The phylogenetic position of the pterobranch hemichordates based on 18S rDNA sequence data. *Molecular Phylogenetics and Evolution*, **4**, 72–76

Halanych, K. M., Bacheller, J. D., Aguinaldo, A. M. A., Liva, S. M., Hillis, D. M. and Lake, J. A. (1995) Evidence from 18S ribosomal DNA that the lophophorates are protostome animals. *Science*, **267**, 1641–1643

Hall, B. K. (1987) Tissue interactions in the development and evolution of the vertebrate head, in *Developmental and Evolutionary Aspects of the Neural Crest*, (ed. P. F. A. Maderson), John Wiley, New York, pp.215–219

Hall, B. K. (1988) *The Neural Crest*, Oxford University Press, Oxford

Hall, B. K. (1992) *Evolutionary Developmental Biology*. Chapman & Hall, London

Halstead, L. B. (1969) Calcified tissues in the earliest vertebrates. *Calcified Tissue Research*, **3**, 107–124

Halstead, L. B. (1985) The vertebrate invasion of fresh water. *Philosophical Transactions of the Royal Society of London*, B **309**, 243–258

Halstead, L. B. (1987) Evolutionary aspects of neural crest-derived skeletogenic cells in the earliest vertebrates, in *Developmental and Evolutionary Aspects of the Neural Crest*, (ed. P. F. A. Maderson, P. F. A.), John Wiley, New York, pp.339–358

Hanken, J. and Hall, B. K. (eds) (1993) *The Skull* (in 3 vols), University of Chicago Press, Chicago

Hanken, J. and Thorogood, P. (1993) Evolution and development of the vertebrate skull: the role of pattern formation. *Trends in Ecology and Evolution*, **8**, 9–15

Harada, Y., Yasuo, H. and Satoh, N. (1995) A sea urchin homologue of the chordate *Brachyury* (*T*) gene is expressed in the secondary mesenchyme founder cells, *Development*, **121**, 2747–2754

Harvey, L. A. (1961) New speculations on the origin of the chordates. *Science Progress*, **49**, 507–514

Hertwig, R. (1909) *A Manual of Zoology*, 2nd American edn from the 5th German edn, trans. J. S. Kingsley. Henry Holt, New York

Higgins, A. (1983) The conodont animal. *Nature*, **302**, 107

Hirakow, R. and Kajita, N. (1994) Electron microscopic study of the development of amphioxus, *Branchiostoma belcheri tsingtauense*: the neurula and larva. *Acta Anatomica Nipponica*, **69**, 1–13

Holland, L. Z., Gorsky, G. and Fenaux, R. (1988) Fertilization in *Oikopleura dioica* (Tunicata, Appendicularia): acrosome reaction, cortical reaction and sperm–egg fusion. *Zoomorphology*, **108**, 229–243

Holland, P. W. H. (1988) Homeobox genes and the vertebrate head. *Development*, **103** (suppl), 25–30

Holland, P. W. H. (1990) Homeobox genes and segmentation: co-option, co-evolution, and convergence. *Seminars in Developmental Biology*, **1**, 135–145

Holland, P. W. H. (1992) Homeobox genes in vertebrate evolution. *BioEssays*, **14**, 267–273

Holland, P. W. H. (1993) Evolution comes to a head. *Trends in Ecology and Evolution*, **9**, 75–76

Holland, P. W. H. and Hogan, B. L. M. (1988) Expression of homeobox genes during mouse development: a review. *Genes and Development*, **2**, 773–782

Holland, P. W. H., Hacker, A. M. and Williams, N. A. (1991) A molecular analysis of the phylogenetic affinities of *Saccoglossus cambrensis* Brambell and Cole (Hemichordata). *Philosophical Transactions of the Royal Society of London* B, **332**, 185–189

Holland, P. W. H., Holland, L.Z., Williams N.A. and Holland, N.D. (1992a) An amphioxus homeobox gene: sequence conservation, spatial expression during development and insights into vertebrate evolution. *Development*, **116**, 653–661

Holland, P. W. H., Ingham, P. and Krauss, S. (1992b) Mice and flies head to head. *Nature*, **358**, 627–628

Holland, P. W. H., Garcia-Fernàndez, J., Holland, L. Z., Williams, N. A. and Holland, N. D. (1994a) The molecular control of spatial patterning in amphioxus. *Journal of the Marine Biological Association of the United Kingdom*, **74**, 49–60

Holland, P. W. H., Garcia–Fernàndez, J., Williams, N. A. and Sidow, A. (1994b) Gene duplications and the origins of vertebrate development. *Development*, suppl., 125–133

Holland, P. W. H., Koschorz, B., Holland, L. Z. and Herrmann, B. G. (1995) Conservation of *Brachyury* (*T*) genes in amphioxus and vertebrates: developmental and evolutionary implications, *Development*, **121**, 4283–4291

Holley, S.A., Jackson, P. D., Sasai, Y., Lu, B., De Robertis, E. M., Hoffmann, F. M. and Ferguson, E. L., (1995) A conserved system for dorsal–ventral patterning in insects and vertebrates involving *sog* and *chordin*. *Nature*, **376**, 249–253

Hou, X. and Bergström, J. (1995) Cambrian lobopodians – ancestors of extant lobopodians? *Zoological Journal of the Linnean Society*, **114**, 3–19

Hubrecht, A. A. W. (1883) On the ancestral form of the chordate. *Quarterly Journal of Microscopical Science*, **23**, 349–367

Hunt, P. and Krumlauf, R. (1991a) *Hox* genes coming to a head. *Current Biology*, **1**, 304–306

Hunt, P. and Krumlauf, R. (1991b) Deciphering the *Hox* code: clues to patterning branchial regions of the head. *Cell*, **66**, 1075–1078

Hunt, P., Gulisano, M., Cook, M., Sham, M.-H., Faiella, A., Wilkinson, D., Boncinelli, E. and Krumlauf, R. (1991a) A distinct *Hox* code for the branchial region of the vertebrate head. *Nature*, **353**, 861–864

Hunt, P., Wilkinson, and Krumlauf, R. (1991b) Patterning the vertebrate head: murine *Hox2* genes mark distinct subpopulations of premigratory and migrating cranial neural crest. *Development*, **112**, 43–50

Hyman, L. H. (1959) *The Invertebrates V: Smaller Coelmate Groups; Chaetognatha, Hemichordata, Pogonophora, Phoromida, Ectoprocta, Brachipoda, Sipunculida, The Coelomate Bilateria*, McGraw-Hill, New York

Ingham, P. W. (1988) The molecular genetics of embryonic pattern formation in *Drosophila*. *Nature*, **335**, 25–34

Ingham, P. W. (1990) The X, Y, Z of head development. *Nature*, **346**, 412–413

Ingham, P. W. and Martinez-Arias, A. (1992) Boundaries and fields in early embryos. *Cell*, **68**, 221–235

Ivanova-Kazas O. M. (1990) Current state of the problem of the origin of chordates. *Soviet Journal of Marine Biology*, **15**, 223–224 [translated from the original Russian]

Jacobs, D. K. and De Salle, R. (1994) Engrailed: homology of metameric units, molluscan phylogeny and relationship to other homeodomains. *Developmental Biology*, **163**(2), 536

Jacobson, A. G. (1987) Determination and morphogenesis of axial structures, in *Development and Evolutionary Aspects of the Neural Crest*, (ed. P. F. A. Maderson), John Wiley, New York

Jacobson, A. G. (1988) Somitomeres: mesodermal segments of vertebrate embryos. *Development*, **104** (suppl.), 209–220

Jaekel, O. (1918) Phylogenie und System der Pelmatozoen. *Palaontogische Zeitschrift*, **3**, 1–128

Janvier, P. (1981) The phylogeny of the Craniata, with particular reference to the significance of the fossil 'agnathans'. *Journal of Vertebrate Paleontology*, **1**, 121–159

Janvier, P. (1995) Conodonts join the club. *Nature*, **374**, 761–762

Janvier, P. and Lund, R. (1983) *Hardistiella montanensis*, n. gen. et sp. (Petromyzontida) from the Lower Carboniferous of Montana, with remarks on the affinities of the lampreys. *Journal of Vertebrate Paleontology*, **2**, 407–413

Jefferies, R. P. S. (1967) Some fossil chordates with echinoderm affinities. *Symposia of the Zoological Society of London*, **20**, 163–208

Jefferies, R. P. S. (1968a) Fossil chordates with echinoderm affinities. *Proceedings of the Geological Society of London*, **1968(1649)**, 128–140

Jefferies, R. P. S. (1968b) The subphylum Calcichordata (Jefferies, 1967) – primitive fossil chordates with echinoderm affinities. *Bulletin of the British Museum, Natural History (Geology)*, **16**, 243–339

Jefferies, R. P. S. (1969) *Ceratocystis perneri* Jaekel – a middle Cambrian chordate with echinoderm affinities. *Palaeontology*, **12**, 494–535

Jefferies, R. P. S. (1971) Some comments on the origin of chordates. *Journal of Palaeontology*, **45**, 910–912

Jefferies, R. P. S. (1973) The Ordovician fossil *Lagynocystis pyramidalis* (Barrande) and the ancestry of amphioxus. *Philosophical Transactions of the Royal Society of London* B, **265**, 409–469

Jefferies, R. P. S. (1975) Fossil evidence concerning the origin of the chordates. *Symposia of the Zoological Society of London*, **36**, 253–318

Jefferies, R. P. S. (1979) The origin of chordates – a methodological essay, in *The Origin of Major Invertebrate Groups* (ed. M. R. House), *Systematics Association Special Volume* **12**, Academic Press, London and New York, pp.443–447

Jefferies, R. P. S. (1980) Zur Fossilgeschichte des Ursprungs der Chordaten und der Echinodermen, *Zoologische Jahrbuch*, **103**, 285–353

Jefferies, R. P. S. (1981a) In defence of calcichordates. *Zoological Journal of the Linnean Society*, **73**, 351–396

Jefferies, R. P. S. (1981b) Fossil evidence on the origin of chordates and echinoderms. *Atti Conv. Accad. naz. Lincei*, **49**, 487–561

Jefferies, R. P. S. (1982) The calcichordate controversy – comments on *Notocarpos garratti* Philip. *Alcheringa*, **6**, 78

Jefferies, R. P. S. (1984) Locomotion, shape, ornament, and external ontogeny in some mitrate calcichordates. *Journal of Vertebrate Paleontology*, **4**, 292–319

Jefferies, R. P. S. (1986) *The Ancestry of the Vertebrates*, British Museum (Natural History), London

Jefferies, R. P. S. (1988a) How to characterize the Echinodermata – some implications of the sister-group relationship between echinoderms and chordates, *Echinoderm Phylogeny and Evolutionary Biology*, (eds C. R. C. Paul and A. B. Smith), Liverpool Geological Society and Clarendon Press, Oxford

Jefferies, R. P. S. (1988b) Our boot-shaped ancestors in Wales. *Geology Today* (November–December 1988), 211–213

Jefferies, R. P. S. (1990) The solute *Dendrocystoides scoticus* from the Upper Ordovician of Scotland and the ancestry of chordates and echinoderms. *Palaeontology*, **33**, 631–679

Jefferies, R. P. S. (1991) Two types of bilateral symmetry in the Metazoa: chordate and bilaterian, in *Biological Asymmetry and Handedness*, (eds G. R. Bock and J. Marsh), Wiley, Chichester (Ciba Foundation Symposium **162**)

Jefferies, R. P. S. and Brown, N. A. (1995) Dorsoventral axis inversion? *Nature*, **374**, 22

Jefferies, R. P. S. and Lewis, D. N. (1978) The English Silurian fossil *Placocystites forbesianus* and the ancestry of the vertebrates. *Philosophical Transactions of the Royal Society of London* B, **282**, 205–323

Jefferies, R. P. S. and Prokop, R. J. (1972) A new calcichordate from the Ordovician of Bohemia and its anatomy, adaptations and relationships. *Biological Journal of the Linnean Society*, **4**, 69–115

Jefferies, R. P. S., Joysey, K. A., Paul, C. R. C. and Ramsbottom, W. H. C. (1967) Echinodermata: Pelmatozoa, in *The Fossil Record*, Geological Society, London, pp.565–581

Jefferies, R. P. S., Lewis, M. and Donovan, S. K. (1987) *Protocystites menevensis* – a stem-group chordate (Cornuta) from the Middle Cambrian of South Wales. *Palaeontology*, **30**, 429–484

Jefferies, R. P. S., Brown, N. and Daley, P. E. J. (1995), *Acta Zoologica, Stockholm*

Jeffery, W. R. (1994) A model for ascidian development and developmental modifications during evolution. *Journal of the Marine Biological Association of the United Kingdom*, **74**, 35–48

Jeffery, W. R. and Swalla, B. J. (1992) Factors necessary for restoring an evolutionary change in an anural ascidian embryo. *Developmental Biology*, **153**, 194–205

Jegalian, B. G. and De Robertis, E. M. (1992) Homeotic transformations in the mouse induced by overexpression of a human *Hox3.3* transgene. *Cell*, **71**, 901–910

Jensen, D. D. (1960) Hoplonemertines, myxinoids, and deuterostome origins. *Nature*, **187**, 649–650

Jensen, D. D. (1963) Hoplonemertines, myxinoids and vertebrate origins, in *The Lower Metazoa: Comparative Biology and Phylogeny*, (eds E. C. Dougherty *et al.*), University of California Press, Berkeley, pp.113–126

Jessell, T. M. and Melton, D. A. (1992) Diffusible factors in vertebrate embryonic induction. *Cell*, **68**, 257–270

Jollie, M. (1973) The origin of chordates. *Acta Zoologica, Stockholm*, **54**, 81–100

Jollie, M. (1982) What are the 'Calcichordata'? and the larger question of the origin of chordates. *Zoological Journal of the Linnean Society*, **75**, 167–188

Katz, M. J. (1983) Comparative anatomy of the tunicate tadpole, *Ciona intestinalis*. *Biological Bulletin*, **164**, 1–27

Kemp, T. S. (1988) Haemothermia or Archosauria? The interrelationships of mammals, birds and crocodiles. *Zoological Journal of the Linnean Society*, **92**, 67–104

Kenyon, C. (1994) If birds can fly, why can't we? Homeotic genes and evolution. *Cell*, **78**, 175–180

Kenyon, C. and Wang, B. (1991) A cluster of *Antennapedia*–class homeobox genes in a nonsegmented animal. *Science*, **253**, 516–517

Kessel, M. and Gruss, P. (1990) Murine developmental control genes. *Science*, **249**, 374–379

Kessel, M. and Gruss, P. (1991) Homeotic transformations of murine vertebrae and concomitant alteration of *Hox* codes induced by retinoic acid. *Cell*, **67**, 89–104

Keynes, R. and Lumsden, A. (1990) Segmentation and the origin of regional diversity in the vertebrate central nervous system. *Neuron*, **4**, 1–9

Keynes, R. and Stern, C. (1988) Mechanisms of vertebrate segmentation. *Development*, **103** (suppl.), 413–429

Knoll, A. H. (1992) The early evolution of eukaryotes: a geological perspective. *Science*, **256**, 622–627

Knoll, A. H. and Walter, M. R. (1992) Latest Proterozoic stratigraphy and Earth history. *Nature*, **356**, 673–678

Kolata, D. R. and Guensberg, T. E. (1979) *Diamphidiocystis*, a new mitrate 'carpoid' from the Cincinnatian (Upper Ordovician) Maquoketa Group in southern Illinois. *Journal of Paleontology*, **53**, 1121–1135

Kolata, D. R., Frest, T. J. and Mapes, R. H. (1991) The youngest carpoid: occurrence, affinities and life mode of a Pennsylvanian mitrate from Oklahoma. *Journal of Paleontology*, **65**, 844–854

Komai, T. (1951) The homology of the 'notochord' found in pterobranchs and enteropneusts. *American Naturalist*, **85**, 270–271

Kowalevsky, A. (1866a) Entwickelungsgeschichte der einfachen Ascidien. *Mémoires de l'Académie des Sciences de St Pétersbourg*, **(7)10**, no. 15, 1–19

Kowalevsky, A. (1866b) Entwicklungsgeschichte des Amphioxus lanceolatus. *Mémoires de l'Académie des Sciences de St Pétersbourg*, **(7)11**, no. 4, 1–17

Kowalevsky, A. (1871) Weitere studien über die Entwicklung der einfachen Ascidien. *Archiv fur mikroskopische Anatatomie (Bonn)*, **7**, 101–130

Kozlowski, R. (1947) Les affinités des graptolithes. *Biological Reviews*, **22**, 93–108

Krauss, S., Johansen, T., Korzh, V., Moens, U., Ericson, J. H. and Fjose, A. (1991a) Zebrafish *pax[zf-a]*: a paired box-containing gene expressed in the neural tube. *EMBO Journal*, **10**, 3609–3619

Krauss, S., Johansen, T., Korzh, V. and Fjose, A. (1991b) Expression patterns of zebrafish *pax* genes suggests a role in early brain regionalization. *Nature*, **353**, 267–270

Krumlauf, R. (1992a) Transforming the *Hox* code. *Current Biology*, **2**, 641–643

Krumlauf, R. (1992b) evolution of the vertebrate *Hox* homeobox genes. *BioEssays*, **14**, 245–252

Krumlauf, R. (1994) *Hox* genes in vertebrate development. *Cell*, **78**, 191–201

Lacalli, T. (1995) Dorsoventral axis inversion. *Nature*, **373**, 110–111

Lacalli, T. C., Holland, N. D. and West, J. E. (1994) Landmarks in the anterior central nervous system of amphioxus larvae. *Philosophical Transactions of the Royal Society of London B*, **344**, 165–185

Lake, J. A. (1990) Origin of the metazoa. *Proceedings of the National Academy of Science of the United States of America*, **87**, 763–766

Lamarck, J. B. (1809) *Philosophie Zoologique, ou Exposition des Considérations Relatives à l'Histoire Naturelles des Animaux*, Dentu, Paris, 487pp. Republished by The University of Chicago Press, Chicago (1984)

Lankester, E. R. (1882) The vertebration of the tail of Appendiculariae, *Quarterly Journal of Microscopical Science*, **22**, 387–390

Lankester, E. R. and Willey, A. (1890) The development of the atrial chamber of amphioxus. *Quarterly Journal of Microscopical Science*, **31**, 445–466

Lans, D., Wedeen, C. J. and Weisblat, D. A. (1993) Cell lineage analysis of the expression of an *engrailed* homolog in leech embryos. *Development*, **117**, 857–871

Lawrence, P. A. (1990) Compartments in vertebrates? *Nature*, **344**, 382–383

Lawrence, P. A. and Morata, G. (1994) Homeobox genes: their function in *Drosophila* segmentation and pattern formation. *Cell*, **78**, 181–189

Lawrence, P. A. and Sampedro, J. (1993) *Drosophila* segmentation: after the first three hours. *Development*, **119**, 971–976

Le Douarin, N. (1983) *The Neural Crest*, Cambridge University Press, Cambridge

Le Mouellic, H., Lallemand, Y. and Brûlet, P. (1992) Homeosis in the mouse induced by a null mutation in the *Hox-3.1* gene. *Cell*, **69**, 251–264

Lewis, E. B. (1978) A gene complex controlling segmentation in *Drosophila*. *Nature*, **276**, 565–570

Lewis, J. (1989) Genes and segmentation. *Nature*, **341**, 382–383

Lim, T.M., Jaques, K. F., Stern, C. D. and Keynes, R. J. (1991) An evaluation of myelomeres and segmentation of the chick embryo spinal cord. *Development*, **113**, 227–238

Lipps, J. H. and Signor, P. W. (eds) (1992) *Origin and Early Evolution of the Metazoa*, Plenum, New York

Lonai, P. and Orr-Urtreger, A. (1990) Homeogenes in mammalian development and the evolution of the cranium and central nervous system. *FASEB Journal*, **4**, 1436–1443

Løvtrup, S. (1977) *The Phylogeny of Vertebrata*, Wiley, New York

Lufkin, T., Dierich, A., LeMeur, M., Mark, M. and Chambon. P. (1991) Disruption of the *Hox–1.6* homeobox gene results in defects corresponding to its rostral domain of expression. *Cell*, **66**, 1105–1119

Lufkin, T., Mark, M., Hart, C., Dollé, P. LeMeur, M. and Chambon, P. (1992) Homeotic transformation of the occipital bones of the skull by ectopic expression of a homeobox gene in transgenic mice. *Nature*, **359**, 835–841

Lumsden, A. (1987) The neural crest contribution to tooth development in the mammalian embryo, in *Developmental and Evolutionary Aspects of the Neural Crest*, (ed. P. F. A. Maderson), John Wiley, New York, pp.261–300

Lumsden, A. (1990) The cellular basis of segmentation in the developing hindbrain. *Trends in Neurosciences*, **13**, 329–335

Lumsden, A. and Keynes, R. (1989) Segmental patterns of neuronal development in the chick hindbrain. *Nature*, **337**, 424–428

Lundin, L. G. (1993) Evolution of the vertebrate genome as reflected in paralogous chromosomal regions in Man and the house mouse. *Genomics*, **16**, 1–19

Lyons, K. M., Jones, C. M. and Hogan, B. L. M. (1991) The *DVR* gene family in embryonic development. *Trends In Genetics*, **7**, 408–412

McGinnis, W. and Krumlauf, R. (1992) Homeobox genes and axial patterning. *Cell*, **68**, 283–302

McGinnis, W., Garber, R. L., Wirz, J., Kuroiwa, A. and Gehring, W. J. (1984a) A homologous protein-coding sequence in *Drosophila* homeotic genes and its conservation in other metazoans. *Cell*, **37**, 403–408

McGinnis, W., Levine, M. S., Hafen, E., Kuroiwa, A. and Gehring, W. J. (1984b) A conserved DNA sequence in homoeotic genes of the *Drosophila* Antennapedia and bithorax complexes. *Nature*, **308**, 428–433

McMahon, A. P. (1992) The *Wnt* family of developmental regulators. *Trends In Genetics*, **8**, 236–243

McMahon, A. P. and Bradley, A. (1990) The *Wnt-1* (*int-1*)-proto-oncogene is required for development of a large region of the mouse brain. *Cell*, **62**, 1073–1085

McMahon, A. P., Joyner, A. L., Bradley, A. and J. A. McMahon (1992) The midbrain–hindbrain phenotype of *Wnt-1*$^-$/Wnt-1$^-$ mice results from stepwise deletion of *engrailed*-expressing cells by 9.5 days postcoitum. *Cell*, **69**, 581–595

Maisey, J. G. (1986) Heads and tails: a chordate phylogeny. *Cladistics*, **2**, 201–256

Malicki, J., Schughart, K. and McGinnis, W. (1990) Mouse *Hox 2.2* specifies thoracic segmental identity in *Drosophila* embryos and larvae. *Cell*, **63**, 961–967

Malicki, J., Cianetti, L. C., Peschle, C. and McGinnis, W. (1992) A human *HOX4B* regulatory element provides head–specific expression in *Drosophila* embryos. *Nature*, **358**, 345–347

Manak, J. R. and Scott, M. P. (1994) A class act: conservation of homeodomain protein functions. *Development*, suppl., 61–71

Marshall, C. R. (1990) The fossil record and estimating divergence times between lineages: maximum divergence times and the importance of reliable phylogenies. *Journal of Molecular Evolution*, **30**, 400–408

Martindale, M. Q., Meier, S. and Jacobson, A. G. (1987) Mesodermal metamerism in the teleost *Oryzias latipes* (the medaka). *Journal of Morphology*, **193**, 241–252

Marx, J. (1992) Homeoboxes go evolutionary. *Science*, **255**, 399–401

Matsumoto, H. (1929) Outline of a classification of the echinodermata. *Science Reports of Tohoku University, Sendai (Geology)*, **13**, 27–33

Meier, S. and Packard, D. S. (1984) Morphogenesis of the cranial segments and distribution of neural crest in the embryos of the snapping turtle, *Chelydra serpentina*. *Developmental Biology*, **102**, 309–323

Melton, D. A. (1991) Pattern formation during animal development. *Science*, **252**, 234–241

Miles, A. and Miller, D. J. (1992) Genomes of diploblastic organisms contain homeoboxes: sequence of *eveC*, an *even-skipped* homologue from the cnidarian *Acropora formosa*. *Proceedings of the Royal Society of London B*, **248**, 159–161

Monge-Najera, J. (1995) Phylogeny, biogeography and reproductive trends in the Onychophora. *Zoological Journal of the Linnean Society*, **114**, 21–60

Morgan, M. J. (1992) On the evolutionary origin of right handedness. *Current Biology*, **2**, 15–17

Morris, R. (1965) Studies on salt and water balance in *Myxine glutinosa*. *Journal of Experimental Biology*, **42**, 359–371

Murphy, P., Davidson, D. R. and Hill, R. E. (1989) Segment-specific expression of a homoeobox-containing gene in the mouse hindbrain. *Nature*, **341**, 156–159

Nichols, D. (1967a) The origin of echinoderms. *Symposia of the Zoological Society of London*, **20**, 209–229

Nichols, D. (1967b) *Echinoderms*, 3rd edn, Hutchinson University Library

Nichols, D. (1972) The water–vascular system in living and fossil echinoderms. *Palaeontology*, **15**, 519–538

Nielsen, C. (1987) Structure and function of metazoan ciliary bands and their phylogenetic significance. *Acta Zoologica (Stockholm)*, **68**, 205–262

Noden, D. M. and Van De Water, T. R. (1992) Genetic analysis of mammalian ear development. *Trends In Neurosciences*, **15**, 235–237

Northcutt, R. G. and Gans, C. (1983) The genesis of neural crest and epidermal placodes: a reinterpretation of vertebrate origins. *Quarterly Review of Biology*, **58**, 1–28

Novacek, M. J. (1994) Whales leave the beach. *Nature*, **368**, 807

Nübler-Jung, K. and Arendt, D. (1994) Is ventral in insects dorsal in vertebrates? *Roux's Archives in Developmental Biology*, **203**, 357–366

Nusse, R. and Varmus, H. E. (1982) Many tumours induced by the mouse mammary tumour virus contain a provirus integrated in the same region of the host genome. *Cell*, **31**, 99–109

Nusse, R. and Varmus, H. E. (1992) *Wnt* genes. *Cell*, **69**, 1073–1087

Ohno, S. (1970) *Evolution by Gene Duplication*. Springer Verlag

Ortner, H., Koschorz, B., Holland, L. Z., Herrmann, B. G. and Holland, P. W. H., Conservation of *Brachyury (T)* genes in amphioxus and vertebrates: developmental and evolutionary implications, (Manuscript submitted).

Orton, J. H. (1913) The ciliary mechanism of the gills and the mode of feeding in amphioxus, ascidians and *Solenomya togata*. *Journal of the Marine Biological Association of the United Kingdom*, **10**, 19–49

Palmer, D. A. and Rickards, B. A. (1991) *Graptolites*, Boydell & Brewer

Parr, B. A., Shea, M. J., Vassileva, G. and McMahon, A. P. (1993) Mouse *Wnt* genes exhibit discrete domains of expression in the early embryonic CNS and limb buds. *Development*, **119**, 247–261

Patel, N. H., Kornberg, T. B. and Goodman, C. S. (1989a) Expression of *engrailed* during segmentation in grasshopper and crayfish. *Development*, **107**, 201–212

Patel, N. H., Martin-Blanco, E., Coleman, K. G., Poole, S. J., Ellis, M. C., Kornberg, T. B. and Goodman, C. S. (1989b) Expression of engrailed protein in arthropods, annelids and chordates. *Cell*, **58**, 955–968

Patten, W. (1890) On the origin of vertebrates from arachnids, *Quarterly Journal of Microscopical Science*, **31**, 317–378

Patten, W. (1912) *The Evolution of the Vertebrates and Their Kin*, Blakiston, Phildaelphia

Patterson, C. (1981) Significance of fossils in determining evolutionary relationships. *Annual Review of Ecology and Systematics*, **12**, 194–223

Patterson, C. and Rosen, D. E. (1977) Review of ichthyodectiform and other Mesozoic teleost fishes and the theory and practice of classifying fossils. *Bulletin of the American Museum of Natural History*, **158**, 85–172

Paul, C. R. C. (1990) Thereby hangs a tail. *Nature*, **348**, 680–681

Pendleton, J. W., Nagai, B. K., Murtha, M. T. and Ruddle, F. H. (1993) Expansion of the *Hox* gene family and the evolution of chordates. *Proceedings of the National Academy of Sciences of the USA*, **90**, 6300–6304

Peterson, K. J. (1994) The origin and early evolution of the Craniata, in *Major Features of Vertebrate Evolution*, (eds D. R. Prothero and R. M. Schoch), Short courses in Paleontology **7**, The Paleontological Society, pp.14–37

Peterson, K. J. (1995a) A phylogenetic test of the calcichordate scenario. *Lethaia*, **28**, 25–38

Peterson, K. J. (1995b) Dorsoventral axis inversion. *Nature*, **373**, 111–112

Philip, G. M. (1979) Carpoids – echinoderms or chordates? *Biological Reviews*, **54**, 439–471

Philippe, H., Chenuil, A. and Adoutte, A. (1994) Can the Cambrian Explosion be inferred through molecular phylogeny? *Development*, suppl. 15–25

Pollock, R. A., Jay, G. and Bieberich, C. J. (1992) Altering the boundaries of *Hox3.1* expression: evidence for antipodal gene regulation. *Cell*, **71**, 911–923

Price, M., Lemaistre, M., Pischetola, M., Di Lauro, R. and Duboule, D. (1991) A mouse gene related to *Distal-less* shows a restricted expression in the developing forebrain. *Nature*, **351**, 748–751

Prince, V. and Lumsden, A. (1994) *Hoxa-2* expression in normal and transposed rhombomeres: independent regulation in the neural tube and neural crest. *Development*, **120**, 911–923

Purnell, M. A. (1995) Microwear on conodont elements and macrophagy in the first vertebrates. *Nature*, **374**, 798–800

Purnell, M. A. Large eyes and vision in conodonts, *Lethaia* (in press)

Ramírez-Solis, R., Hui Zheng, Whiting, J., Krumlauf, R. and Bradley, A. (1993) *Hoxb-4* (*Hox-2.6*) mutant mice show homeotic transformation of a cervical vertebra and defects in the closure of the sternal rudiments. *Cell*, **73**, 279–294

Raw, F. (1960) Outline of a theory of the origin of the Vertebrates. *Journal of Palaeontology*, **34**, 497–539

Regnéll, G. (1975) Review of recent research on 'Pelmatozoans'. *Palaontologische Zeitschrift*, **49**, 530–564

Repetski, J. E. (1978) A fish from the Upper Cambrian of North America. *Science*, **200**, 529–531

Richards, R. J. (1992) *The Meaning of Evolution*, University of Chicago Press, Chicago, 205pp.

Riddihough, G. (1992) Homing in on the homeobox. *Nature*, **357**, 643–644

Riddle, R. D., Johnson, R. L., Laufer, E. and Tabin, C. (1993) *Sonic hedgehog* mediates the polarizing activity of the ZPA. *Cell*, **75**, 1401–1416

Rigby, J. K., Jr. (1983) Conodonts and the early evolution of the vertebrates. *Geological Society of America Abstr. Progr.*, **15**, 671

Rigby, S. (1993) Graptolites come to life. *Nature*, **362**, 209–210

Ritter, W. E. (1894) On a new Balanoglossus larva from the coast of California, and its possession of an endostyle. *Zoologische Anzeiger*, **17**, 24–30

Robertson, J. D. (1954) The chemical composition of the blood of some aquatic chordates, including members of the Tunicata, Cyclostomata and Osteichthyes. *Journal of Experimental Biology*, **31**, 424–442

Robertson, J. D. (1957) The habitat of the early vertebrates. *Biological Reviews*, **32**, 156–187

Robertson, J. D. (1959) The origin of vertebrates – marine or freshwater? *The Advancement of Science*, **61**, 516–520

Robison, R. A. (1965) Middle Cambrian eocrinoids from Western North America. *Journal of Paleontology*, **39**, 355–364

Romer, A. S. (1933) Eurypterid influence on vertebrate history. *Science*, **78**, 114–117

Romer, A. S. (1955) Fish origins – fresh or salt water? in *Papers in Marine Biology and Oceanography*, Pergamon Press, London, pp.261–280

Romer, A. S. (1970) *The Vertebrate Body*, 4th edn, Saunders, Philadelphia

Romer, A. S. (1972) The vertebrate as a dual animal – somatic and visceral. *Evolutionary Biology*, **6**, 121–156

Romer, A. S. and Grove, B. H. (1935) Environment of the early vertebrates. *American Midland Naturalist*, **16**, 805–862

Rosen, D. E., Forey, P. L., Gardiner, B. G. and Patterson, C. (1981) Lungfishes, tetrapods, palaeontology and plesiomorphy. *Bulletin of the American Museum of Natural History*, **167**, 154–276

Rosenfeld, M. G. (1991) POU-domain transcription factors: pou-er-ful developmental regulators. *Genes and Development*, **5**, 897–907

Ruddle, F. H., Hart, C. P. and McGinnis, W. (1985) Structural and functional aspects of the mammalian homeobox sequences. *Trends in Genetics*, **1**, 48–51

Ruddle, F. H., Bentley, K. L., Murtha, M. T. and Risch, N. (1994) Gene loss and gain in the evolution of the vertebrates. *Development*, suppl., 155–161

Ruiz i Altaba, A. (1991) Vertebrate development: an emerging synthesis. *Trends in Genetics*, **7**, 276–279

Ruiz i Altaba, A. and Melton, D. A. (1989a) Bimodal and graded expression of the *Xenopus* homeobox gene *Xhox3* during embryonic development. *Development*, **106**, 173–183

Ruiz i Altaba, A. and Melton, D. A. (1989b) Involvement of the *Xenopus* homeobox gene *Xhox3* in pattern formation along the anterior-posterior axis. *Cell*, **57**, 317–326

Runnegar, B. (1989) Rates and modes of evolution in the Mollusca, in *Rates of Evolution* (eds K. S. W. Campbell and M. F. Day), Allen and Unwin, London

Ruppert, E. E. (1991) Introduction to the aschelminth phyla: a consideration of mesoderm, body cavities and cuticle, in *Microscopic Anatomy of Invertebrates Vol. 4 Aschelminths* (eds F. W. Harrison and E. E. Ruppert), Wiley-Liss, Inc., New York, pp. 1–17

Ruppert, E. E. (1994) Evolutionary origin of the vertebrate nephron. *American Zoologist*, **34**, 542–553

St Johnston, D. and Nüsslein-Volhard, C. (1992) The origin of pattern and polarity in the *Drosophila* embryo. *Cell*, **68**, 201–219

Sansom, I. J., Smith, M. P., Armstrong, H. A. and Smith, M. M. (1992) Presence of the earliest vertebrate hard tissues in conodonts. *Science*, **256**, 1308–1311

Sansom, I. J, Smith, M. P. and Smith, M. M. (1994) Dentine in conodonts. *Nature*, **368**, 591

Sansom, I. J., Smith, M. M. and Smith, M. P. (1996) Scales of thelodont and shark-like fishes from the Ordovician of Colorado. *Nature*, **379**, 628–630

Sasai, Y., Bin Lu, Steinbeisser, H., Geissert, D., Gont, L. K. and De Robertis, E. (1994) Xenopus *chordin*: a novel dorsalizing factor activated by organizer-specific homeobox genes. *Cell*, **79**, 779–790

Schaeffer, B. (1987) Deuterostome monophyly and phylogeny. *Evolutionary Biology*, **21**, 179–235

Schubert, F. R., Nieselt-Struwe, K. and Gruss, P. (1993) The Antennapedia-type homeobox genes have evolved from three precursors separated early in metazoan evolution. *Proceedings of the National Academy of Sciences of the United States of America*, **90**, 143–147

Schummer, M., Scheurlen, I., Schaller, C. and Galliot, B. (1992) HOM/HOX homeobox genes are present in hydra (*Chlorohydra viridissima*) and are differentially expressed during regeneration. *EMBO Journal*, **11**, 1815–1823

Scott, M. P. (1992) Vertebrate homeobox gene nomenclature. *Cell*, **71**, 551–553

Sedgwick, A. (1905) *A Student's Textbook of Zoology*, Swan Sonnenschein, London

Seeliger, O. (1894) Die Bedeutung der 'Segmentation' des Ruderschwanzes der Appedicularien. *Zoologische Anzeiger*, **17**, 162–165

Shawlot, W. and Behringer, R. R. (1995) Requirement for *Lim1* in head-organizer function. *Nature*, **374**, 425–430

Shimell, M. J., Ferguson, E. L., Childs, S. R. and O'Connor, M. B. (1991) The *Drosophila* dorsal–ventral patterning gene *tolloid* is related to human bone morphogenetic protein 1. *Cell*, **67**, 469–481

Shin, G. L. (1994) Epithelial origin of mesodermal structures in arrow-worms (Phylum Chaetognatha). *American Zoologist*, **34**, 523–532

Shinn, G. L. and Roberts, M. E. (1994) Ultrastructure of hatchling chaetognaths (*Ferosagitta hispida*): epithelial arrangement of the mesoderm and its phylogenetic implications. *Journal of Morphology*, **219**, 143–163

Shu, D., Zhang, X. and Chen, L. (1996) Reinterpretation of Yunnanozoon as the earliest known hemichordate. *Nature*, **380**, 428–430

Sidow, A. (1992) Diversification of the *Wnt* gene family on the ancestral lineage of vertebrates. *Proceedings of the National Academy of Sciences of the United States of America*, **89**, 5098–5102

Sillman, L. R. (1960) The origin of the vertebrates. *Journal of Paleontology*, **34**, 540–544

Simeone, A., Acampora, D., Gulisano, M., Stornaiuolo, A. and Boncinelli, E. (1992) Nested expression domains of four homeobox genes in developing rostral brain. *Nature*, **358**, 687–690

Slack, J. (1984) A Rosetta Stone for pattern formation in animals? *Nature*, **310**, 364–365

Slack, J. (1992) The turn of the screw. *Nature*, **355**, 217

Slack, J. M. W., Holland, P. W. H. and Graham, C. F. (1993) The zootype and the phylotypic stage. *Nature*, **361**, 490–492

Smith, A. B. (1984) Classification of the Echinodermata. *Palaeontology*, **27**, 431–459

Smith, A. B. (1990) Evolutionary diversification of echinoderms during the early Palaeozoic, in *Major Evolutionary Radiations*, (eds P. D. Taylor and G. P. Larwood), Systematics Association Special Volume No. 42, Clarendon Press, Oxford

Smith, J. C. (1994) Hedgehog, the floor plate, and the zone of polarizing activity. *Cell*, **76**, 193–196

Smith, M. M., Sansom, I. J. and Smith, M. P. (in press) Teeth before armour: the earliest vertebrate mineralized tissues. Preprint, *IGCP 8 Palaeozoic Microvertebrates-SDS: Gross Symposium (Göttingen, 31 July–6 August 1993)*

Sokol, S., Christian, J. L., Moon, R. T. and Melton, D. A. (1991) Injected *Wnt* RNA induces a complete body axis in *Xenopus* embryos. *Cell*, **67**, 741–752

Southward, E. C. (1975) Fine structure and phylogeny of the Pogonophora. *Symposia of the Zoological Society of London*, **36**, 235–251

Sprinkle, J. (1983) Patterns and problems in echinoderm evolution. *Echinoderm Studies*, **1**, 1–18

Sprinkle, J. (1992) Radiation of the Echinodermata, origin and early evolution of the Metazoa. *Topics in Geobiology*, **10**, 375–398

Stebbing, A. R. D. (1970) The status and ecology of *Rhabdopleura compacta* (Hemichordata) from Plymouth. *Journal of the Marine Biological Association of the United Kingdom*, **50**, 209–221

Storey, K. G., Crossley, J. M., De Robertis, E. M., Norris, W. E. and Stern, C. D. (1992) Neural induction and regionalization in the chick embryo, *Development*, **114**, 729–741

Struhl, G. (1984) A universal genetic key to body plan? *Nature*, **310**, 10–11

Sulston, J. *et al.* (1992) The *C. elegans* genome sequencing project: a beginning. *Nature*, **356**, 37–41

Swalla, B. J., Makabe, K. W., Satoh, N. and Jeffery, W. R. (1993) Novel genes expressed differentially in ascidians with alternate modes of development. *Development*, **119**, 307–318

Sweet, W. C. (1988) *The Conodonta – Morphology, Taxonomy, Paleoecology and Evolutionary History of a Long-extinct Animal Phylum*, Oxford University Press, New York

Telford, M. J. amd Holland, P. W. H. (1993) The phylogenetic affinities of the chaetognaths: a molecular analysis. *Molecular Biology and Evolution*, **10**, 660–676

Thomson, K. S. (1987a) The neural crest and the morphogenesis and evolution of the dermal skeleton in vertebrates, in *Developmental and Evolutionary Aspects of the Neural Crest*, (ed P. F. A. Maderson), John Wiley, New York, pp.301–338

Thomson, K. S. (1987b) Spinal discord. *Nature* **327**, 196–197

Thomson, K. S. (1991) Segmentation, the adult skull and the problem of homology, in *The Vertebrate Skull: Volume 1, Development*, (eds. J. Hanken and B. K. Hall), University of Chicago Press, Chicago

Thorogood, P. and Hanken, J. (1992) Body building exercises. *Current Biology*, **2**, 83–85

Turbeville, J. McC., Field, K. G. and Raff, R. A. (1992) Phylogenetic position of Phylum Nemertini, inferred from 18S rRNA sequences: molecular data as a test of morphological character homology. *Molecular Biology and Evolution*, **9**, 235–249

Turbeville, J. McC., Schulz, J. R. and Raff, R. A. (1994) Deuterostome phylogeny and the Sister Group of the Chordates: Evidence from Molecules and Morphology. *Molecular Biology and Evolution*, **11**, 648–655

Ubaghs, G. (1961) Sur la nature de l'organe appelé tige ou pédoncule chez les carpoides Cornuta et Mitrata, *Comptes Rendus Hebdomadaires des Séances de l'Academie des Sciences Paris*, **253**, 2738–2740

Ubaghs, G. (1963) *Cothurnocystis* Bather, *Phyllocystis* Thoral and an undetermined member of the order Soluta (Echinodermata, Carpoidea) in the uppermost Cambrian of Nevada. *Journal of Paleontology*, **37**, 1133–1142

Ubaghs, G. (1967) Stylophora, *Treatise on Invertebrate Paleontology Part S, Echinodermata I*, 2 vols, (ed. R. C. Moore), Geological Society of America and University of Kansas Press, Lawrence, Kansas, pp.495–565

Urbanek, A. (1986) The enigma of graptolite ancestry: lesson from a phylogenetic debate, *Problematic Fossil Taxa*, (eds A. Hoffman and M. H. Nitecki), Oxford University Press, New York/Clarendon Press, Oxford

Vacelet, J. and Boury-Esnault, N. (1995) Carnivorous sponges. *Nature*, **373**, 333–335

Van Wijhe, J. W. (1913) On the metamorphosis of *Amphioxus lanceolatus*. *Proceedings van de Koninglijk Akademie van Wetenschappen, Amsterdam*, **16**, 574–583

Wada, H. and Satoh, N. (1994) Details of the evolutionary history from invertebrates to vertebrates, as deduced from the sequences of 18S rDNA. *Proceedings of the National Academy of Sciences of the United States of America*, **91**, 1801–1804

Wells, H. G., Huxley, J. and Wells, G. P. (1931) *The Science of Life*, Cassell, London

Whitear, M. (1957) Some remarks on the ascidian affinities of vertebrates. *Annals and Magazine of Natural History*, **(12) 10**, 338–347

Whiting, J., Marshall, H., Cook, M., Krumlauf, R., Rigby, P. W. J., Stott, D. and Allemann, R. J. (1991) Multiple spatially specific enhancers are required to reconstruct the pattern of *Hox-2.6* gene expression. *Genes and Development*, **5**, 2048–2059

Wilkinson, D. G., Bhatt, S., Chavrier, P., Bravo, R. and Charnay, P. (1989a) Segment-specific expression of a zinc-finger gene in the developing nervous system of the mouse. *Nature*, **337**, 461–464

Wilkinson, D. G., Bhatt, S., Cook, M., Boncinelli, E. and Krumlauf, R. (1989b) Segmental expression of Hox-2 homoeobox-containing genes in the developing mouse hindbrain. *Nature*, **341**, 405–409

Willey, A. (1894) *Amphioxus* and the ancestry of the vertebrates, *Columbia University Biol. Ser. II*, Macmillan, London and New York

Williams, A., Cusack, M. and Mackay, S. (1994) Collagenous chitinophosphatic shell of the brachiopod *Lingula*. *Philosophical Transactions of the Royal Society of London B*, **346**, 233–266

Willmer, E. N. (1974) Nemertines as possible ancestors of the vertebrates. *Biological Reviews*, **49**, 321–363

Willmer, E. N. (1975) The possible contribution of the nemertines to the problem of the phylogeny of the protochordates. *Protochordates, Symposia of the Zoological Society of London*, **36**, 319–346

Willmer, P. G. (1990) *Invertebrate Relationships: Patterns in Animal Evolution*, Cambridge University Press, Cambridge

Willmer, P. G. and Holland, P. W. H. (1991) Modern approaches to metazoan relationships. *Journal of Zoology*, **224**, 689–694

Wills, M. A., Briggs, D. E. G. and Fortey, R. A. (1994) Disparity as an evolutionary index: a comparison of Cambrian and Recent arthropods. *Paleobiology*, **20**, 93–130

Woods, I. S. and Jefferies, R. P. S. (1992) A new stem-group chordate from the Lower Ordovician of South Wales, and the problem of locomotion in boot-shaped cornutes. *Palaeontology*, **35**, 1–25

Wright, C. and Hogan, B. (1991) Another hit for gene targeting. *Nature*, **350**, 458–459

Wright, C. V. E. (1991) Vertebrate homeobox genes. *Current Opinions in Cell Biology*, **3**, 976–982

Yamada, T. (1994) Caudalization by the amphibian organizer: *brachyury*, convergent extension and retinoic acid. *Development*, **120**, 3051–3062

Yasuo, H. and Satoh, N. (1993) Function of vertebrate *T* gene. *Nature*, **364**, 582–583

Yasuo, H. and Satoh, N. (1994) An ascidian homolog of the mouse *Brachyury* (*T*) gene is expressed exclusively in notochord cells at the fate restricted stage. *Development, Growth and Differentiation*, **36**, 9–18

Yokoyama, T., Copeland, N. G., Jenkins, N. A., Montgomery, C. A., Elder, F. F. B. and Overbeek, P. A. (1993) Reversal of left–right asymmetry: a *situs inversus* mutation. *Science*, **260**, 679–682

Yost, H. J. (1992) Regulation of vertebrate left–right asymmetries by extracellular matrix. *Nature*, **257**, 158–161

Young, J. Z. (1981) *The Life of Vertebrates*, 3rd edn, Clarendon Press, Oxford

Index

Page references in **bold** refer to figures.